S0-BQZ-894

Sławomir Wiak, Andrzej Krawczyk and Ivo Dolezel (Eds.)

Intelligent Computer Techniques in Applied Electromagnetics

Studies in Computational Intelligence, Volume 119

Editor-in-chief
Prof. Janusz Kacprzyk
Systems Research Institute
Polish Academy of Sciences
ul. Newelska 6
01-447 Warsaw
Poland
E-mail: kacprzyk@ibspan.waw.pl

Further volumes of this series can be found on our homepage: springer.com

Vol. 98. Ashish Ghosh, Satchidananda Dehuri and Susmita Ghosh (Eds.)
Multi-Objective Evolutionary Algorithms for Knowledge Discovery from Databases, 2008
ISBN 978-3-540-77466-2

Vol. 99. George Meghabghab and Abraham Kandel
Search Engines, Link Analysis, and User's Web Behavior, 2008
ISBN 978-3-540-77468-6

Vol. 100. Anthony Brabazon and Michael O'Neill (Eds.)
Natural Computing in Computational Finance, 2008
ISBN 978-3-540-77476-1

Vol. 101. Michael Granitzer, Mathias Lux and Marc Spaniol (Eds.)
Multimedia Semantics - The Role of Metadata, 2008
ISBN 978-3-540-77472-3

Vol. 102. Carlos Cotta, Simeon Reich, Robert Schaefer and Antoni Ligeza (Eds.)
Knowledge-Driven Computing, 2008
ISBN 978-3-540-77474-7

Vol. 103. Devendra K. Chaturvedi
Soft Computing Techniques and its Applications in Electrical Engineering, 2008
ISBN 978-3-540-77480-8

Vol. 104. Maria Virvou and Lakhmi C. Jain (Eds.)
Intelligent Interactive Systems in Knowledge-Based Environment, 2008
ISBN 978-3-540-77470-9

Vol. 105. Wolfgang Guenthner
Enhancing Cognitive Assistance Systems with Inertial Measurement Units, 2008
ISBN 978-3-540-76996-5

Vol. 106. Jacqueline Jarvis, Dennis Jarvis, Ralph Rönnquist and Lakhmi C. Jain (Eds.)
Holonic Execution: A BDI Approach, 2008
ISBN 978-3-540-77478-5

Vol. 107. Margarita Sordo, Sachin Vaidya and Lakhmi C. Jain (Eds.)
Advanced Computational Intelligence Paradigms in Healthcare - 3, 2008
ISBN 978-3-540-77661-1

Vol. 108. Vito Trianni
Evolutionary Swarm Robotics, 2008
ISBN 978-3-540-77611-6

Vol. 109. Panagiotis Chountas, Ilias Petrounias and Janusz Kacprzyk (Eds.)
Intelligent Techniques and Tools for Novel System Architectures, 2008
ISBN 978-3-540-77621-5

Vol. 110. Makoto Yokoo, Takayuki Ito, Minjie Zhang, Juhnyoung Lee and Tokuro Matsuo (Eds.)
Electronic Commerce, 2008
ISBN 978-3-540-77808-0

Vol. 111. David Elmakias (Ed.)
New Computational Methods in Power System Reliability, 2008
ISBN 978-3-540-77810-3

Vol. 112. Edgar N. Sanchez, Alma Y. Alanís and Alexander G. Loukianov
Discrete-Time High Order Neural Control: Trained with Kalman Filtering, 2008
ISBN 978-3-540-78288-9

Vol. 113. Gemma Bel-Enguix, M. Dolores Jiménez-López and Carlos Martín-Vide (Eds.)
New Developments in Formal Languages and Applications, 2008
ISBN 978-3-540-78290-2

Vol. 114. Christian Blum, Maria José Blesa Aguilera, Andrea Roli and Michael Sampels (Eds.)
Hybrid Metaheuristics, 2008
ISBN 978-3-540-78294-0

Vol. 115. John Fulcher and Lakhmi C. Jain (Eds.)
Computational Intelligence: A Compendium, 2008
ISBN 978-3-540-78292-6

Vol. 116. Ying Liu, Aixin Sun, Han Tong Loh, Wen Feng Lu and Ee-Peng Lim (Eds.)
Advances of Computational Intelligence in Industrial Systems, 2008
ISBN 978-3-540-78296-4

Vol. 117. Da Ruan, Frank Hardeman and Klaas van der Meer (Eds.)
Intelligent Decision and Policy Making Support Systems, 2008
ISBN 978-3-540-78306-0

Vol. 118. Tsau Young Lin, Ying Xie, Anita Wasilewska and Churn-Jung Liau (Eds.)
Data Mining: Foundations and Practice, 2008
ISBN 978-3-540-78487-6

Vol. 119. Sławomir Wiak, Andrzej Krawczyk and Ivo Dolezel (Eds.)
Intelligent Computer Techniques in Applied Electromagnetics, 2008
ISBN 978-3-540-78489-0

Sławomir Wiak
Andrzej Krawczyk
Ivo Dolezel
(Eds.)

Intelligent Computer Techniques in Applied Electromagnetics

With 201 Figures and 24 Tables

Sławomir Wiak
Institute of Mechatronics and Information Systems
Technical University of Lodz
ul. Stefanowskiego 18/22
90-924 Lodz
Poland
wiakslaw@p.lodz.pl

Andrzej Krawczyk
Central Institute for Labour Protection
ul. Czerniakowska 16
00-701 Warsaw
Poland
ankra@ciop.pl

Ivo Dolezel
Czech Technical University in Prague
Faculty of Electrical Engineering
Department of Electrical Power Engineering
Technická 2
166 27 Praha 6
Czech Republic
dolezel@fel.cvut.cz

ISBN 978-3-540-78489-0 e-ISBN 978-3-540-78490-6

Studies in Computational Intelligence ISSN 1860-949X

Library of Congress Control Number: 2008923177

Cover design: Deblik, Berlin, Germany

Printed on acid-free paper

9 8 7 6 5 4 3 2 1

springer.com

Preface

This book contains papers presented at the International Symposium on Electromagnetic Fields in Mechatronics, Electrical and Electronic Engineering ISEF'07 which was held in Prague, the Czech Republic, from September 13 to 15, 2007. ISEF conferences have been organized since 1985 and from the very beginning it was a common initiative of Polish and other European researchers who have dealt with electromagnetic field in electrical engineering. The conference travels through Europe and is organized in various academic centres. Relatively often, it was held in some Polish city as the initiative was on the part of Polish scientists. Now ISEF is much more international and successive events take place in different European academic centres renowned for electromagnetic research. This time it was Prague, famous for its beauty and historical background, as it is the place where many cultures mingle. The venue of the conference was the historical building of Charles University, placed just in the centre of Prague. The Technical University of Prague, in turn, constituted the logistic centre of the conference.

It is the tradition of the ISEF meetings that they try to tackle quite a vast area of computational and applied electromagnetics. Moreover, the ISEF symposia aim at combining theory and practice; therefore the majority of papers are deeply rooted in engineering problems, being simultaneously of a high theoretical level. The profile of the conference changes, however, year-by-year and one can find more and more contributions dealing with applied electromagnetics coupled with hardware and software technologies. That is why, for the first time, the organizers decided to use the Springer Lecture Notes in Artificial Intelligence as the proper place for publishing some of the papers. Generally speaking, one can observe the trend of a decreasing number of papers which deal with classical electrical engineering in preference to information technology and biomedical applications. This direction seems to be comprehensible in the light of modern industry. Nevertheless, even beyond electrical engineering, we do touch the heart of the matter in electromagnetism.

The ISEF'07 Proceedings cover over 240 papers and after a selection process, 24 papers were accepted for publication in this volume, while the others were directed to COMPEL: The International Journal for Computation and Mathematics in Electrical and Electronic Engineering and to another book to be published by IOS Press.

The methods of evaluation and control of the electromagnetic field, based on Artificial Intelligence (AI), constitute the core of the papers presented at the ISEF'07 conference. Indeed, nowadays it would be hard to imagine dealing with electromagnetic field analysis without computer simulation. The papers which have been selected for this volume have strong links with AI theory and methodology, i.e. they do not use the numerical method automatically but they attempt to contribute some new AI tools to computer modelling. We hope that such an approach would be developed further in future ISEF conferences.

The papers selected for this volume have been grouped into three parts, which represent the following topics:

- Algorithms and Intelligent Computer Methods
- Computer Methods and Engineering Software
- Applications of Computer Methods

The first part consists of papers dealing with the algorithms which are more general and concern the class of more general problems, like the problems of optimization and/or inverse problems. Thus, the first part is focused on information techniques rather than modelling particular devices. The reader may find here such problems as multi-agent algorithms, neural networks, genetic algorithms, wavelet transformation, paralleling of computation problems, monitoring and others.

The second part directs the reader's attention to similar problems but referring to more technical aspects. Indeed, one can find here the papers which deal with numerical modelling of such devices as high voltage power lines, actuators, NMR and induction tomography, adaptive plasma, electrochemical machining and others. It is clearly seen that the attention here is focused on the device rather than the methodology.

The third part consists of papers dealing with the application of advanced algorithms and methods to selected technical engineering problems. The reader can focus attention on such problems as **optimization of a drive system**, **combined electromagnetic and thermal analysis by means of approximate methods**, combined finite element method (FEM) and finite volume method (FVM), algorithms based on stochastic methodsand others.

At the end of these remarks let us, the Editors of the volume, be allowed to express our thanks to our colleagues who have contributed to the book by peer-reviewing the papers at the conference as well as in the publishing process. We also convey our thanks to Springer-Verlag Publisher for their effective collaboration in shaping this editorial enterprise. As ISEF conferences are organised biannually we do hope to maintain our strong links with Springer-Verlag Publisher in the future.

Poland *Sławomir Wiak*
Poland *Andrzej Krawczyk*
Czech Republic *Ivo Dolezel*
May, 2008

Contents

Part I
Algorithms and Intelligent Computer Methods

Comparison of Two Multi-Agent Algorithms: ACO and PSO for the Optimization of a Brushless DC Wheel Motor

Fouzia Moussouni, Stéphane Brisset, and Pascal Brochet

Abstract Particle swarm optimization (PSO) and ant-colony optimization (ACO) are novel multi-agent algorithms able to solve complex problem. By consequence, it would seem wise to compare their performances for solving such problems. For this purpose, both algorithms are compared together and with Matlab's GA in term of accuracy of the solution and computation time. In this paper the optimization is applied on the design of a brushless DC wheel motor that is known as a nonlinear multimodal benchmark.

1 Introduction

Often, to shown the practical interest of any optimization metaheuristic it is necessary to test them through more and more difficult experiments like solving hard optimization problems. For improving its applicability, it is important to answer how and why the methods work. This paper tackles the comparison of the two nature-inspired metaheuristics particle swarm optimization (PSO) and ant-colony optimization (ACO) in order to adapt them for handling hard optimization problems.

Indeed, PSO and ACO are novel techniques created for solving hard optimization problems. As they are multi-agent methods, they can be used to handle multi-level optimization problems. Unfortunately, the tuning of the control parameters for these algorithms is not acquired. So, to do a successful selection of their control parameters, a brushless DC wheel motor benchmark is used. In this study, the GA algorithm implemented in the Matlab Genetic Algorithm and Direct Search Toolbox is used to evaluate both multi-agents algorithms.

The comparison among the both algorithms and the brushless DC wheel motor benchmark are presented in Sects. 1 and 2, respectively. In Sect. 3, the mixed-integer

F. Moussouni, S. Brisset, and P. Brochet
L2EP – Ecole Centrale de Lille, Cité Scientifique, BP 48, 59651 Villeneuve d'Ascq Cedex, France
fouzia.moussouni@ec-lille.fr

F. Moussouni et al.: *Comparison of Two Multi-Agent Algorithms: ACO and PSO for the Optimization of a Brushless DC Wheel Motor*, Studies in Computational Intelligence (SCI) **119**, 3–10 (2008)
www.springerlink.com

NSGA-II is applied to tune ACO and PSO parameters in order to reduce the Euclidean distance between the known optimal point and the solution found by these algorithms, and the number of evaluations. Then, both methods (ACO and PSO) are evaluated in terms of convergence speed, and quality of the results in Sect. 4. Thereafter Sect. 5 tackles how the both algorithms are behaving to obtain effective solutions to a multi-level optimization problem. This is one open question among many with a certain interest of being solved in the near future. The final section, gives some concluding remarks.

2 Comparing PSO and ACO Methods

PSO and ACO are considered as a global optimization method. In particular, these algorithms manage very well combinatorial and mixed problems. Unlike gradient search methods, they are less susceptible to be trapped in local optima. When GA mimics the natural biological evolution, ACO and PSO draw inspiration from the interactive behavior of social species [1, 2].

GA has a relatively old history since the first work of its author John Holland backs to 1962. ACO was proposed by Dorigo et al. in 1990 [1]. It is inspired by the collaborative behavior of the ants [1]. PSO is the most recent one of both. It was developed by Eberhart and Kennedy in 1995 [2]. PSO is inspired by social behavior of bird flocking. Both algorithms (ACO, PSO) are population based stochastic optimization techniques.

PSO and ACO have many similarities with GA. All are initialized with a randomly generated population and updating solution following the fitness values at each generation. However, unlike GA, ACO and PSO have no evolution operators like crossover and mutation. But both have memory, which is important to the algorithms [1, 2]. Compared with GA where individuals are in rivalry, the efficiency of PSO and ACO is in the cooperative behavior of agents (particles or ants).

In the following, a version of ACO called Generalize Ant Colony Optimization (GACO) algorithm [3], and the algorithm implemented in Matlab Genetic Algorithms and Direct Search Toolbox [4] are used for comparison. But first, PSO and ACO are briefly presented below.

2.1 Particle Swarm Optimization

The PSO algorithm is based on two rules [2, 5, 6]: (a) each particle has a memory that enables him to memorize its best position found in the past and it tends to be attracted by this point. (b) Each particle is informed about the best position find in its vicinity and it tends to go towards this point.

In other words, starting from some information, particles must be able to choose its next movement, i.e., to calculate its new velocity which is an updating operator for its position.

Therefore, a particle i combines, linearly, three components:

- Its current velocity: v_i
- Its best position: p_i
- The best position of its neighbors: G_i.

Using three parameters which balance three tendencies:

- c_1: tendency to follow its own way
- c_2: tendency to reconsider its steps (preserving)
- c_3: tendency to follow the best neighbor.

To resume, in PSO each single solution is a particle (an agent) that has fitness value, and a velocity to direct his flying. The particles show a tendency towards current optimum particles. In other terms, the PSO algorithm consists on the calculation of the velocity:

$$v_i(k+1) = c_{1i} \cdot v_i(k) + c_{2i} \cdot rand \cdot (p_i - x_i(k)) + c_{3i} \cdot rand \cdot (G_i - x_i(k)) \tag{1}$$

and the position:

$$x_i(k+1) = x_i(k) + v_i(k+1), \tag{2}$$

where i is a particle index, v, x are velocity and current position of ith particle, respectively. p is the best position found by ith particle and G is the best position found by swarm. rand is a random number $\in [0,\ 1]$. c_2, c_3 are learning factors. c_1 is the inertia of ith particle.

2.2 Ant Colony Optimization

Contrary to a natural ant, an artificial ant is not completely blind and can see a little further that its direct entourage. It has a certain memory enabling him to adopt the solution which it built. According to the quality of the solution an artificial ant put down a quantity of pheromone [1,7,8]. To resume, at each point requiring a choice, a decision table produces a probability distribution on which the movements of the ants are based. Thus, by considering that the ant is on a point i, the probability p_{ij} which it moves towards a point j will be given by the following:

$$p_{ij} = \frac{[\tau_{ij}(t)]^{\alpha}[\eta_{ij}(t)]^{\beta}}{\sum_{k\in V(i)}[\tau_{ik}(t)]^{\alpha}T[\eta_{ik}(t)]^{\beta}}, \tag{3}$$

where at iteration t, τ_{ij} is the concentration of pheromone associated with the chosen solution, η_{ij} is the attraction of this solution. $V(i)$ represents the vicinity of the point i. α and β are two parameters giving a relative importance to the pheromone and the attraction values of each ant.

The concentration of pheromone is updating at each iteration t according to the following:

$$\tau_{ij}(t+1) = \rho\tau_{ij}(t) + \sum_{k=1}^{m} \Delta\tau_{ij}^{k}(t+1), \tag{4}$$

where $\rho \in [0, 1]$ is the pheromone persistence such that $(1-\rho)$ represents the evaporation of trail between iteration $t+1$ and t. $\Delta\tau_{ij}$ is the pheromone addition when ant k moves from the decision point i to the solution j.

3 Benchmark

In [9] an original motor is proposed as a benchmark to compare optimization methods. As it is sufficiently detailed to be adapted to different design contexts, this benchmark is considered as a reference for the optimal design of electrical machines optimization problems.

It consists on an analytical model for the design of a brushless DC wheel motor, where the aim is to design a motor with the best efficiency η. In this study, only five design variables are considered. This analytical model has been validated via prototype measures (Fig. 1). All materials useful to deal with this benchmark can be found in [10].

As it is highlighted in [9], this constraint optimization problem is multimodal which makes optimization process difficult.

The optimization problem was defined in [9] as follows:

$$\begin{aligned} & \text{minimize } 1-\eta \\ \text{with} \quad & 150mm \le D_s \le 330mm,\ 0.9\,T \le B_d \le 1.8\,T \\ & 2.0A/mm^2 \le \delta \le 5.0A/mm^2,\ 0.5T \le B_e \le 0.76T, \\ & 0.6T \le B_{cs} \le 1.6T \\ \text{s.t.} \quad & M_{tot} \le 15kg,\ D_{ext} \le 340mm,\ D_{int} \ge 76mm,\ I_{max} \ge 125A, \\ & discr(D_s, \delta, B_d, B_e) \ge 0,\ T_a \le 120^{\circ}C, \end{aligned} \tag{5}$$

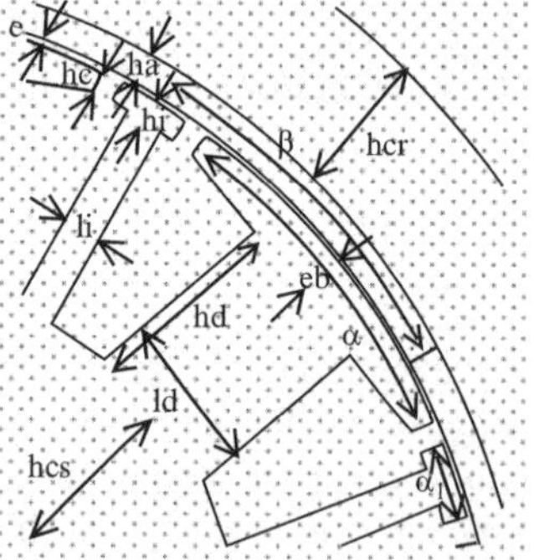

Fig. 1 Prototype of the brushless DC wheel motor

where η, M_{tot}, D_{ext}, D_{int}, I_{max}, and T_a are results of the analytical model and $discr(D_s, \delta, B_d, B_e)$ is the determinant used for the calculation of the slot height [9].

As optimization of the DC wheel motor is a non-linear constrained problem the external penalty approach is used. The objective function and non-linear constraints are combined to formulate the following sub-problem:

$$\phi(X,\mu) = f(X) + \mu \left[\sum_{i=1}^{m} (\max[0, g_i(X)])^2 \right] + \sum_{i=1}^{l} (h_i(X))^2, \tag{6}$$

where $g(X)$, $h(X)$ are the l equality and m inequality constraints respectively. The parameter μ is given by the users and must be positive and very large. $f(X) = (1-\eta)$ is the original objective function to be minimized.

This benchmark is used for comparison in case of single objective and for the optimal tuning of the PSO and ACO parameter control.

4 Optimal Tuning of the Parameters

Unfortunately, to now the tuning of the control parameters for ACO PSO and GA algorithms is not acquired. However, there are not many parameter need to be tuned in PSO contrarily to GA and ACO.

In order to obtain a maximum of interesting information on the optimal tuning of the parameters of each algorithm, multiple parametric configurations are applied. A multi-objective optimization method that is called mixed-integer NSGA-II [11] is used. Particularly, it evaluates the three algorithms in term of the Euclidean distance between the known optimal point and the solution found, and also the number of evaluations.

To draw on the experiences of the other researchers [12–15] and the state of art on the swarm intelligence domain, the PSO and ACO parameters to be tuning by NSGA-II are chosen. In this study, only six GA parameters, and four PSO/ACO parameters are tuned by the cited method. The GA parameters are: rate of the elite individuals and probability of crossover which are continuous parameters, and four discrete parameters that are: population size, fitness scaling type, selection scheme, and type of crossover. For ACO, the parameters are number of ants, two parameters α, β that allow a user control on the relative importance of trail vs. visibility, and the pheromone's persistence factor ρ. Regarding to PSO the tuned parameters are number of particles, learning factors c_2, c_3 and the inertia factor c_1.

Figure 2 shows the Pareto's fronts found by the Mixed-integer NSGA-II. As GA, PSO and ACO are stochastic methods, the optimization is performed 20 times, and then the Pareto's fronts are built with the average values of the two objectives (Fig. 2). Some interesting points are highlighted. The optimal values of the PSO parameters founded by Mixed-integer NSGA-II are: particle number is $N = 20$, $c2 = 1.6$, $c3 = 1.8$ and the inertia factor $c_1 \in [0.4,\ 1]$ is decreasing in time. For ACO the optimal parameters are: ant number is 45, $\alpha = \beta = 1$. Pheromone's

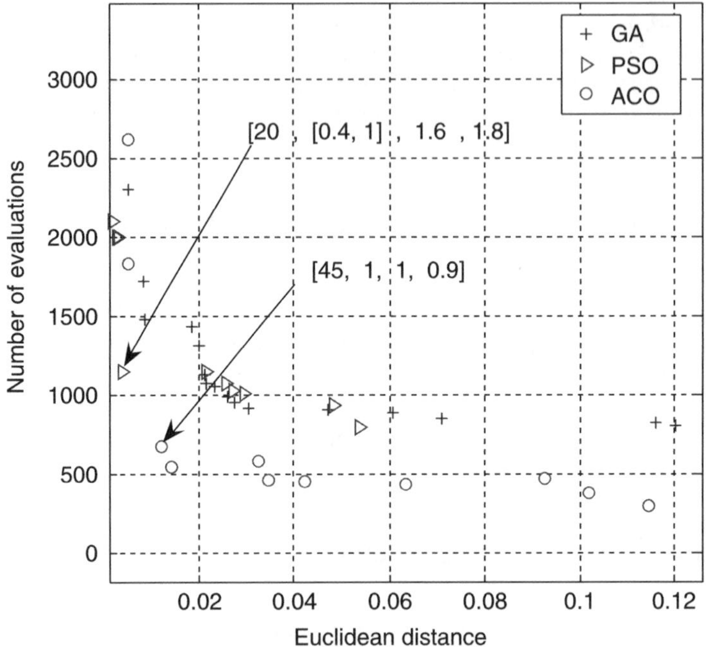

Fig. 2 Pareto fronts of GA, ACO and PSO

persistence factor is $\rho = 0.9$. As shown in Fig. 1, ACO is the most efficient method whereas PSO and GA have partially the same solutions.

In principle, to assess the performance of two multi-objective optimizers, two basic approaches exist in the literature; the attainment function approach [7], and the indicator approach [5, 8]. To assess just one of criteria such as (a) convergence to the Pareto-optimal front, (b) having an uniform distribution of the Pareto front and (c) having a better coverage of the objective space, many quality indicators (metrics) have been proposed. The trade-off solutions between the two objectives are in the bend of the Pareto front. Then, the use of the quality indicators is not required in this study.

5 Mono-Objective Optimization

In this section, GA, ACO and PSO are compared in the case of a mono-objective optimization. According to [11], the best values of GA parameters for this benchmark are: *20* individuals per generation, *10%* of elite individuals, crossover probability is *0.5*, rank fitness scaling, roulette selection, and scattered crossover operators. Following the previous study, PSO parameters are chosen as follow: particle number is *20*, $c_2 = 1.6$, $c_3 = 1.8$. The inertia factor $c_1 \in [0.4,\ 1]$ is decreasing in time. And a

Table 1 Results of the mono-objective optimization

Symbol (unit)	GA	PSO	ACO
M_{tot} (kg)	15	15.0011	15.00
I_{max} (A)	125.0128	124.9996	125.00
D_{int} (mm)	76.9	79.2	76.0
D_{ext} (mm)	239.2	239.8	238.9
T_a (°C)	95.2077	94.9793	95.3464
D_s (mm)	201.5	202.1	201.2
B_e (T)	0.6480	0.6476	0.6481
delta (A mm^{-2})	2.0602	2.0417	2.0437
B_d (T)	1.7991	1.8	1.8
B_{cs} (T)	0.8817	0.9298	0.8959
$\eta(\%)$	95.3112	95.32	95.32
Evaluation number	3,380	1,600	1,200

complete neighborhood is used. For ACO, ant number is *45*, $\alpha = \beta = 1$, and $\rho = 0.9$. All algorithms stop when the maximum number of iterations or the minimum error requirement is reached. It can be seen in Table 1 that all methods produce a numerical solution which well approximates the exact solution. However, the ACO method obtains a more accurate solution in a smaller number of objective function evaluations. The optimal configurations given by the different methods are very close; the Euclidian distance is less than 0.01. So, all solutions are effectives. ACO algorithm is faster.

6 Conclusion

In this paper, the multi-agent algorithms PSO and ACO have proved to be effective to handle hard optimization problems. For this purpose both algorithms are compared together and with Matlab's GA on the optimization of the brushless DC wheel motor with constraints and a single objective. The accuracy of the solution and the computation time are compared. ACO is faster and more accurate. The mixed-integer NSGA-II is employed to find the optimal tuning of ACO and PSO taking into account the accuracy of the solution and the computation time. On this benchmark, ACO and PSO are more computationally effective than Matlab's GA. PSO is the less powerful but research about it is still ongoing. However, compared to GA and ACO, the advantages of PSO are its easy implementation and the few parameters to tune.

Acknowledgements The work presented in this paper was done within OSCAR project of "Centre National de Recherche Technologique en Génie Electrique", with the support of ERDF, French Government and Région Nord – Pas de Calais.

References

1. M. Dorigo, V. Maniezzo, and A. Colorni, Ant system: optimization by a colony of cooperating agents, IEEE Trans. Systems Man Cybernet – Part B26 (1), Italy, 1996.
2. Kennedy and R.C. Eberhat, Particle swarm optimization, IEEE International Conference on Neural Networks, Perth, Australia. IEEE Service Center, Piscataway, NJ, 1995.
3. Y. Hou, Y. Wu, L. Lu, and X. Xiong, Generalized ant colony optimization for economic dispatch of power system, Power System Technology, Proceedings PowerCon, vol. 1, pp. 225–229, 2002.
4. P. Venkataraman, Applied Optimization with Matlab® Programming. Wiley, New York, 2001.
5. C. Erbas, S. Cerav-Erbas, and D. Pimentel, Multiobjective optimization and evolutionary algorithms for the application mapping problem in multiprocessor system-on-chip design, IEEE Trans. Evol. Comp. 10(3): 358–374, Jun. 2006.
6. F. Moussouni, S. Brisset, and P. Brochet, Some results on the design of the brushless DC wheel motor using SQP and GA, International Journal of Applied Electromagnetics and Mechanics-IJAEM, vol. 3, 4-2007. http://12ep.univ-lillel.fr/come/benchmark-wheel-motor.htm.
7. D. Joshua, D. Knowles, L. Thiele, and E. Zitzler, A tutorial on the performance assessment of stochastic multi-objective optimizers, TIK-Report No. 214, Computer Engineering and Networks Laboratory, ETH Zurich, February 2006.
8. Y. Colette and P. Siarry, Three new metrics to measure the convergence of meta-heuristics towards the Pareto frontier and the aesthetic of a set of solutions in bi-objective optimization, available online at http://www.sciencedirect.com.
9. S. Brisset and P. Brochet, Analytical model for the optimal design of a brushless DC wheel motor, COMPEL 24(3), 2005.
10. http://l2ep.univ-lille1.fr/come/benchmark-wheel-motor.htm.
11. F. Moussouni, S. Brisset, and P. Brochet, Some results on the design of the brushless DC wheel motor using SQP and GA, Int. J. Appl. Electromagn. Mech.–IJAEM, 3: 4, 2007.
12. M. Dorigo and C. Blum, Ant colony optimization theory: a survey, Theor. Comp. Sci. 344: 243–278, 2005.
13. A. Colorni, M. Dorigo, and M. Maniezzo, Investigation of some properties of an ant algorithm, Proceedings of the Second Conference on the Parallel Solving from Nature, Bruxelles, pp. 509–520, 1992.
14. M. Clerc and J. Kennedy, The particle swarm: explosion, stability, and convergence in a multidimensional complex space, Evol. Comput. IEEE Trans. 6(1): 58–73, Feb 2002.
15. I.C. Trelea, The particle swarm optimization algorithm: convergence analysis and parameter selection, Inf. Proceed. Lett. 85: 317–325, 2003.

Generalized RBF Neural Network and FEM for Material Characterization Through Inverse Analysis

Tarik Hacib, Mohammed Rachid Mekideche, Fouzia Moussouni, Nassira Ferkha, and Stéphane Brisset

Abstract This paper describes a new methodology for using artificial neural networks (ANN) and finite element method (FEM) in an electromagnetic inverse problem (IP) of parameters identification. The approach is used to identify unknown parameters of ferromagnetic materials. The methodology used in this study consists in the simulation of a large number of parameters in a material under test, using the FEM. Both variations in relative magnetic permeability and electric conductivity of the material under test are considered. Then, the obtained results are used to generate a set of vectors for the training of generalized radial basis function neural networks (RBFNN). Finally, the obtained neural network (NN) is used to evaluate a group of new materials, simulated by the FEM, but not belonging to the original dataset. The reached results demonstrate the efficiency of the proposed approach.

1 Introduction

IPs in electromagnetic are usually formulated and solved as optimization problems, so iterative methods are commonly used approaches to solve this kind of problems [1]. These methods involve solving well behaved forward problem in a feedback loop. The numerical models such as FE model are used to represent the forward process. However, iterative methods using the numerical based forward models are computationally expensive. Parameters identification using NNs can be recast as a problem in multidimensional interpolation, which consists of finding the unknown nonlinear relationship between inputs and outputs in a space spanned

T. Hacib, M. R. Mekideche, and N. Ferkha
Laboratoire L.A.M.E.L, Université de Jijel, BP 98 Ouled Aissa Jijel, Algeria
tarik_hacib@mail.univ-jijel.dz

F. Moussouni and S. Brisset
Laboratoire L2EP – Ecole Centrale de Lille, Cité Scientifique – BP 48
F59651 Villeneuve d'Ascq cedex, France

T. Hacib et al.: *Generalized RBF Neural Network and FEM for Material Characterization Through Inverse Analysis*, Studies in Computational Intelligence (SCI) **119**, 11–19 (2008)
www.springerlink.com

by the activation functions associated with the NN nodes [2, 3]. The input space corresponds to the signal generated by sensors and the output corresponds to the electromagnetic parameters such as relative magnetic permeability and electrical conductivity.

In this paper, we propose a new method for the robust identification of electromagnetic properties. The method is based on the FEM and a RBFNN scheme. We present a comparison of the results obtained using the proposed method with those obtained from a multilayer perceptron (MLP). It is shown that the generalized RBFNN is faster both in training as well as identification of parameters.

2 Neural Networks

NNs are connectionist models proposed in an attempt to mimic the function of the human brain. A NN consists of a large number of simple processing elements called neurons [1]. Neurons implement simple functions on their inputs and are massively interconnected by means of weighted interconnections. These weights, estimated by means of a training process, determine the functionality of the NN. The training process uses a training database to determine the network parameters (weights). NNs have been widely used for function approximation and multidimensional interpolation [4]. Given a set of p ordered pairs $(x_i,\ d_i), i = 1,2,\ldots,p$ with $x_i \in R^N$ and $d_i \in R$, the problem of interpolation is to find a function $F : R^N \rightarrow R^1$ that satisfies the interpolation condition

$$F(x_i) = d_i,\ i = 1,\ldots,p. \tag{1}$$

For strict interpolation, the function F is constrained to pass through all the p data points. The definition can be easily extended to the case where the output is M-dimensional. The desired function is then $F : R^N \rightarrow R^M$. In practice, the function F is unknown and must be determined from the given data $(x_i, d_i), i = 1,2,\ldots,\ p$. A typical NN implementation of this problem is a two step process: training, where the NN learns the function F given the training data $\{x_i, d_i\}$, and generalization, where the NN predicts the output for a test input. Networks using two different types of basis functions (BFs) are described in the following sections.

2.1 Radial Basis Function Neural Networks

An RBFNN consists of an input and output layer of nodes and a single hidden layer (Fig. 1). Each node in the hidden layer implements a BF $G(x,x_i)$ and the number of hidden nodes is equal to the number of data points in the training database. The RBFNN approximates the unknown function that maps the input to the output in terms of a BF expansion, with the functions $G(x,x_i)$ as the BFs. The I/O relation for RBFNN is given by

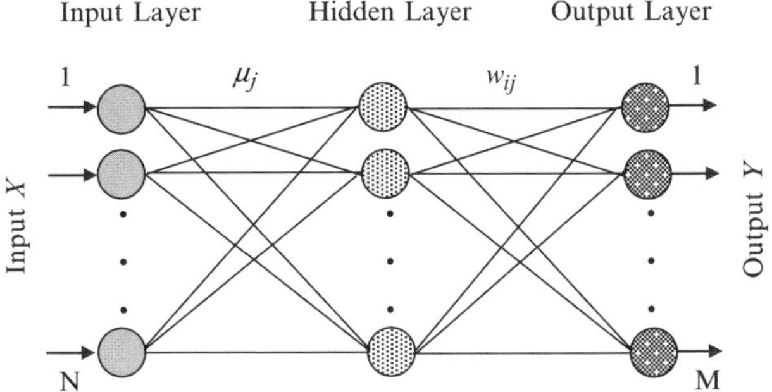

Fig. 1 Three layer feed-forward network

$$y_l = \sum_{j=1}^{N} w_{lj} G(x, x_j),\ l = 1, 2 \ldots, M. \tag{2}$$

Where N is the number of BFs used, $\mathrm{y} = (y_1, y_2, \ldots, y_M)^T$ is the output of the RBFNN, x is the test input, x_j is the center of the BF, and w_{lj} are the expansion coefficients or weights. Each training data sample is selected as the center of a BF. BFs $G(x, x_i)$ that are radially symmetric are called radial BFs. Commonly used radial BFs include the Gaussian and inverse multiquadrics [2].

The network described above is called an exact RBFNN, since each training data point is used as a basis center. The storage costs of an exact RBFNN can be enormous, especially when the training database is large. An alternative to an exact RBFNN is a generalized RBFNN, where the number of BFs is less than the number of training data points. The problem then changes from strict interpolation (in an exact RBFNN) to an approximation, where certain error constraints are to be satisfied. The operation of the generalized RBFNN is summarized in the following steps.

2.2 Center Selection

This is achieved by using either the k-means clustering algorithm [5] or other optimization techniques that select the BF locations by minimizing the error in the approximation. The I/O relation for a generalized RBFNN using Gaussian BFs is given by

$$y_l = \sum_{j=1}^{H} w_{lj} \exp\left(-\frac{\left\|x - c_j\right\|^2}{2\sigma_j^2}\right), \tag{3}$$

where H is the total number of BFs used, c_j is the center of the Gaussian BF, and σ_j is the width of the Gaussian. The NN architecture is then selected by setting the number of input nodes equal to the input dimension, the number of hidden nodes to the number of centers obtained in this step, and the number of output nodes equal to the output dimension.

2.3 Training of Generalised RBF Neural Network

Training of the NN involves determining the weights w_{lj}, in addition to the centers and widths of the BFs. Writing (2) in matrix-vector form as

$$\mathbf{Y} = \mathbf{G}\mathbf{W}, \tag{4}$$

Equation (4) can be solved for $\mathbf{W}$ as:

$$\mathbf{W} = \mathbf{G}^{+}\mathbf{Y}, \tag{5}$$

where $\mathbf{G}^{+}$ is the pseudo-inverse.

2.4 Generalization

In the test phase, the unknown pattern x is mapped using the relation

$$F(x) = \sum_{j=1}^{H} w_{lj}\, exp\left(-\frac{\left\|x - c_j\right\|^2}{2\sigma_j^2}\right). \tag{6}$$

3 Electromagnetic Field Computation

In this study, the magnetic field is calculated using the FEM. This method is based on the A representation of the magnetic field. The calculations are performed in two steps. First, the magnetic field intensity is calculated by solving the equation:

$$rot\left(\frac{1}{\mu} rot(A)\right) + \mathrm{j}\omega\sigma A = J, \tag{7}$$

where μ is the magnetic permeability, σ the electric conductivity and $\boldsymbol{J}$ the electric current density.

Equation (7) is discretized using the Galerkin FEM, which leads to the following algebraic matrix equation:

$$([\boldsymbol{K}] + \mathrm{j}\omega[\boldsymbol{C}])\,[\boldsymbol{A}] = [\boldsymbol{F}]. \tag{8}$$

In the second step, the field solution is used to calculate the magnetic induction $\boldsymbol{B}$. More details about the FE theory can be found in [6].

4 Methodology for Parameters Identification

First of all, an electromagnetic device was idealized to be used as an electromagnetic field exciter (Fig. 2). In this paper, we have considered direct current in the coils. To increase the sensitivity of the electromagnetic device a magnetic core with a high permeability is used and the air gap between the core and the metallic wall is reduced to a minimum. Deviations of the magnetic induction (difference in magnetic induction without and with material under test) at equally stepped points in the external surface of the material under test are taken.

Figure 3 show the steps of the methodology used in this work. Steps 1–4 correspond to the FE analysis.

The simulations were done for a hypothetic metallic wall with 1 mm height and 15 mm width. The material of the metallic wall is 1006 Steel (a magnetic material). The relative magnetic permeability of the core is supposed to be 2,500 and the air gap is 0.1 mm.

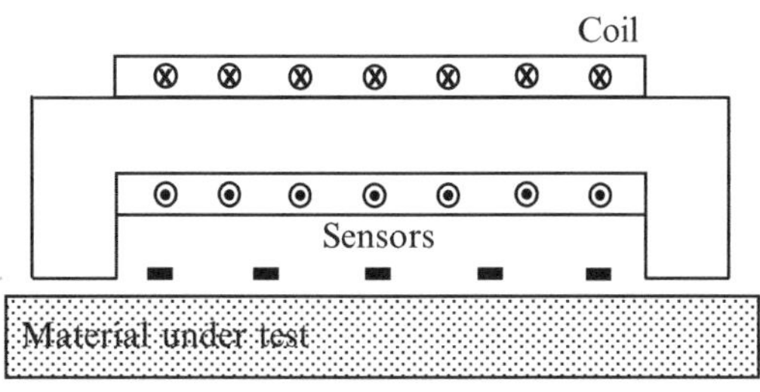

Fig. 2 Arrangement for the measurements

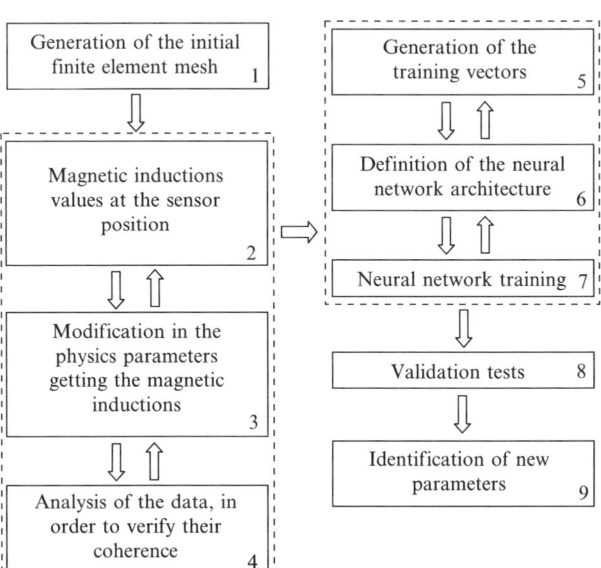

Fig. 3 Flowchart of the used methodology

During the phase of FEs simulations, errors can appear, due to it's massively nature. So, the results of the simulations must be carefully analyzed. This can be done, for instance, plotting in the same graphic the magnetic induction deviations for a set of parameters. Figure 4 shows the magnetic induction deviation at the sensor position for three materials having the same electrical conductivity (10^3 [S m^{-1}]), and relative magnetic permeability ranging from 50 to 300. Figure 5 shows the graphics for a fixed magnetic relative permeability (240), and three different electrical conductivity ranging from 6×10^5 [S m^{-1}] to 10^8 [S m^{-1}].

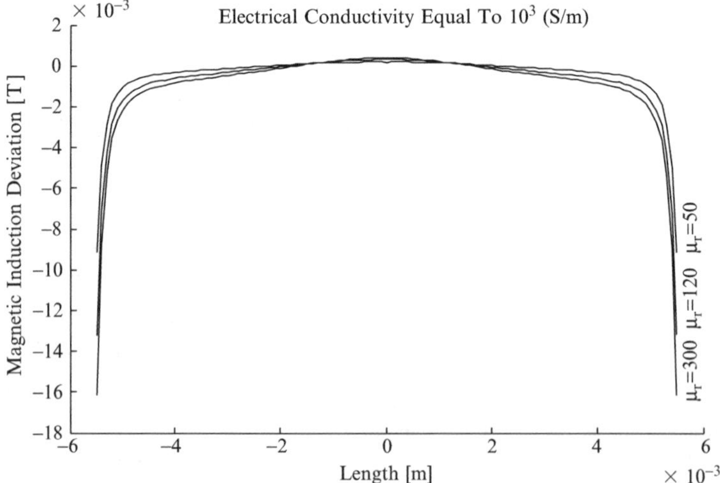

Fig. 4 Magnetic induction deviation for three values of magnetic relative. Permeability and electrical conductivity equal to 10^3 (S m^{-1})

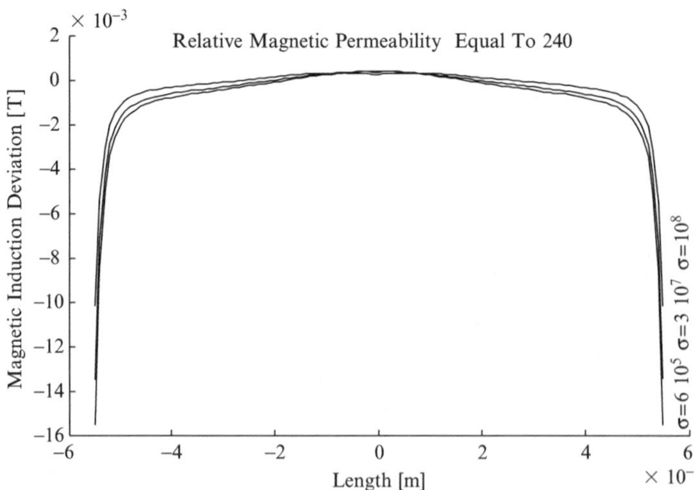

Fig. 5 Magnetic induction deviation for three values of electrical conductivity and magnetic relative permeability equal to 240

5 Formulation of Network Models for Parameters Identifications

In the step 5, we generate the training vectors for NNs. In this work, we generated 300 vectors for NNs training. Each of the vectors consists of 11 input values, which represent the deviation of magnetic induction, and two output values, which represent the relative magnetic permeability and electrical conductivity. Of the 300 vectors, a random sample of 225 cases (75%) was used as training, 75 (25%) for validation.

To show stability of the proposed approach, the measured values, which intrinsically contains errors in the real word, is obtained by adding a random perturbation to the exact inputs values

$$\tilde{In} = In_{exact} + \delta\lambda, \tag{9}$$

where δ is the standard deviation and λ is a random variable taken from a Gaussian distribution.

MLP network architecture considered for this application was a single hidden layer with sigmoid activation function. A back-propagation algorithm based on Levenberg–Marquardt optimization technique [7] was used to train the MLP network for the above data. The MLP architecture had 11 input variables, one hidden layer with 24 hidden nodes and two output nodes. Total number of weights present in the model was 338. The best MLP was obtained at lowest mean square error (MSE) of 10^{-6}. Percentage correct prediction of the MLP model was 95.6%. Generalized RBFNN performed best at 140 centres and Gaussian BFs. MSE using the best centres and Gaussian BFs was 5×10^{-6}. Percentage correct prediction of the generalized RBFNN model was 98.2%. Figure 6 shows the performance of the generalized RBFNN during a training session. Table 1 shows some results for the validation of the network by using 5% of noise ($\delta = 0.05$), for this session.

As we can see, the results obtained in the validation are very close to the expected ones. The worse identification parameter was obtained with MLP network.

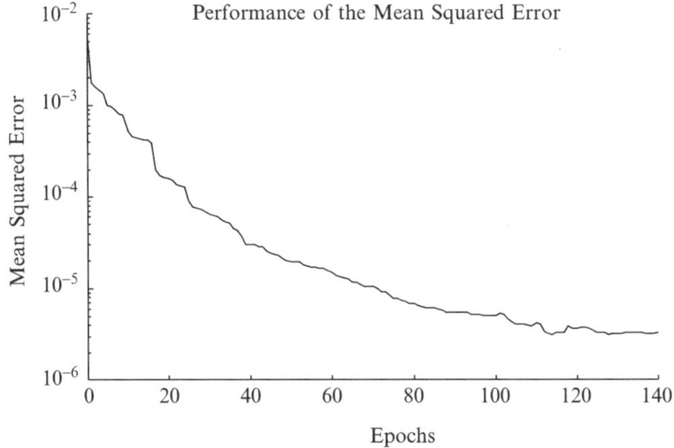

Fig. 6 Performance of the generalized RBFNN during a training session

Table 1 Expected and obtained values during a training session

Relative magnetic permeability, μ_r			Electric conductivity, σ (S m^{-1})		
Expected	Generalized RBF	MLP	Expected	Generalized RBF	MLP
83.34	83.329	83.321	7.020 e006	7.018 e006	7.017 e006
287.55	287.54	287.53	17.99 e003	18.01 e003	18.02 e003
311.34	311.35	311.35	4.500 e002	4.506 e002	4.519 e002
446.21	446.21	446.24	22.50 e006	22.49 e006	22.46 e006
527.07	527.04	527.33	3.010 e005	3.011 e005	3.014 e005

Table 2 Simulation results, for new parameters

Material no.	Relative magnetic permeability, μ_r			Electric conductivity, σ (S m^{-1})		
	Expected	Obtained		Expected	Obtained	
		Generalized RBF	MLP		Generalized RBF	MLP
1	89	89.001	88.933	65.00	65.02	65.03
2	212	211.99	211.98	2.500 e002	2.501 e002	2.502 e002
3	360	360.02	359.95	2.200 e006	2.199 e006	2.201 e006
4	472	472.01	472.03	4.100 e005	4.099 e005	4.101 e005

6 New Parameter Identification

After the NNs training and respective validations, new electromagnetic parameters were simulated by the FEM, for posteriori identification by the networks. Table 2 shows the parameters values of material under test, and the obtained values, by the NNs.

As we can see, the results obtained in the identification of new parameters, obtained by the NNs agree very well with the expected ones.

7 Conclusion

In this paper we presented an investigation on the use of the FEM and Generalized RBFNN for the identification of metallic walls parameters, present in industrial plants. The obtained weights of the network can be embedded in electronic devices in order to identify electromagnetic parameters of real metallic walls. The proposed approach was found to be highly effective in identification of parameters in electromagnetic devices. Comparison of the result indicates that Generalized RBF is trained and identifies the electromagnetic parameters faster than MLP.

References

1. S. R. H. Hoole, Artificial neural networks in the solution of inverse electromagnetic field problems, *IEEE Trans. Magn.*, 29(2), pp. 1931–1934, 1993.
2. S. Haykin, Neural networks: a comprehensive foundation, Englewood Cliffs, NJ: Prentice-Hall, 1999.
3. A. Fanni and A. Montisci, A neural inverse problem approach for optimal design, *IEEE Trans. Magn.*, 39(3), pp. 1305–1308, 2003.
4. P. Ramuhalli, L. Udpa and S.S. Udpa, Finite element neural networks for solving differential equations, *IEEE Trans. Neural Netw.*, 16(6), pp. 1381–1392, 2005.
5. J. A. Hartigan, Clustering Algorithms, John Wiley and Sons, Agnetism, CA: Academic, 2000.
6. P. P. Silvester and R. L. Ferrari, Finite Elements for Electrical Engineers. Cambridge, University Press, 1996.
7. M. T. Hagan and M. Menhaj, Training feed-forward networks with the Levenberg–Marquardt algorithm, *IEEE Trans. Neural Netw.*, 5(6), pp. 989–993, 1994.

Optimization of Complex Lighting Systems in Interiors with the Use of Genetic Algorithm and Elements of Paralleling of the Computation Process

Leszek Kasprzyk, Ryszard Nawrowski, and Andrzej Tomczewski

Abstract The paper presents the application of a genetic algorithm to optimization of the costs related to complex lighting systems. A form of objective function, a group of constraints, and a method of their implementing into the computation algorithm are proposed. For purposes of analysis of the electromagnetic field in visible range the Boundary Elements Method is used. Some elements of the computation algorithm are proposed that may be organized in parallel, as well as a method of communication between the computer issuing an order for the computation process and the cluster-supervising computer. Example optimization calculation is carried out and discussed.

1 Introduction

Industrial objects, athletic areas, cities, communication paths, and millions of households are reckoned as the main sites of numerous receivers of electric energy serving as light sources. Reduction of electric power consumption for lighting purposes, with maintained standard requirements, might result in significant energy savings. This, however, is feasible by optimization of lighting equipment design, the use of indirect component of the luminous flux as an element improving basic standard parameters, and optimization of proper choice and arrangement of light fittings inside the objects. While solving the optimization tasks the economic aspects of the optimization are in many cases neglected. Nevertheless, the to-day state of the knowledge, computer simulation and analysis means, and manufacturing technology

Leszek Kasprzyk, Ryszard Nawrowski, and Andrzej Tomczewski
Poznan University of Technology, Institute of Electrical Engineering and Electronics
60-965 Poznan, ul. Piotrowo 3A, Poland
Leszek.Kasprzyk@put.poznan.pl, Ryszard.Nawrowski@put.poznan.pl
Andrzej.Tomczewski@put.poznan.pl

L. Kasprzyk et al.: *Optimization of Complex Lighting Systems in Interiors with the Use of Genetic Algorithm and Elements of Paralleling of the Computation Process*, Studies in Computational Intelligence (SCI) **119**, 21–29 (2008)
www.springerlink.com

enable designing and sale of the solutions that meet, at the same time, economical and technological criteria of the optimization process.

In many technical tasks the optimal solutions are searched with the use of numerical deterministic or random optimization methods [1, 2]. The optimization calculation performed at personal computers is usually highly time-consuming, lasting even up to tens hours. This results from the need of many times repeated numerical analysis of the electromagnetic field in the visible range. The use of the Boundary Elements Method for an accurate mapping of the geometrical and physical structure (differentiated reflective parameters of the system) enables obtaining good computation accuracy.

2 Optimization of Lighting Systems

In case of the objects illuminated with artificial light sources a conception of a lighting system may be defined, considered as a system composed of light fittings arranged at an Ω area, light sources, and the parts of electric wiring system. The Ω area is open in the case of outside illumination or constrained with an S surface in the case of an inside one. This affects the choice of the method applied in order to calculate the luminous flux distribution and the luminance [3].

Lighting systems of large objects may be of complex features, including K separate sub-areas $(A_1,\ A_2,\ldots A_K)$. The manner of definition of the sub-areas should take into consideration both the geometrical structure of the object and appropriation of its particular parts. Each of the system parts is characterized by a set of lighting parameters (average illuminance of the working plane E_{av}, exploitation illuminance E_m, uniformity $\mathrm{E_{min}/E_{av}}$, colour rendering factor R_a, the value of Unified Glare Rating (UGR), and the type, number M_i (for $i = 1, 2, \ldots, K$) and the manner of arrangement of the light fittings. So-called transient areas are located between the above mentioned sub-areas, in which, first of all, the requirements resulting from physiology of human vision should be considered. Total number of the fittings of a complex lighting system is equal to M, being a sum of the numbers of the fittings located in each of the sub-areas. Light sources of one of the sub-areas my affect illumination of another, in result of direct and indirect light fluxes. Numerical algorithm of the analysis of the electromagnetic field in the visible range for the case of an interior lighting system shown in Chap. 3 of the present paper. The optimization task is defined as aimed at searching such a point x' for which the function $J(\mathbf{x})$ $(\mathbf{x} = \{\mathrm{x}_1,\ \mathrm{x}_2,\ldots,\ \mathrm{x}_n\}$ being a vector of independent variables), called an objective function, takes its optimal value (maximal or minimal, according to the job type). The searching process undergoes in a set X of admissible solutions that may be of continuous or discrete character. Values of the independent variables may be constrained, for example, by technological or standard requirements, or other needs. In order to solve the task it must be previously converted into a problem free of constraints, with the us of one of known methods. A method of penalty function is frequently used for this purpose. It requires defining a modified form of the objective function, including a so-called penalty term related to transgressing a group of constraints of equality (N_R) or inequality (N_N) type [1,2]:

$$\hat{J}(\mathbf{x}) = J(\mathbf{x}) + \sum_{j=1}^{N} F_j(\mathbf{x}) \tag{1}$$

where $\hat{J}(\mathbf{x})$ – a modified objective function, F_j – the penalty functions for $j = 1,2,..,N$ $(N = N_R + N_N)$. A form of the penalty function F_j may differ according to selected particular method [2].

An optimization task requires definition of a criterion of solution quality and, in consequence, a form of the objective function used for purpose of the solving procedure. It seems that a main assessment criterion is related to financial effects, e.g. power or materials savings, reduction of maintenance expenses, etc. Therefore, the objective function selected for purpose of optimization of complex lighting systems is of economic character and defines total cost of the lighting system (i.e. a sum of investment $J^0(\mathbf{x})$ and operation $J^t(\mathbf{x})$ costs). The costs may be written in the form of the sums:

$$J^0(\mathbf{x}) = K_{fit} + K_{src} + K_{fix} + K_{ass} + K_{pro} \text{ and } J^t(\mathbf{x}) = K_{pow} + K_{main} + K_{srcexch} + K_{util}, \tag{2}$$

where: K_{fit} – the cost of lighting fittings, K_{src} – the cost of light sources, K_{fix} – the cost of fixtures, K_{ass} – the cost of assembling (as a per cent part of the investment costs), K_{pro} – the cost of the lighting system project, K_{pow} – the cost of electric power, K_{main} – maintenance cost, $K_{srcexch}$ – the cost of periodic exchange of the light sources, K_{util} – the cost of utilization of the light sources.

From among the parameters determining the structure of the lighting system a group of decisive variables was separated. In each of the sub-areas the following parameters are distinguished: the lighting fitting and reflector types, the kind of the light source ignition system, the parameters related to distribution and number of the fittings (inclusive of their angular positions). In case of K selected sub-areas the length of the vector of independent variables $\mathbf{x}$ (i.e. total number of the decisive variables) amounts to $14K$ and in practice is equal at least to tens. The objective function is proposed in the following form:

$$J(\mathbf{x}) = r \cdot k_{ass} \left[\sum_{i=1}^{K} \sum_{m=1}^{M_i} (C_{fit_{m_i}} + w_{mi} \cdot C_{src_{m_i}} + C_{fix_{m_i}}) \right] + \sum_{i=1}^{K} \sum_{m=1}^{M_i} \left[P_{fit_{m_i}} \cdot T_w \cdot C^*{}_{pow} \cdot Y + C^*{}_{main_{mi}} \cdot Y + w_{mi} \cdot (C^*{}_{utilsrc_{mi}} + C^*{}_{srcexch_{mi}}) \cdot \frac{T_w \cdot Y}{T_{src_{mi}}} \right], \tag{3}$$

where: k_{ass} – a factor of the assembling cost, M – the number of fittings, $_{mi}$ – index of the mth fitting belonging to the ith lighting subsystem ($i = 1,2,\ldots,K$; $m = 1,2,M_i$), P_{fitmi} – power of the mith fitting [kW], T_w – operating time within one calendar year [h], T_{srcmi} – life of the source located in the mith fitting [h], w_{mi} – number of the sources of the mith fitting, C_{pow} – the price of 1 kWh of electric power [€], C_{mainmi} – yearly maintenance cost of the *mi*th fitting [€], C_{fitmi} – price of the *mi*th fitting [€], C_{srcmi}– price of the source in the *mi*th fitting [€], C_{fixmi} – price of the fixtures of the *mi*th fitting [€], $C_{srcexchmi}$ – price of exchange of the sources of the *mi*th fitting, Y – life-time of the lighting system [years], r – the rate of extended reproduction; * – taking into account the inflation factor. Most of the decisive variables occur inexplicitly in the objective function – they affect values of other parameters.

An important element of the optimization process is consideration of the groups of equality and inequality constraints related to the decisive variables, control of standardized parameters, and general design of the object. They include standardizing requirements defined separately for particular sub-areas (maintained illuminance E_m, uniformity E_{min}/E_{av}, and the UGR and R_a factors), the fitting and source types admissible in considered objects, the ranges of source assembling height, etc. In order to consider the above mentioned constraints the algorithm developed for this purpose uses the method of external penalty function. Prevailing part of the constraints imposed on the function (3) is of inequality type.

Considering analysis of the problem and the form of the objective function (3) proposed in order to find optimal costs of complex lighting systems a random method of genetic algorithm has been selected. The network of genetic algorithm operation is of deterministic type, nevertheless, the operations performed for its purpose (inclusive of genetic selection, crossover, and mutation operators) are of random character. Such a concept gives good results even in case of the tasks the deterministic solution [1] of which appears to be difficult. In the considered task of optimization of the costs of lighting systems a modified genetic algorithm is used (Fig. 1) of the following features and parameters: size of a generation – 100 individuals, binary encoding of the decisive variables, haploidal chromosome model, a single chromosome determines an individual as a whole, linear scaling of the adaptation function, thus ensuring suitable competition level in the beginning and in the end of the computation, two-point crossover with probability $p_k = 0.6$, the mutation operator with assumed probability $p_m = 0.005$.

An original part of the modified genetic algorithm consists in the use of tournament selection method and the elements of exclusive strategy (maintaining the best individual in the generation). Code length of particular decisive variables are so selected as to cover appropriate range of their variability and required resolution [2].

In case of the optimization with a genetic algorithm the important role of the individuals of lower values of the adaptation function for the reproduction process should be noticed. An interesting solution to this problem is the penalty functions the value of which depends not only on the amount of the excess but also on the iteration number. This is conducive to smaller penalties in the initial stage of the algorithm (possibly leaving the weaker individuals) and higher ones in the final stage (resulting in higher differentiation of the fitness factor). Taking the tests and the paper [2] into account the square penalty functions accounting for the current iteration number have been used.

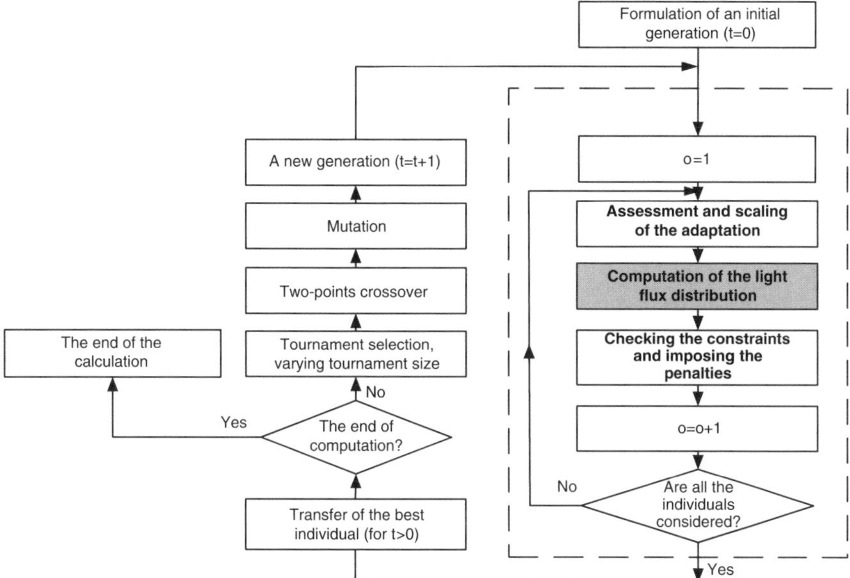

Fig. 1 Diagram of the modified genetic algorithm (t – a generation, o – an individual, the block depicted in gray – a paralleled algorithm of the light field analysis)

3 Analysis of the Electromagnetic Field in Visible Range

The electromagnetic field in visible range (i.e. the light) includes the radiation of the wavelengths from 380 to 780 nm. For such type of the field the analysis of the radiation power propagation becomes interesting, due to the process of energy transfer. In case of the visible radiation the radiation stream is called the luminous flux, while its space distribution (depending on the excitation types and distribution) determines the values of basic lighting parameters [4].

For purposes of numerical computation of the luminous flux distribution in interiors accounting for multiple reflections the Boundary Elements Method was used [2, 5]. It enables accurate determining the lighting parameters in the analyzed object. The S surface delimiting the range is to be divided into N elementary surfaces $dS_i(\text{dla } i = 1,2,\ldots,N)$. Inside the S surface the L light sources are located, considered as the excitations of the electromagnetic field (luminous flux). Additionally, only the artificial light sources are assumed.

The dS_i elements may take various values of the reflection coefficient $\rho \in <0,1>$. Each of the dS_i surfaces of the coefficient $\rho > 0$ reflects a part of the luminous flux Φ' incident to it directly from the sources. In result of multiple reflections between the parts of the S surface the total luminous flux Φ incident to this surface equals to the sum of the direct Φ' and indirect Φ'' components.

The quotient of the luminous flux reflected from the kth element of the S surface and incident on the ith element (Φ_{ki}) by the flow emitted into the entire space from the kth element (Φ_k) is called the usage factor or coupling factor (f_{ki}) of the ith and kth elements of the S surface. The values of the usage factors may be also determined on the grounds of geometric parameters of the system subject to the analysis [3]. Consideration of usage factor enables determining the total luminous flux incident on the ith element coming from all the other elements:

$$\Phi_i = \Phi_i' + \sum_{k=1}^{N} \rho_k f_{ki} \Phi_k \text{ for } (k \neq i). \tag{4}$$

Equation (4) for the index $i = 1, 2, \ldots, N$ may be formulated in the form of a system of linear equations with unknown total luminous fluxes at elementary surfaces dS_i. Once the luminous fluxes are arranged and the condition of equality of direct flows Φ'_i $(i = 1, 2, \ldots, \mathrm{N})$ and luminous fluxes of all L light sources included inside the considered interior Φ_{src} activated, the task may be written in a matrix form, thus enabling determining total flows at the elementary surfaces dS_i:

$$\begin{bmatrix} 1+\rho_1 \frac{S_{10}}{S_1} f_{1N} & -\frac{S_{20}}{S_2}(\rho_2 f_{21} - \rho_2 f_{2N}) & .. & -\rho_N \frac{S_{N0}}{S_N} f_{N1} - 1 \\ -\frac{S_{10}}{S_1}(\rho_1 f_{12} - \rho_1 f_{1N}) & 1+\rho_2 \frac{S_{20}}{S_2} f_{2N} & .. & -\rho_N \frac{S_{N0}}{S_N} f_{N2} - 1 \\ \cdots & \cdots & .. & \cdots \\ 1-\rho_1 \frac{S_{10}}{S_1} & 1-\rho_2 \frac{S_{20}}{S_2} & .. & 1-\rho_N \frac{S_{N0}}{S_N} \end{bmatrix} \begin{bmatrix} \Phi_1 \\ \Phi_2 \\ \cdots \\ \Phi_N \end{bmatrix} = \begin{bmatrix} \Phi_1' - \Phi_N' \\ \Phi_2' - \Phi_N' \\ \cdots \\ \Phi_{zr} \end{bmatrix}, \tag{5}$$

where: i, k – indexes of surface elements; ρ_i – reflection coefficient of the ith surface element; f_{ki}– the usage factor of the ith and kth element dS_i, Φ_{scr} – the total luminous flux of the light sources of the analyzed object, S_{kO} – the surface closing the kth concave element, S_i – total surface area of the kth concave element.

Solution of the system (5) preceded by determination of the usage factors enables determining the distribution of the luminous flux in the object. This, in consequence, allows for finding a set of the lighting parameters (i.e. the constraints of the optimization process). In order to achieve the results of proper accuracy the S surface should be precisely divided, with all additional elements of the Ω space.

4 Computation Example

Example optimization calculation was carried out for an industrial object composed of three sub-areas of various illumination requirements: total length 100 m (50 m, 30 m, 20 m), width 30 m and the height varying from 5 to 7 m (Fig. 2). Boundary sub-areas (A_1 i A_3) are production bays of electrical industry, the middle one is a material store. According to the European Standard EN 12464-1:2002 the basic standard requirements are formulated for the above mentioned areas and operations performed therein. Location of the separated sub-areas of the considered lighting system, its maximal height H, the levels of calculation planes H_r and illumination parameters defined on the grounds of the standard are marked on the object draft (Fig. 2).

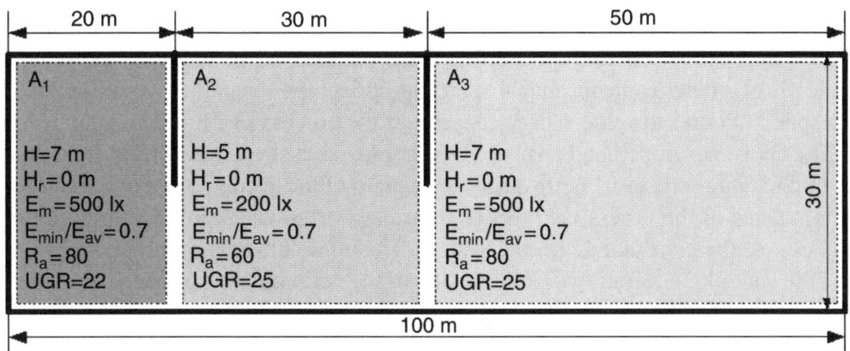

Fig. 2 Geometry and required lighting parameters in the sub-areas of the lighting system subject to the optimization

The optimal solution was found for a group of the fittings and light sources (meeting the required conditions related, for example, to the fitting type, source type, and unit power of a lighting point, etc.) of various values of the luminous fluxes and efficiency factors, produced by leading European manufacturers. It was assumed that life-time of a lighting system amounts to 10 years, with the discount rate equal to 4%.

The most time-consuming part of the genetic algorithm shown in Fig. 1 is many times repeated determination of luminous flux distribution (the block marked in gray). This provides a basis for determination of lighting parameters of particular sub-areas, necessary for the control of basic constraints group – i.e. normative constraints.

In case of optimization of complex lighting systems one of two versions of paralleling of the computation algorithm elements may be applied: i.e. parallel analysis of an individual (each of the processors performs calculation of luminous flux distribution coming from a single source) or parallel analysis of a generation (each of the processors performs calculation of luminous flux of all the sources for a single individual).

In the research presented in earlier works [3, 5] the authors applied the first of the algorithm paralleling methods for computation of complex lighting systems. Taking into account linear character of the system, computation of the direct flow distribution caused by all the excitations was arranged in parallel. Evident reduction of numerical analysis time was additionally achieved in result of dissipation of the computation process of the coefficients of linear equations system among many processors. Similar dissipation was introduced in computation of the usage factors f_{ik} (numerical calculation of many double integrals), total area values and the penetration areas of the concave elements, and solving procedure of a large system of linear equations [5].

The algorithm together with the homogeneous machine (10 Pentium P4 2.6 GHz computers) enable attaining the capacity of paralleling the luminous field determination from 80 to 95%, that is equivalent to significant shortening of numerical analysis time. High capacity of the calculation process results from specific features of the computation task and transferring small numbers of small messages. In case of the use of a heterogeneous machine an appropriate job scheduling algorithm must be applied in order to achieve high capacity of the process [3, 5].

The above developed and characterized computer system served for purposes of the optimization process. Figure 3 shows a pattern of the average (Avg) and minimal (Min) values of the fitness factor of the optimization process aimed at minimizing total cost of the considered lighting system. The minimal value was found between the 20th and 40th generations. Continuation of the iterations was no more conducive to further improving of the optimization point, resulting only in longer duration of the computation process.

The optimization calculation allowed for defining the types, number, and the manner of arranging the lighting fittings in the sub-areas ensuring minimal total cost of the lighting system. Table 1 specifies the results of optimal and other technically correct solutions that satisfy the standard requirements. For the non-optimal solutions presented in the Table 1 (i.e. technically correct and satisfying the standards requirements and constraints) the per cent increase in total cost is determined with regard to the optimal solution. "The other solutions" include the admissible ones the lighting parameters of which exceed the levels determined for the optimal solution no more than by 5%.

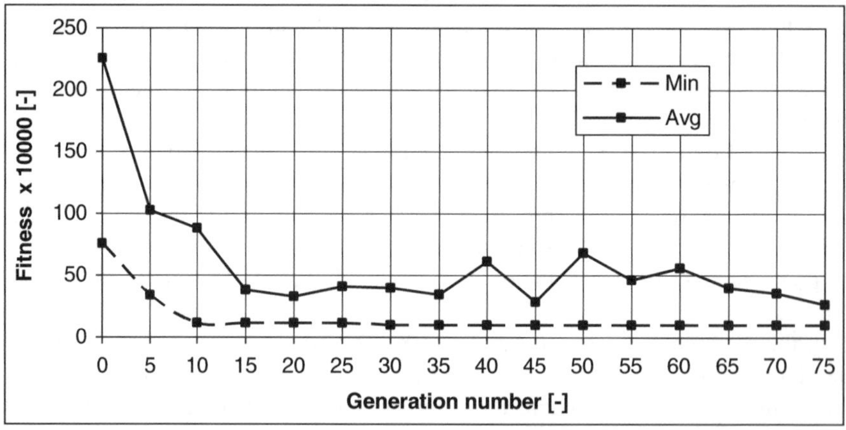

Fig. 3 The pattern of the average (Avg) and minimal (Min) values of the fitness factor of the optimization process for the case of the example lighting system

Table 1 Comparison of total costs of the lighting system of an optimal solution and the others ones satisfying standard requirements and constraints

Item	Type of the solution	Total cost [€]	Cost increase [per cent]
1	Optimal	68,020.2	–
2	Non-optimal 1	77,981.9	14.6
3	Non-optimal 2	79,784.8	17.3

5 Conclusions

The optimal solutions found in the presented example are distinguished by reduction of total expenses of the lighting system down to 20%, as compared to the others options meeting the normative requirements. In case of large objects this may result in considerable financial advantages.

Optimization of complex lighting systems is a task in which a genetic algorithm is doing very well. This is due to large number of decisive variables in the objective function defined for this purpose and possible occurrence of many local extreme points. The use of a tournament selection together with exclusive strategy, a two-points crossover, and scaling of the adaptation function enable obtaining the optimization results usually after a cycle of tens generations.

The algorithm used for the light field analysis (the Boundary Elements Method) allows for accurate defining the values of lighting parameters required by appropriate standards in selected sub-areas. They make an important part of the whole optimization process, as they belong to one of the constraint groups and for each individual a full computation cycle of the luminous flux distribution must be carried out.

Computation complexity of the task is so elevated that application of the methods allowing for shortening total computation time becomes important. In the computation system developed by the authors it was achieved by the use of a paralleled algorithm of determining the luminous flux distribution. Its high efficiency reaches even 90%. In result of the paralleling procedure an 8.8 times reduction of the average total duration of the optimization computation is observed in the considered example, as compared to duration of the analysis using a sequential algorithm (performed on a single processor).

References

1. M. Corcione, L. Fontana, Optimal design of outdoor lighting systems by genetic algorithms, Lighting Res. Technol., 35(3): 261–280, 2003
2. Z. Michalkiewicz, Genetic Algorithms + Data Structures = Evolution Programs. Springer, Berlin Heidelberg New York, 1992
3. R. Nawrowski, The use of multiple reflection method for calculation of luminous flux in interiors, J. Light Vis. Environ., 24(2): 44–48, 2000
4. P. Flesch, Light and light sources, High-Intensity Discharge Lamps. Springer, Berlin Heidelberg New York, 2006
5. L. Kasprzyk, R. Nawrowski, A. Tomczewski, Application of a Parallel Virtual Machine for the Analysis of a Luminous Field, LNCS, Vol. 2474. Springer-Verlag, Berlin Heidelberg New York, pp. 122–129, 2002

Transformer Monitoring System Taking Advantage of Hybrid Wavelet Fourier Transform

Piotr Lipinski, Dariusz Puchala, Agnieszka Wosiak, and Liliana Byczkowska-Lipinska

Abstract In this paper, the power transformer monitoring system (PTMS), which uses expert database to detect failures and hybrid wavelet-fourier transform (HWFT) to compress data is introduced. The HWFT takes advantage of both: discrete wavelet transform (DWT) and discrete cosine transform (DCT), to compress data. It has been proved that HWFT outperforms DCT and DWT, which are used in JPEG and JPEG2000 compression standards, in terms of compression ratio while preserves the same accuracy of decompressed signal.

1 Introduction

The concurrence on industrial market exacts constant reduction in operation costs of power transformer. New power transformers as well as those which has already been installed in transmission lines should be cost-effective and reliable. This can be achieved through constant system assessment, which leads to optimized power transformer exploitation. The assessment can be made basing on series of measurements, which are strongly related to each other, and experts knowledge. Similar approaches have become popular for recent 20 years, however they do not have the expert knowledge database implemented into the system [1]. Power transformer monitoring systems (PTMS) are widely used in Japan [2], USA [3], Germany [4–8], Switzerland [3], France [8], Swiss and Canada [6]. In all those systems several hundreds of parameters are captured at the frequency of about 1 Hz. As a result, throughout

Piotr Lipinski
Technical University of Lodz, Division of Computer Networks, ul. Stefanowskiego 18/22, Poland
piter@amuz.lodz.pl

Dariusz Puchala, Agnieszka Wosiak, and Liliana Byczkowska-Lipinska
Technical University of Lodz, Institute of Computer Science, ul. Wolczanska 215, Poland
dpuchala@ics.p.lodz.pl, awosiak@ics.p.lodz.pl, lilip@ics.p.lodz.pl

P. Lipinski et al.: *Transformer Monitoring System Taking Advantage of Hybrid Wavelet Fourier Transform*, Studies in Computational Intelligence (SCI) **119**, 31–36 (2008)
www.springerlink.com

years the databases become very large and standard compression techniques do not yield sufficient results. Therefore, advanced data compression techniques must be developed to compress data in PTMS.

2 Transformer Monitoring System Architecture

The PTMS which uses expert database to detect failures and hybrid wavelet-fourier transform (HWFT) to compress data has been developed at the Faculty of Technical Physics, Computer Science and Applied Mathematics in cooperation with Institute of Power Engineering. It captures 196 signals from sensors and actuators and saves them in a database. According to the requirements of Institute of Power Engineering the signals from sensors and actuators are captured in remote test station and transmitted through fiber optic connection to a system server, from which data are transmitted to data acquisition system. The system architecture allows to capture data from several remote systems simultaneously and assures high safety standards. Data acquisition system has been implemented in Microsoft SQL server, while the expert knowledge together with the Monitoring System Console (MSC), the data assessment, the system analysis, the failure detection and the decision modules has been implemented in Microsoft Visual Studio. Net and can be executed on a remote computer. Experts knowledge is represented as a set of rules, which describe the state of the power transformer basing on input signal values. The choice of appropriate fitness evaluation rule, failure localization rule and suggested procedure is made based on suitable decision tree. Final information transmitted to MSC includes failure localization, reason and suggested procedure. This helps the system operators to make the appropriate decisions fast. System operators can also analyze the input data through MSC in form of diagrams, which is very comfortable, because they can access the parameter values in real time as well as analyze historical data. The number of parameters, and as a result the number of samples, which are stored in database is large (about 2 GB currently) and increases 1 GB per year per transformer. This generates high costs of storage and data transmission (to analyze data remotely). Therefore, to reduce those costs we have developed advanced compression algorithms. Furthermore, data which are stored in database are often plotted in form of graphs, which cover long time periods. As a result, large number of samples are first selected from database, next transmitted to remote MCS and plotted on very small area, which is ineffective. To solve this problem, we have developed advanced compression techniques including progressive signal decompression (Fig. 1).

3 Data Compression in Power Transformer Monitoring System

The data compression system has been developed under the following assumptions/constraints:

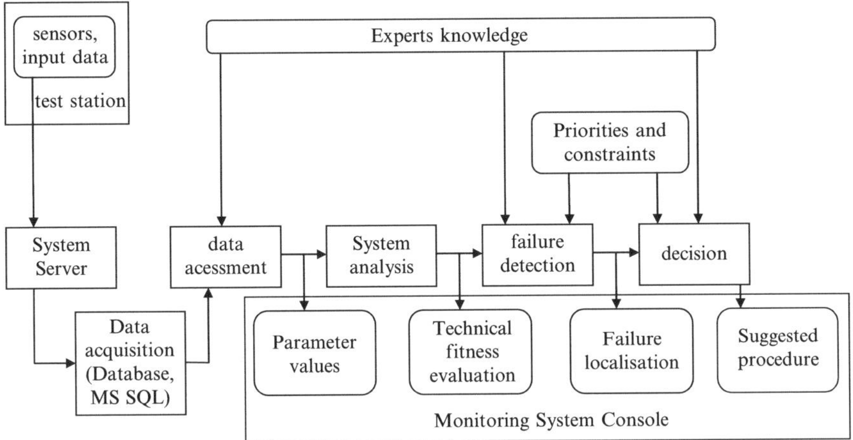

Fig. 1 Logic structure of the power transformer monitoring system

- Compression algorithm is lossless, however little losses are accepted
- Progressive decompression can be performed
- Data are transmitted to MSC in compressed form
- All historical data must be stored in database
- All data must be stored in one database
- Input signal is the sum of low frequency stationary signal and high frequency non-stationary signal.

Data can be compressed in databases in multiple ways: before inserting samples it into database, tables can be compressed using internal database compression tools, database files can be compressed [9]. In the PTMS described here, input samples are grouped into blocks, which are compressed and inserted into database. The compression is performed in three steps. First, the input data are divided into blocks (e.g. 1 year). Second they are transformed using HWFT. Third they are compressed using lossless compression algorithm. This leads to high compression ratio and allows to decompress data progressively. It has been proved in Sect. 5 that HWFT outperforms up to 15% known transforms such as discrete cosine transform (DCT) and discrete wavelet transform (DWT), which are used in JPEG and JPEG2000 standards [10], while preserves the same accuracy of decompressed signal. Lossless compression, which is performed after HWFT can be in general performed using two techniques: dictionary and statistical [11]. The research which has been carried out proved that signals which are stored in PTMS can be compressed with up to 50% higher ratio using dictionary techniques.

In next section the HWFT, used in PTMS data compression algorithm is described in detail.

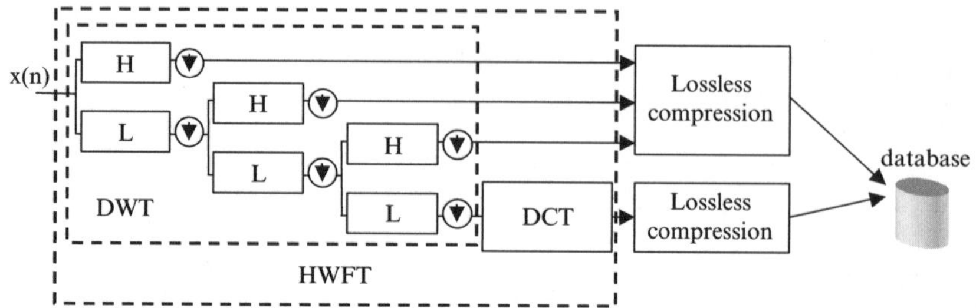

Fig. 2 The compression scheme including HWFT algorithm

4 Compression Scheme Taking Advantage of Hybrid Wavelet-Fourier Transform

In the considered system, signals subjected to data compression display steady (i.e. low oscillatory) as well as non-steady (i.e. highly oscillatory) characteristics. Hence in order to obtain good results in reduction of data volumes it is advantageous to exploit two different transforms. For steady and low oscillatory signals DCT yields compact spectral data representation while non-steady and highly oscillatory signals find good representatives in DWT domain. Therefore, in this paper the authors proposed an HWFT that uses the experimentally chosen, optimal operational scope for DCT and DWT. In DWT Haar mother wavelet [12] has been chosen as it displays the best correlation coefficient with compressed signal of all Daubechies mother wavelets [13].

The HWFT structure is shown in Fig. 2. First, we calculate three steps of DWT, using Mallat's algorithm [14]. The outputs of high-pass filters (H) are downsampled and compressed using lossless compression algorithm (see Fig. 2). Second the output from the third low-pass filter (L) is transformed using DCT and also compressed using lossless compression algorithm. In both abovementioned cases the LZW compression algorithm has been chosen to perform lossless compression as it has displayed the best compression ratio results of all tested lossless compression algorithms (LZW, Huffman, RLE) [15, 16]. All compressed data are stored in one table in the database.

The analogous structure to the one given in Fig. 2, is used to perform decompression, but compressed data are transformed in reverse direction. As a result, a decompressed signal can display different accuracy depending on the required scale which is used to plot the decompressed signal.

5 Experimental Results

In this section we compare the compression results of three sample signals taken from the transformer monitoring system: current, power and temperature. Here we compare the compression scheme described in previous section, taking advantage

Table 1 Comparison of compression ratio, mse, psnr of three sample signals from the transformer monitoring system

Input sequence	Algorithm	Compression ratio (%)	mse	psnr
Current	DWT	19	35	81
	DCT	22	183	74
	HWFT	16	46	80
Power	DWT	20	36	81
	DCT	25	152	75
	HWFT	16	40	80
Temperature	DWT	19	37	81
	DCT	21	165	74
	HWFT	15	48	80

of HWFT, with well known compression algorithms, which use DCT (JPEG) and DWT (JPEG2000) in terms of compression ratio, MSE and SNR [6]. We perform lossless compression, however we allow signal distorsion at the level of about SNR = 90 dB, which is caused by multiplication accuracy constraints. The comparison of three sample signals is shown in Table 1.

6 Conclusions

In this paper, the PTMS, which uses expert database to detect failures and HWFT to compress data has been introduced. The main novelties of the system are: the HWFT algorithm and the experts knowledge database which are embedded into the monitoring system. The use of HWFT yields better compression ratio up to 15% than well known compression algorithms while the expert database improves power transformer fitness assessment, transformer data analysis and failure detection. The experimental result which has been presented here has proved that HWFT outperforms well known compression algorithms based on DCT (JPEG) and DWT (JPEG2000) algorithms used separately, in terms of compression ratio, while preserves the same accuracy of decompressed signal, when compressed signals are composed of two separate signals: low frequency stationary signal and high frequency nonstationary signal. The example given here proves that hybrid transforms give promising results and should be developed and improved in future.

References

1. Provanzana J.H., Gattens P.R., Hagman W.H., Moore H.R., Harley J.W., Triner J.E.: Transformer Condition Monitoring – Realizing an Integrated Adaptive Analysis System, CIGRE 1992, Rep. No. 12–105

2. Kawamura T., Fushimi Y., Shimano T., Amano N., Ebisawa Y., Hosokawa N.: Improvement in Maintenance and Inspection and Pursuit of Economical Effectiveness of Transformers in Japan, CIGRE 2002, Rep No. 12–107
3. Boss P., Lorin P., Viscardi A., Harley J.W., Isecke J.: Economical aspects and practical experiences of power transformer on-line monitoring, CIGRE 2000, Rep. No. 12–202
4. Aschwanden T., Hässig M., Fuhr J., Lorin P., Der Houhanessian V., Zaengl W., Schenk A., Zweiacker P., Piras A., Dutoid J.: Development and Application of New Condition Assessment Methods for Power Transformers, CIGRE 1998, Rep. No. 12–207
5. Liebfried T., Knorr W., Viereck K., Dohnal D., Kosmata A., Sundermann U., Breitenbauch B.: On-line Monitoring of Power Transformers – Trends, New Developments and First Experiences, CIGRE 1998, Rep. No. 12–211
6. Bengtsson T., Kols H., Foata M., Léonard F.: Monitoring Tap Changer Operations, CIGRE 1998, Rep. No. 12–209
7. Poittevin J., Tenbohlen S., Uhde D., Sundermann U., Borsi H., Werle P., Matthes H.: Enhanced Diagnosis of Power Transformers Using On- and Off-Line Methods: Results, Examples and Future trends, CIGRE 2000, Rep No. 12–204
8. Tenbohlen S., Stirl T., Stach M., Breitenbauch B., Huber R., Bastos G., Baldauf J., Mayer P.: Experienced-Based Evaluation of Economic Benefits of On-line Monitoring Systems for Power Transformers. CIGRE 2002, Rep. No. 12–110
9. Bassiouni, M.A.: Data compression in scientific and statistical databases. IEEE Trans. Software Eng. SE 11(10): 1047–1058 (October 1985)
10. Walker J.S.: Fast Fourier Transforms. CRC Press, West Palm Beach, FL, 2nd edition, 1996
11. Burrows M., Wheeler D.J.: A block-sorting lossless data compression algorithm. Digital Systems Res. Ctr., Palo Alto, CA, 1994
12. Haar A., Zur Theorie der orthogonalen Funktionen-Systeme, Math. Ann., 69, pp. 331–371
13. Daubechies I., Ten Lectures on Wavelets, SIAM, Philadelphia, Pennsylvania, 1992
14. Mallat S., A theory for multiresolution signal decomposition: The wavelet representation. IEEE Trans. Pattern Anal. Machine Intell., vol. 11, pp. 674–693, July 1989
15. Skarbek W., Mulitmedia Algorytmy i Standardy Kompresji, Akademicka Oficyna Wydawnicza, Warszawa 1998
16. Huffman D.A., A method for the construction of minimum redundancy codes, Proc. IRE, no. 40, pp. 1098–1101, Sept. 1952

SALHE-EA: A New Evolutionary Algorithm for Multi-Objective Optimization of Electromagnetic Devices

Emonuele Dilettoso, Santi Agatino Rizzo, and Nunzio Salerno

Abstract This paper presents a new evolutionary algorithm (EA) for the optimization of electromagnetic devices called SALHE-EA (self-adaptive low-high evaluations – evolutionary algorithm). Its main aspects are identification of the optima of the objective function and evaluation of their sensitivity. Moreover, SALHE-EA works well if combined with the deterministic pattern search (PS) method, forming a good hybrid method. It performs well in the design of electromagnetic devices when the optimization of a multimodal function is required.

1 Introduction

In industrial applications optimized design is often problematic because of the simultaneous occurrence of many conflicting targets. There are different methods to solve this kind of problem [1–3], such as finding multiple Pareto-Optimal solutions or optimizing a single multi-objective function obtained by a weighted sum of targets.

Whichever method one adopts, real-world optimization problems often exhibit multiple optima: i.e. it is necessary to find a set of Pareto optimal solutions or several local optima of a single multimodal function. In the paper we deal with the latter case, in which deterministic methods do not perform well, while standard stochastic methods tend to find a single global optimum. To explore a multimodal function correctly, the EA must maintain population diversity: several methods have been developed to adapt standard EAs for multimodal function optimization, such as the niching genetic algorithm (NGA) [4–6]. Unfortunately, they can be extremely expensive, because the exploitation of different peaks of the search space requires a

E. Dilettoso, S.A. Rizzo, and N. Salerno
Dipartimento di Ingegneria Elettrica, Elettronica e dei Sistemi (DIEES), Università di Catania
Viale A. Doria, 6, I-95125 Catania, Italy
alfo@diees.unict.it

E. Dilettoso et al.: *SALHE-EA: A New Evolutionary Algorithm for Multi-Objective Optimization of Electromagnetic Devices*, Studies in Computational Intelligence (SCI) **119**, 37–45 (2008)
www.springerlink.com

great number of evaluations of the objective function. This is especially harmful in electromagnetic problems when each estimation of the objective function calls for a numerical solution by means of the finite element method (FEM) [7]. Better results could be obtained by performing the recognition of subpopulations by means of NGAs, and then identifying the best point in every niche by means of a deterministic zero-order method. In practice this approach is nontrivial: the identification of niches and the attribution of each individual (a point in the search space) to its niche is very hard due to the unknown behaviour of the objective function. In order to give good results, they require a crucial a priori specification of a dissimilarity measure, corresponding to the "niche radius".

In the new EA proposed in this paper, a self-adaptive mechanism is implemented in order to preserve diversity in the population. The SALHE algorithm correctly identifies different local optima and estimates their sensitivity to variations in the design parameters.

2 The SALHE Evolutionary Algorithm

Without loss of generality in the following optimization will refer to maximization.

The proposed evolutionary algorithm is a coupled stochastic-deterministic optimization algorithm. The stochastic section consists of five fundamental steps: (1) selection; (2) mutation; (3) elimination of useless individuals; (4) identification of new hypothetical maxima (optima) and new hypothetical minima (they are "hypothetical" because they could differ from the true maxima or minima); (5) evaluation of niche radii. These steps are performed for a fixed number of times, n_g, at the end of which the doublets are deleted. At the end of the stochastic section, a deterministic method, pattern search (PS) [8], is applied to the hypothetical maxima in order to improve their value and to delete doublets.

The pseudo-code of SALHE-EA is shown in Fig. 1. In the following section the steps of the algorithm are described in detail.

- begin
    - generates random population
    - do n_g times
        - evaluation of fitness functions, f_H and f_L, for each individual in the population
        - selection
        - mutation
        - elimination of useless individuals
        - identification of new hypothetical maxima and new hypothetical minima
        - updating of niche radii
    - end do
    - deletion of doublets
    - starting of PS algorithm from each hypothetical maxima
    - updating of niche radii
    - deletion of doublets
- end

Fig. 1 Pseudo-code of SALHE evolutionary algorithm

2.1 Niche Radius and Sensitivity

Indicating a point of maximum with the vector of design parameters $\mathbf{x_{jM}}$ and its closest point of minimum with $\mathbf{x_{jm}}$, we define the niche radius of the *j*th maximum as the Euclidean distance:

$$r_j = \left\| \mathbf{x}_{jM} - \mathbf{x}_{jm} \right\| \tag{1}$$

We consider the niche radius (1) as the sensitivity of the maximum j to parameter variation, and its niche as the region of the search space bounded by the hypersphere with centre $\mathbf{x_{jM}}$ and radius r_j.

2.2 Selection

To search for the maxima, a "modified fitness function" $f_{i,H}$ is evaluated for the *i*th individual:

$$f_{i,H} = (f_i)^{a_H} \left(\frac{d_i}{d_{\max}} \right)^b (\rho_{i,H})^c \tag{2}$$

in this way the probability that the individual i is selected for reproduction depends on three factors:

1. The value of the scalar objective function f_i, i.e. the weighted sum of the targets, calculated for the i-th individual (we assume that f_i is positive);
2. A penalization factor d_i/d_{max}, for the individuals present in crowded zones, where d_i is the average Euclidean distance between the n_c individuals closest to the individual i:

$$d_i = \frac{\sum_{k=1}^{n_c} d_{ik}}{n_c} \tag{3}$$

   (d_{ik} being the Euclidean distance between individual k and individual i) and d_{max} is the largest d_i;
3. A penalization factor $\rho_{i,H}$, for the individuals present in discovered niches, defined as:

$$\rho_{i,H} = \prod_{j=1}^{m_H} \frac{d_{ij}}{r_j} \tag{4}$$

   where m_H is the number of niches, with a hypothetical maximum the individual *i* belongs to, d_{ij} is the Euclidean distance between the individual *i* and the maximum in niche *j* and r_j is the niche's radius.

In (2), a_H, b and c are scaling powers greater than or equal to one.

To obtain niche radii by means of (1) it is essential to identify the minima. To search for the minima, a "modified fitness function" $f_{i,L}$ is evaluated for the *i*th individual:

$$f_{i,L} = (f_i)^{a_L} \left(\frac{d_i}{d_{\max}} \right)^b (\rho_{i,L})^c \tag{5}$$

where

$$\rho_{i,L} = \prod_{w=1}^{m_L} \frac{d_{iw}}{r_w} \tag{6}$$

in which m_L is the number of niches with a hypothetical minimum the individual i belongs to, d_{iw} is the Euclidean distance between the individual i and the minimum in niche w and r_w is the niche's radius, computed as the Euclidian distance between the minimum in niche w and its closest maximum. The scaling power a_L is less than or equal to minus one.

After evaluation of the modified fitness functions (2) and (5) for all individuals in the population, two individuals are selected for reproduction by means of the well-known "roulette wheel selection" operator [6]: two wheels are created, one with slots sized according to f_H and the other with slots proportional to f_L.

Intuitively, since we are interested in finding maxima, the search for minima may appear to be useless. It is, however, fundamental to identify the niches, as it restricts the search space (and thus reduces the overall computational cost), gives us an automatic estimate of the sensitivity of solutions and avoids the a priori specification of a fixed niche radius by means of a heuristic criterion, which is the principal problem of niching GAs [5].

2.3 Mutation

A selected individual $\mathbf{x_{par}}$, termed "parent", generates an offspring $\mathbf{x_{off}}$ by means of the mutation operator, i.e. adding a random vector to $\mathbf{x_{par}}$. The value of the *k*th parameter is computed as:

$$x_{k.off} = x_{k,par} + sign(r_1 - 0.5) r_2 \; step_k, \tag{7}$$

where r_1 and r_2 are [0,1] random numbers, and $step_k$ is computed as follows:

$$step_k = \frac{\Delta_k}{N} \sqrt{\frac{N}{n_k}} \left(1 + \frac{1 - \sqrt{n_k}}{\sqrt{n_k}} \frac{f_{par} - f_{\min}}{f_{\max} - f_{\min}} \right) (2 - \rho_{par,H})^e, \tag{8}$$

where *N* is the number of individuals in the population, Δ_k is the domain of the *k*th design parameter, f_{par}, f_{min} and f_{max} are the values of the scalar objective function for the parent, the worst and the best individual, respectively. Considering *N* regular subdivisions of Δ_k, n_k is the number of intervals in which at least one individual is present. Parameter e is a scaling power greater than or equal to one. In (8), $\rho_{par,H}$ is substituted by $\rho_{par,L}$ for the individuals selected by the roulette with slots sized according to f_L.

Applying the mutation operator twice, each selected individual generates two offspring. After reproduction, the individual with the intermediate fitness value among a parent and its two offspring is chosen and eliminated because it is less able to identify local optima or local minima, as shown in Fig. 2.

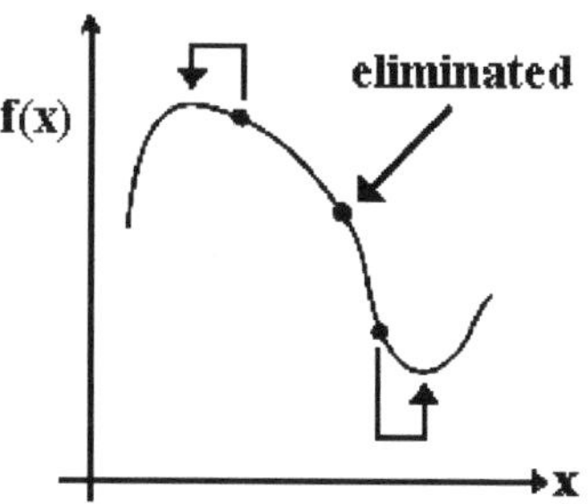

Fig. 2 Elimination of an individual with intermediate fitness value in a mono-dimensional search space

Equation (8) controls the balance between *exploration* of unknown zones of the search space and *exploitation* of attractive areas of the search space. If the number N of individuals in the population is sufficiently large to explore the search space, the factor Δ_k/N in (8) is small and forces new individuals to search locally. Otherwise the term Δ_k/N is larger and the new individuals are spaced out, favouring exploration of the whole space. The third factor in (8) also has an influence on the balance between exploration and exploitation, getting individuals with high fitness values to search locally and individuals with low fitness values to search globally. The term $\sqrt{N/n_k}$ in (8) maintains diversity in the population: if individuals tend to crowd, the number n_k becomes small and $step_k$ goes up. Finally, the last factor in (8) increases $step_k$ if the parent belongs to a discovered niche, thus preventing unnecessary computations of the objective function.

2.4 Elimination

Individuals not selected for reproduction for many generations are called "non-active" individuals. A threshold parameter n_a is introduced: if an individual is not selected for n_a generations it becomes a non-active individual.

Individuals that have an intermediate fitness value in comparison to their neighbours are called "local middle" individuals.

While the algorithm proceeds, the number N of individuals tends to increase, so, at the end of the mutation step, some individuals are eliminated in order to recondition the initial number N_{pop} of individuals in the population. Non-active and local-middle individuals are mainly suitable to be eliminated. The pseudo-code of the elimination mechanism is shown in Fig. 3.

2.5 Identification of Hypothetical Maxima and Hypothetical Minima

If, for several consecutive times, an individual selected for reproduction generates two offspring with fitness values lower than its own, then we assume that it is probably

```
- begin elimination
    - number of exceeding individuals: Ne = N – Npop
    - number of non-active individuals: Nna
    - number of local-middle individuals: Nlm
    - if Ne> Nna
        - elimination of all non-active individuals
        - Ne = Ne – Nna
        - if Ne> Nlm
            - elimination of all local-middle individuals
            - Ne = Ne – Nlm
        - else
            - elimination of the Ne worst local-middle individuals
            - Ne = 0
        - endif
    - else
        - elimination of the Ne worst non-active individuals
        - Ne = 0
    - endif
- end elimination
```

Fig. 3 Pseudo-code of the elimination mechanism

a maximum and we call it a "hypothetical maximum". A minimum can be identified analogously: we call an individual that generates, several consecutive times, offspring with fitness values higher than its own a "hypothetical minimum". Two numbers n_H and n_L are fixed in order to quantify the meaning of "several times" for the identification of hypothetical maxima and minima, respectively.

2.6 Coupling with the Deterministic Pattern Search Algorithm

Two hypothetical maxima belong to the same niche if their Euclidean distance is lower than both their niche radii. If only one hypothetical maximum has a niche radius r_1 greater than their Euclidian distance, this niche radius is reduced:

$$r_{1,new} = r_{1,old} - r_2, \tag{9}$$

where r_2 is the niche radius of the other hypothetical maximum.

We call hypothetical maxima that belong to the same niche and have fitness values lower than that of the best hypothetical maximum in the niche "doublets". At the end of the main loop all doublets are deleted.

A deterministic search is then started from each hypothetical maximum and carried out by means of the PS algorithm. At the end of the PS, the niche radii are updated and new doublets are deleted, if they occur. The remaining hypothetical maxima are the optimal solutions obtained and their niche radii provide a measure of the sensitivity of the maxima to variations in the design parameters.

3 Results

3.1 Mathematical Functions

An accurate comparison was performed between the stochastic section of the proposed method and an NGA [9] and the artificial immune system (AIS) [10]. Performance was evaluated on both newly designed and existing test functions, typically used for NGA benchmarking [6].

All the comparisons were performed setting the following values for the SALHE parameters: $N_{\text{pop}} = 10$, $a_{\text{H}} = 2.5$, $a_{\text{L}} = -1$, $b = 1.5$, $c = 5$, $e = 2.5$, $n_{\text{c}} = 2$, $n_{\text{H}} = 5$, $n_{\text{L}} = 4$ and $n_{\text{a}} = 30$.

Some results are shown in the following tables where n_{v} is the number of the fitness evaluations while $n_{90\%}$ is the percentage of authentic maxima found. An authentic maximum is considered found if there is a hypothetical maximum close to it with a fitness value higher than 90% of that of the authentic maximum. The results are averaged over 100 trials.

Table 1 shows the results obtained on Shekel's Foxholes function. Using the same number of evaluations SALHE-EA works better than both NGA and AIS. Moreover, SALHE-EA approaches the performance of the other algorithms with a smaller number of objective function evaluations, showing that it is well-suited when the containment of computational cost is a priority, as in electromagnetic optimization.

Table 2 shows the results obtained on the function F [11]. Once again, with unequally spaced and non-uniform peaks, SALHE-EA works better than both NGA and AIS.

Figure 4 shows the final hypothetical maxima with their radii in a solution found by means of the SALHE algorithm for the Foxholes function.

Table 1 Foxholes function (comparison between SALHE, NGA and AIS)

n_{v}	$n_{90\%}$		
	SALHE (%)	NGA (%)	AIS (%)
1,000	50.30	25.92	26.44
2,500	87.00	45.60	38.10
5,000	97.76	56.10	56.38

Table 2 F function (comparison between SALHE, NGA and AIS)

n_{v}	$n_{90\%}$		
	SALHE (%)	NGA (%)	AIS (%)
1,000	47.15	9.90	13.73
2,500	64.10	11.92	22.46
5,000	75.83	11.75	35.71

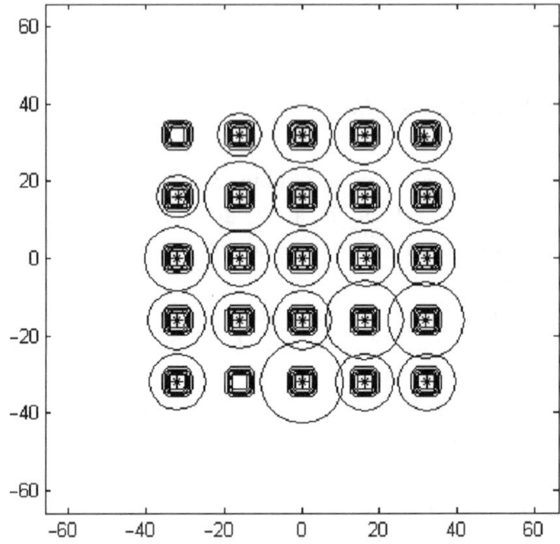

Fig. 4 Final hypothetical maxima with their radii for the foxholes function

Table 3 Optima for the SMES benchmark

	MAX 1	MAX 2	MAX 3
R_2	3.213	3.176	3.097
$h_2/2$	0.742	0.382	0.266
d_2	0.117	0.231	0.35
Radius	0.293	0.18	0.204

3.2 Electromagnetic Device

The SALHE algorithm was applied to solve the TEAM Workshop Problem 22 (SMES), for the case with three parameters [12]. The field problem was solved by means of the FEM-DBCI method [13]. The fictitious boundary coincides with the line where the stray field is calculated. The mesh is constituted by 1,728 second-order triangles and 3,575 nodes. Optimization of the device required 1,233 objective function evaluations (900 for the stochastic section, 333 for PS). The method identified three niches: Table 3 shows the optima values in each niche, together with the estimated normalized niche radius.

4 Conclusions

In this paper the SALHE algorithm for the optimization of electromagnetic devices is presented. SALHE is a new evolutionary algorithm able to discover niches in the objective function domain and estimate their radius. The new idea is that the niche radius is not assigned a priori, but is estimated taking advantage of the information on the positions of maxima and minima. The tests performed on mathematical functions and electromagnetic devices show that SALHE-EA, coupled with the deterministic pattern search algorithm, is a very efficient hybrid optimization method which works with a low number of objective function evaluations and is thus particularly suitable for use in the design of electromagnetic devices. Comparisons between SALHE-EA and other methods able to find multiple optima, like standard fitness sharing and AISs with the clonal selection algorithm, confirmed the effectiveness of the proposed algorithm.

References

1. K. Deb, *Evolutionary Algorithms for Multi-Criterion Optimization in Engineering Design*, in Evolutionary Algorithms in Engineering and Computer Science, Wiley, Chichester, UK, 1999, pp. 135–161.
2. D. A. Van Veldhuizen, G. B. Lamont, *Multiobjective Evolutionary Algorithms: Analyzing the State-of-the-Art*, Evolutionary Computation, vol. 8, no. 2, 2000, pp. 125–147.
3. A. Ghosh, S. Dehuri, *Evolutionary Algorithms for Multi-Criterion Optimization: A Survey*, International Journal of Computing & Information Sciences, vol. 2, no. 1, 2004, pp. 38–57.
4. D. E. Goldberg, J. Richardson, *Genetic Algorithms with Sharing for Multimodal Function Optimization*, Proceedings of the Second International Conference on Genetic Algorithms, 1987, pp. 41–49.
5. B. Sareni, L. Krahenbuhl, *Fitness Sharing and Niching Method Revisited*, IEEE Transactions on Evolutionary Computation, vol. 2, no. 3, 1998, pp. 97–106.
6. S. W. Mahfoud, *Niching Methods for Genetic Algorithms*, Ph.D. Dissertation, University of Illinois, Urbana Champaign, 1995.
7. P. P. Silvester, R. L. Ferrari, *Finite Elements for Electrical Engineers*, 3a ed., Cambridge University Press, Cambridge, 1996.
8. R. Hooke, T. A. Jeeves, *Direct Search Solution of Numerical and Statistical Problems*, Journal of the Association for Computing Machinery, vol. 8, 1961, pp. 212–229.
9. D. L. Carroll, *Fortran GA Driver*, http://cuaerospacee.com/carroll/ga.html, ver. 1.6.4, 1997.
10. L. N. de Castro, F. J. Von Zuben, *Learning and Optimization Using the Clonal Selection Principle*, IEEE Transactions on Evolutionary Computation, vol. 6, no. 3, June 2002, pp. 239–251.
11. E. Dilettoso, N. Salerno, *A Self-Adaptive Niching Genetic Algorithm for Multimodal Optimization of Electromagnetic Devices*, IEEE Transactions on Magnetics, vol. 42, no. 4, 2006, pp. 1203–1206.
12. Team Workshop Problem 22 [online]. Available: http://www.igte.tugraz.at/archive/team/team3dis.htm
13. G. Aiello, S. Alfonzetti, E. Dilettoso, *Finite Element Solution of Eddy Current Problems in Unbounded Domains*, IEEE Transactions on Magnetics, vol. 39, no. 3, 2003, pp. 1409–1412.

Non-Standard Nodal Boundary Elements for FEM-BEM

Salvatore Alfonzetti and Nunzio Salerno

Abstract This paper presents a novel family of nodal boundary elements to be used in the context of the hybrid FEM-BEM method to deal with unbounded static and quasi-static electromagnetic field problems. In this new type of boundary elements the field variable is developed by means of classical polynomial shape functions of a given order, whereas its normal derivative is developed with lower-order shape functions. A numerical example is provided regarding a simple electrostatic problem.

1 Introduction

In the context of the finite element method (FEM) [1] several auxiliary techniques have been devised to make it able to solve static and quasi-static electromagnetic field problems in unbounded domains, such as infinite elements [2], coordinate transformations [3], the hybrid FEM-BEM (boundary element method) method [4, 5], and the hybrid FEM-DBCI (Dirichlet boundary condition iteration) method proposed by the authors to solve electrostatic [6, 7], time-harmonic skin effect [8, 9] and eddy current [10] problems.

The hybrid FEM-BEM is undoubtedly the most popular of these methods. In this method a fictitious truncation boundary is introduced, so FEM is applied to the interior bounded domain (dealing with non-homogeneities, non-linearities, etc.), whereas BEM is applied to the exterior unbounded domain in order to substitute the unknown Dirichlet or Neumann conditions on the truncation boundary.

In order to combine the two methods, the two meshes need to be suitably chosen. A standard way to do this is to adopt boundary elements which restrict the finite elements on the boundary. This means that both the field variable and its normal

S. Alfonzetti and N. Salerno
Dipartimento di Ingegneria Elettrica, Elettronica e dei Sistemi, Università di Catania, Viale A. Doria 6, I-95125 Catania, Italy
alfo@diees.unict.it

S. Alfonzetti and N. Salerno: *Non-Standard Nodal Boundary Elements for FEM-BEM*, Studies in Computational Intelligence (SCI) **119**, 47–54 (2008)
www.springerlink.com

derivative are developed by employing the same nodes and the same polynomial shape functions of a given order, disregarding the fact that the normal derivative could be better described by a lower-order polynomial. This standard implementation of the BEM equations leads to some difficulties when the truncation boundary is not smooth, because in a corner node the normal derivative is multi-valued.

In this paper a new family of simplex boundary elements (for 2-D applications) is proposed in which the nodes of the field variable are placed in the canonical positions, whereas the nodes of the normal derivative are placed in between them and internally to the element, in such a way that the fact that one variable is the derivative of the other is fully exploited.

The paper is structured as follows: in the next section the new family of boundary elements is proposed for the hybrid FEM-BEM method; then a brief description of the FEM-BEM method is given for a simple electrostatic problem; an example of application is given and the authors conclusions follow.

2 The New Boundary Elements for FEM-BEM

In order to avoid the above-mentioned difficulties, which arise when the truncation boundary is not smooth, and to take into account the fact that the variable q is the derivative of v, a new family of finite elements has been conceived, in which the nodes of the variable v are placed in the canonical positions, whereas the nodes of the variable q are inserted in between them, as depicted in Fig. 1 for first-, second- and third-order triangles.

On the boundary sides of such elements, the variables v and q are approximated by

$$\mathrm{v} = \sum\nolimits_{n} \mathrm{v}_n \alpha_n \qquad \mathrm{q} = \sum\nolimits_{m} \mathrm{q}_m \beta_m, \tag{1}$$

where v_n and q_m are the nodal values of v and q, and α_n and β_m are the associated shape functions. Note that, if ν is the order of the α_n functions, $\nu - 1$ is the order of the β_m ones, in congruence with the fact that q is the derivative of v. The shape functions α_n are the canonical ones [1], whereas the shape functions β_m are given

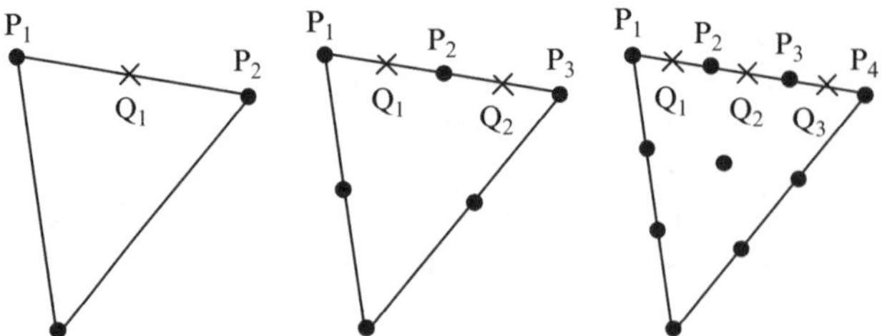

Fig. 1 Finite elements having a side lying on the fictitious boundary. On this boundary, the symbol • denotes the nodes P_n of the potential v, whereas the symbol × denotes the nodes Q_m of its normal derivative q

Table 1 Shape functions β_m associated with the nodes Q_m of the normal derivative q

Element order	Node Q_1	Node Q_2	Node Q_3
1	$\beta_1 = 1$	–	–
2	$\beta_1 = \frac{1}{2}(3–4\xi)$	$\beta_2 = \frac{1}{2}(-1+4\xi)$	–
3	$\beta_1 = \frac{1}{8}(15–48\xi+36\xi^2)$	$\beta_2 = \frac{1}{4}(-5+36\xi-36\xi^2)$	$\beta_3 = \frac{1}{8}(3–24\xi+36\xi^2)$

in Table 1, where ξ is the local coordinate, ranging from $\xi = 0$ at vertex node P_1 to $\xi = 1$ at the other vertex of the boundary side.

The use of these non-standard shape functions makes it necessary to compute some integrals to derive both the FEM and the BEM equations.

Let us assume that the FEM equations are obtained by means of the Galerkin method in which the α_n shape functions are used as weighting functions. The following integrals have to be evaluated:

$$c_{nm} = \int_{S_k} \alpha_n \beta_m ds = L_k \int_0^1 \alpha_n \beta_m d\xi = L_k \eta_{nm}, \tag{2}$$

where S_k is the generic kth boundary side and L_k is its length, and the η_{nm} coefficients are the entries of the following universal matrices:

$$I^{(1)} = \frac{1}{2}\begin{bmatrix}1\\1\end{bmatrix} \quad I^{(2)} = \frac{1}{12}\begin{bmatrix}3 & -1\\4 & 4\\-1 & 3\end{bmatrix} \quad I^{(3)} = \frac{1}{320}\begin{bmatrix}55 & -26 & 11\\81 & 66 & -27\\-27 & 66 & 81\\11 & -26 & 55\end{bmatrix} \tag{3}$$

for 1st-, 2nd- and 3rd-order triangular elements, respectively.

Moreover, in the evaluation of the BEM integral equation, the following integrals have to be computed:

$$h_{in}{}^{(k)} = \int_{S_k} \alpha_n(P) \frac{\partial G(P, Q_i)}{\partial n} ds, \tag{4}$$

$$g_{im}{}^{(k)} = \int_{S_k} \beta_m(P) G(P, Q_i) ds, \tag{5}$$

where S_k is the kth boundary element, Q_i is the ith node of the q unknown, and G is the two-dimensional free-space Green function, given by

$$G(P, Q) = \frac{1}{2\pi} \ln \frac{1}{r}, \tag{6}$$

where r is the distance between points P and Q. Possible symmetries are simply taken into account by modifying the Green function accordingly [1]. If node Q_i does not belong to S_k the integrand functions are regular and a simple Gauss quadrature technique may be used [1]. If, on the contrary, the node Q_i belongs to S_k, the integrand functions are singular and the integrations are performed analytically: all the

Table 2 Analytical formulas for evaluation of the coefficient g_{im} in the BEM equations

Element order	g_{im} coefficients	
1	$g_{11}^{(k)} = \frac{L_k}{2\pi}\left[1+\ln\left(\frac{2}{L_k}\right)\right]$	
2	$g_{11}^{(k)} = g_{22}^{(k)} = \frac{L_k}{4\pi}\left[\frac{3}{2}-\frac{3}{8}\ln(3)+\ln\left(\frac{4}{L_k}\right)\right]$	$g_{12}^{(k)} = g_{21}^{(k)} = \frac{L_k}{4\pi}\left[\frac{1}{2}-\frac{9}{8}\ln(3)+\ln\left(\frac{4}{L_k}\right)\right]$
3	$g_{11}^{(k)} = g_{33}^{(k)} = \frac{L_k}{2\pi}\left[\frac{13}{24}-\frac{5}{36}\ln(5)+\frac{3}{8}\ln\left(\frac{6}{L_k}\right)\right]$	$g_{22}^{(k)} = \frac{L_k}{2\pi}\left[\frac{3}{4}+\frac{1}{4}\ln\left(\frac{2}{L_k}\right)\right]$
	$g_{12}^{(k)} = g_{32}^{(k)} = \frac{L_k}{2\pi}\left[\frac{5}{12}-\frac{25}{72}\ln(5)+\frac{1}{4}\ln\left(\frac{6}{L_k}\right)\right]$	$g_{21}^{(k)} = g_{23}^{(k)} = \frac{L_k}{2\pi}\left[\frac{1}{8}+\frac{3}{8}\ln\left(\frac{2}{L_k}\right)\right]$
	$g_{13}^{(k)} = g_{31}^{(k)} = \frac{L_k}{2\pi}\left[\frac{1}{24}-\frac{25}{72}\ln(5)+\frac{3}{8}\ln\left(\frac{6}{L_k}\right)\right]$	

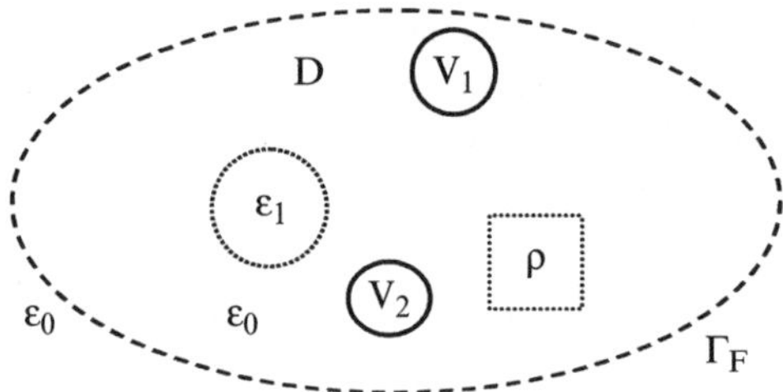

Fig. 2 An electrostatic system (made of voltaged conductors, non-homogeneities and distributed charges) enclosed by a fictitious truncation boundary

coefficients h_{in} vanish, whereas the coefficients g_{im} are calculated by means of the analytical formulas given in Table 2.

3 The Hybrid FEM-BEM for Electrostatic Field Problems

In order to show a concrete application of the boundary elements described in Sect. 2, a simple electrostatic problem will be formulated in this section by the hybrid FEM-BEM.

Consider a system of N_C conductors, voltaged at V_k, $k = 1,\dots,N_C$, embedded in an unbounded free space. Dielectric bodies and/or charge distributions may be present in the proximity of the conductors (see Fig. 2).

A fictitious truncation boundary Γ_F enclosing the whole system is introduced so that a bounded domain D is obtained. In D the Poisson equation holds:

$$\nabla \cdot (\varepsilon \nabla v) = -\rho \tag{7}$$

where ε is the electric permittivity and ρ the charge density. Equation (1) is subject to Dirichlet conditions on the conductors, whereas an unknown Neumann condition $\partial v/\partial n = q$ is assumed to hold on Γ_F, n being the inward unit normal to the boundary.

By dicretizing the domain D by means of Lagrangian simplex finite elements, and applying the Galerkin method using the shape function α_n as weighting functions, the following equations are obtained:

$$\iint_D \varepsilon \nabla v \cdot \nabla \alpha_n dxdy = \iint_D \rho \alpha_n dxdy - \int_{\Gamma_F} \varepsilon \alpha_n q ds. \tag{8}$$

By developing the variables v and q as in (1), these equations read:

$$\sum_h \varepsilon_h \sum_j v_j \iint_{E_h} \nabla \alpha_n \cdot \nabla \alpha_j dxdy = \sum_h \sum_j \rho_j \iint_{E_h} \alpha_n \cdot \alpha_j dxdy - \sum_k q_m \int_{S_k} \alpha_n \beta_m ds, \tag{9}$$

These equations are rewritten in matrix form as:

$$\begin{bmatrix} \mathbf{A} & \mathbf{A}_F \\ \mathbf{A}_F^t & \mathbf{A}_{FF} \end{bmatrix} \begin{bmatrix} \mathbf{V} \\ \mathbf{V}_F \end{bmatrix} = \begin{bmatrix} \mathbf{B}_0 \\ \mathbf{0} \end{bmatrix} - \begin{bmatrix} \mathbf{0} \\ \mathbf{C} \end{bmatrix} [\mathbf{Q}_F], \tag{10}$$

where: $\mathbf{V}$ and $\mathbf{V}_F$ are the vectors of the unknown values of the potential v in the nodes inside the domain and on the boundary Γ_F, respectively, $\mathbf{A}$, $\mathbf{A}_F$ and $\mathbf{A}_{FF}$ are sparse matrices of geometrical coefficients, $\mathbf{B}_0$ is the part of the known term array due to the conductor potentials and sources, $\mathbf{C}$ is a sparse matrix of geometrical coefficients, and $\mathbf{Q}_F$ is the vector of the unknown nodal values of q on Γ_F.

The matrix equation (3) alone is not sufficient to solve the problem because it only allows $\mathbf{V}$ and $\mathbf{V}_F$ to be obtained once the correct $\mathbf{Q}_F$ is known. To solve the unbounded problem, it is thus necessary to derive another matrix equation relating the unknown vectors. These equations are obtained by evaluating the BEM integral equation at the nodes Q_j of the q unknowns:

$$\frac{1}{2} v(Q_i) + \int_{\Gamma_F} v(P) \frac{\partial G(P,Q_i)}{\partial n_P} d\Gamma = \int_{\Gamma_F} q(P) G(P,Q_i) d\Gamma. \tag{11}$$

Note that the factor $1/2$ in the first term of (11) is due to the smoothness of the boundary Γ_F at Q_i. By developing the variable v and q as in (1), (11) reads

$$\frac{1}{2} \sum_n v_n \alpha_n(Q_i) + \sum_k \sum_n v_n \int_{S_k} \alpha_n(P) \frac{\partial G(P,Q_i)}{\partial n} ds = \sum_k \sum_m q_m \int_{S_k} \beta_m(P) G(P,Q_i) ds. \tag{12}$$

This equation is rewritten in matrix form as:

$$\mathbf{H}\,\mathbf{V}_F = \mathbf{G}\,\mathbf{Q}_F, \tag{13}$$

where $\mathbf{H}$ and $\mathbf{G}$ are fully-populated and non-symmetric matrices of geometrical coefficients. It is worth noticing that matrix $\mathbf{G}$ is square by construction, whereas $\mathbf{H}$ may be rectangular.

In order to solve the global system (10)–(13) a simple common approach is the direct use of an iterative conjugate gradient (CG)-like solver for non-symmetric systems of equations [11]. A similar approach is also used in which the array Q_F is derived from the BEM equation and substituted in the FEM one, to obtain a reduced system. Both these approaches suffer from the fact that in each step of the solver the matrix-by-vector multiplication is very costly, due to the presence of dense parts in the matrices, and the number of steps is high (typically a few hundred). In [12] an iterative solving strategy was proposed which takes into account the very different nature of equations (10) and (13). An improvement on this approach was presented in [13], based on the use of the generalized minimal residual (GMRES) solving algorithm [14]. In each step of the GMRES algorithm, the FEM equations are solved by means of the standard CG solver, whereas the BEM equations are not solved but used to perform fast matrix-by-vector multiplications.

4 A Validation Example

In order to validate the proposed method and the relative non-standard boundary element, in this section a simple electrostatic system will be considered, exhibiting an analytical solution. The system is a two-wire transmission line (Fig. 3) constituted by two parallel conducting circular cylinders of radius R whose centres are separated by a distance of $D = 2.4R$, voltaged with opposite potentials $V_0/2$ and $-V_0/2$. This problem exhibits the well-known analytical solution [15]:

$$\mathrm{v}^*(\mathrm{x},\mathrm{y}) = \frac{\mathrm{V}_0}{4\cosh^{-1}(\mathrm{D}/2\mathrm{R})} \ln\left[\frac{(\mathrm{x}+\mathrm{a})^2+\mathrm{y}^2}{(\mathrm{x}-\mathrm{a})^2+\mathrm{y}^2}\right], \tag{14}$$

where:

$$\mathrm{a} = \sqrt{\mathrm{D}^2/4-\mathrm{R}^2} \tag{15}$$

and the exact capacitance per unit length is $c = 5.04785\varepsilon_0$. Although this problem could be solved by means of a pure BEM, an FEM-BEM solution will be pursued in this paper. The fictitious truncation boundary is selected as constituted by two circumferences of radius $1.14R$ centred at the cylinder centres, so the gap between the conductors and the fictitious boundary is 0.14R. The analysis can be restricted to the first quadrant only, by imposing homogeneous Neumann and Dirichlet boundary conditions on the x- and y-axis, respectively. The resulting bounded domain is filled

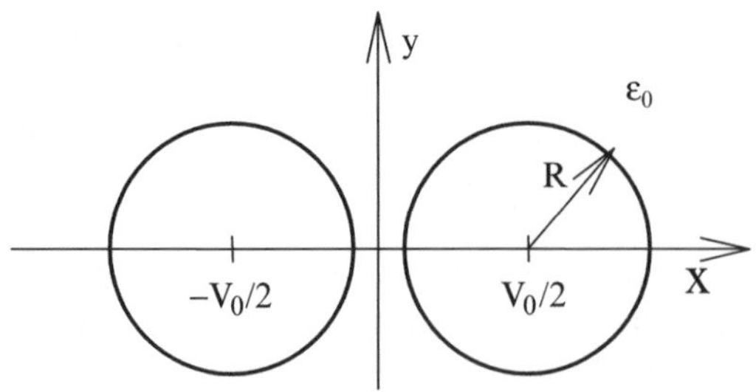

Fig. 3 The two-wire transmission line of the example

with eight layers of elements (the mesh is formed by 2,568 second-order triangular elements with 5,473 nodes). The global system is solved with the GMRES schemes (with restarting parameter $m = 10$).

To test the accuracy of the numerical solutions obtained, an accuracy indicator is defined as:

$$\zeta = 100\sqrt{\frac{\iint_{\mathrm{D}} (\mathrm{v}^* - \mathrm{v})^2 \mathrm{dxdy}}{\iint_{\mathrm{D}} (\mathrm{v}^*)^2 \,\mathrm{dxdy}}}, \tag{16}$$

where D is the domain of FEM analysis (lying in the first quadrant).

Having set an end-iteration tolerance of 0.01% for the GMRES and 0.001% for the CG, convergence was obtained with eight GMRES iterations. The computing time for the solution of the linear system was 0.453 s (on a Pentium IV, 3.2 GHz, 4 Gb RAM).

The accuracy indicator ζ is evaluated by sampling the electric potential v^* given by the analytical solution (14) at the FEM nodes and by computing (16) on a finite element basis: by this algorithm the indicator gives the value 8.52×10^{-3}, which means that the accuracy obtained is very good.

Another indication of the accuracy was obtained by computation of the capacitance per unit length by means of the cross-energy formula:

$$\mathrm{c} = 2\frac{\varepsilon_0}{\mathrm{V}_0}\iint_{\mathrm{D}} \nabla \mathrm{v} \cdot \nabla \mathrm{u} \;\mathrm{dxdy}, \tag{17}$$

where u is an arbitrary function which is 1 on the conductor and 0 on the truncation boundary; this function is very simply built by setting its nodal values to 1 on the nodes of the conductor and 0 in all the other nodes, so that only the finite elements having a node on the conductor surface actually contribute to the computation. By this computing algorithm the capacitance yields $c = 5.04740\varepsilon_0$, with an error of 9×10^{-3} per cent with respect to the value obtained by means of the analytical formula [15].

5 Conclusions

In this paper a new family of non-standard boundary elements has been proposed to be used in the hybrid FEM-BEM for the solution of open-boundary static or quasi-static electromagnetic field problems. The BEM equations have been written in a non-conventional way, by making the nodes for the potential non-coinciding with the nodes for its normal derivative.

The proposed approach has been implemented in ELFIN [16], a large FEM code developed by the authors for research in computational electromagnetism.

References

1. P. P. Silvester and R. L. Ferrari, *Finite Elements for Electrical Engineers*, Cambridge University Press, Cambridge, UK, 1996.
2. P. Bettess, Infinite elements, *Int. J. Numer. Methods Eng.*, vol. 11, pp. 53–64, 1977.
3. D. A. Lowther, E. M. Freeman, and B. Forghani, A sparse matrix open boundary method for finite element analysis, *IEEE Trans. Magn.*, vol. 25, pp. 2810–2812, July 1989.
4. C. A. Brebbia, J. C. F. Telles, and L. C. Wrobel, *Boundary Element Technique*, Springer-Verlag, Berlin Heidelberg New York, 1984.
5. S. J. Salon and J. D'Angelo, Applications of the hybrid finite element – boundary element method in electromagnetics, *IEEE Trans. Magn.*, vol. 24, pp. 80–85, Jan. 1988.
6. G. Aiello, S. Alfonzetti, and S. Coco, Charge iteration: a procedure for the finite element computation of unbounded electrical fields, *Int. J. Numer. Methods Eng.*, vol. 37, pp. 4147–4166, Dec. 1994.
7. G. Aiello, S. Alfonzetti, and G. Borzì, A generalized minimal residual acceleration of the charge iteration procedure, *J. Phys. III*, vol. 7, pp. 1955–1966, Oct. 1997.
8. G. Aiello, S. Alfonzetti, S. Coco, and N. Salerno, Finite element iterative solution to skin effect problems in open boundaries, *Int. J. Numer. Modelling*, Special issue on 'Computational Magnetics', vol. 9, pp. 125–143, Jan.–April 1996.
9. G. Aiello, S. Alfonzetti, G. Borzì, and N. Salerno, An improved solution scheme for open-boundary skin effect problems, *IEEE Trans. Magn.*, vol. 37, pp. 3474–3477, Sept. 2001.
10. G. Aiello, S. Alfonzetti, and E. Dilettoso, Finite element solution of eddy current problems in unbounded domains by means of the hybrid FEM-DBCI method, *IEEE Trans. Magn.*, vol. 39, pp. 1409–1412, May 2003.
11. G. H. Golub and C. F. Van Loan, *Matrix Computations*, J. Hopkins University Press, Baltimore, USA, 1996.
12. G. Aiello, S. Alfonzetti, E. Dilettoso, and N. Salerno, An iterative solution to FEM-BEM algebraic systems for open-boundary electrostatic problems, *IEEE Trans. Magn.*, vol. 43, pp. 1249–1252, April 2007.
13. G. Aiello, S. Alfonzetti, G. Borzì, E. Dilettoso, and N. Salerno, Solution of linear FEM-BEM systems for electrostatic field problems by means of GMRES, *International Symposium on Electric and Magnetic Fields* (EMF), Aussois (F), June 19–22, 2006.
14. Y. Saad and M. H. Schultz, GMRES: a generalized minimal residual algorithm for solving non-symmetric linear systems, *SIAM J. Sci. Stat. Comput.*, vol. 7, pp. 856–869, 1986.
15. E. Durand, *Electrostatique*, Masson Ed., Paris, 1968.
16. G. Aiello, S. Alfonzetti, G. Borzì, and N. Salerno, An overview of the ELFIN code for finite element research in electrical engineering, in *Software for Electrical Engineering Analysis and Design*, A. Konrad and C. A. Brebbia (ed.), WIT Press, Southampton, UK, 1999.

On the Use of Automatic Cuts Algorithm for $\mathbf{T_0 - T - \Phi}$ Formulation in Nondestructive Testing by Eddy Current

Anh-Tuan Phung, Patrice Labie, Olivier Chadebec, Yann Le Floch, and Gérard Meunier

Abstract In this paper, an application of the automatic cut algorithm applied to non-destructive testing (NDT) is proposed. The $T_0 - T - \Phi$ finite elements formulation based on the scalar magnetic potential is used in our application. This formulation requires the creation of cut in order to compute the eddy current density in multiply connected conducting region. However, the generation of these cuts can be difficult for the user, particularly in the case of complex geometries. In order to improve robustness of NDT modeling, we proposed an algorithm generating these cuts automatically.

1 Introduction

Eddy current inspection is a NDT technique using the induced alternating current generated by a emitting coil. Changes in the flow of eddy currents caused by cracks, dimensional variations, or changes in the material's conductivity can be detected by a receiving coil. Let us notice that is some applications, these both coils can be the same. EC-NDT is particularly sensitive to detect small and near surface defects.

This technique has been widely used in inspection and maintenance in the power generation and aircraft industries. Because of the high cost and fault consequences, electromagnetic simulations are intensively used to improve the robustness of NDT device.

In finite elements method, two main formulations can be used to solve eddy currents problems: A–V formulation and $\mathrm{T} - \mathrm{T_0} - \Phi$ formulation [1, 2]. This paper

A.-T. Phung, P. Labie, O. Chadebec, and G. Meunier
Grenoble Electrical Engineering Lab (G2Elab), UMR 5269 INPG-UJF-CNRS, ENSIEG, BP 46, 38402 Saint-Martin-d'Hères Cedex, France
Gerard.Meunier@G2elab.Inpg.Fr

Y.L. Floch
Cedrat 15, Chemin de Malacher, Innovallée, 38246 Meylan, France

A.-T. Phung et al.: *On the Use of Automatic Cuts Algorithm for $T_0 - T - \Phi$ Formulation in Nondestructive Testing by Eddy Current*, Studies in Computational Intelligence (SCI) **119**, 55–62 (2008)
www.springerlink.com

deals with $T - T_0 - \Phi$ formulation which is known as powerful one in term of computation cost and memory requirement. However, its utilization needs some slight precautions. More precisely, in order to compute current distribution in conducting media around a crossing crack, an artificial cut needs to be created. The automatic cut generation algorithm introduced in this paper avoids users to define these cuts manually. Hence it makes the $T - T_0 - \Phi$ formulation easier to use and the evaluation more reliable.

2 T_0-T-Φ Formulation for EC-NDT

We consider a typical example in Fig. 1. In magnetic part (Ω_1) without current source $(Js = 0)$, the Maxwell–Ampère equation is written as:

$$\text{rot}\,\mathbf{H} = 0. \tag{1}$$

It means that the magnetic field can be derived from gradient of a scalar potential Φ:

$$\mathbf{H} = -\text{grad}(\Phi). \tag{2}$$

To describe the current source J_S, a special source potential $\mathbf{T}_0$ is introduced such that:

$$\text{rot}\,\mathbf{T}_0 = J_S. \tag{3}$$

Hence, the magnetic field in air (Ω_0) is called reduced to this $\mathbf{T}_0$:

$$\mathbf{H} = \mathbf{T}_0 - \text{grad}(\Phi). \tag{4}$$

The $\mathbf{T}_0$ calculation must be done in region Ω_{T_0} that encloses excitation coils, to respect Ampere's law. In order to represent induced current in the conductive part (Ω_c), from the current conservation law div $\mathbf{J} = 0$, we have $\mathbf{J} = \text{rot}\,\mathbf{T}$. The field expression is then written as $\mathbf{H} = \mathbf{T} - \text{grad}(\Phi)$. Edge elements are used because their good physical representation of the phenomena. As a result, Maxwell–Ampere

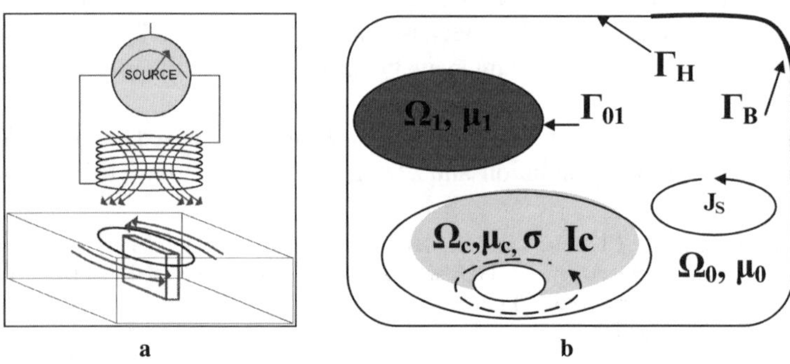

Fig. 1 **a** Typical EC-NDT problem. **b** Typical schematic problem

equation is naturally respected. We have to solve the weak form of Maxwell–Gauss and Maxwell–Faraday equations:

$$\mathrm{div}(-\mu_1 \,\mathrm{grad}(\Phi)) = 0 \text{ in } \Omega_1, \tag{5}$$

$$\mathrm{div}(\mu_0 \mathbf{T}_0 - \mu_0 \,\mathrm{grad}\,(\Phi)) = 0 \text{ in } \Omega_0, \tag{6}$$

$$\mathrm{div}(\mu_C \mathbf{T} - \mu_C \,\mathrm{grad}\,(\Phi)) = 0 \text{ in } \Omega_C, \tag{7}$$

$$\text{and rot}\left(\frac{1}{\sigma} rot\mathbf{T}\right) + \frac{\partial}{\partial t}(\mu_C(\mathbf{T} - \,\mathrm{grad}\,(\Phi))) = 0 \text{ in } \Omega_C. \tag{8}$$

If the conductive parts contain crossing holes, eddy current I_C will develop around them. The conductive parts will become multiply connected or closed solid conductors. Hence, the above formulation must be modified in order to take into account this latter induced current I_C. If we call $\mathbf{j}_{0C}$ an arbitrary current density of a prescribed net current circulating around holes and $\mathbf{t}_{0C}$ a special vector potential which verifies $\mathbf{j}_{0C} = \mathrm{rot}\ \mathbf{t}_{0C}$ and $\mathbf{t}_{0C} \times \mathrm{n} = 0$ on a domain Ω_{0C} surrounding Ω_C. Thus, we can choose Ω_{0C} as the union between Ω_0 and Ω_C. The magnetic field expression in (Ω_c) is written as:

$$\mathbf{H} = I_C \mathbf{t}_{0C} + \mathbf{T} - \mathrm{grad}(\Phi). \tag{9}$$

The current density expression is:

$$\mathbf{J} = \mathrm{rot}(\mathbf{T} + I_C \mathbf{t}_{0C}). \tag{10}$$

In air region (Ω_0), the field expression is reduced to this induced current:

$$\mathbf{H} = I_C \mathbf{t}_{0C} + \mathbf{T}_0 - \mathrm{grad}(\Phi). \tag{11}$$

Let us have a look to the Ampere's theorem. Along any closed path C_1 which is entirely in the air and going through the hole, we suppose there is no source current going through surface enclosed by C_1. We have in this case:

$$\int_{C_1} \mathbf{H} dl = \int_{C_1} (I_C \mathbf{t}_{0C} + \mathbf{T}_0 - \mathrm{grad}(\Phi))\, dl = I_C. \tag{12}$$

The second term in the right-hand side is equal to zero due to the above condition; the third one vanishes thanks to nature of scalar potential. Finally, Ampere's theorem is verified.

If there are k holes, the presented formulation can be generalized to take into account the k currents circulating around holes. In that case:

$$\mathbf{H} = \sum_k I_{Ck} \mathbf{t}_{0Ck} + \mathbf{T} - \mathrm{grad}(\Phi) \text{ in conducting region}(\Omega_c), \tag{13}$$

$$\mathbf{J} = \mathrm{rot}(\mathbf{T} + \sum_k I_{Ck} \mathbf{t}_{C0k}), \tag{14}$$

$$\mathbf{H} = \sum_k I_{Ck} \mathbf{t}_{0Ck} + \mathbf{T}_0 - \mathrm{grad}(\Phi) \text{ in air region}(\Omega_0). \tag{15}$$

Analyzing the field expression in this formulation, we have to clarify some points. Firstly, the induced current I_{Ck} is an unknown and must be linked to the total current in the solid conductor by an additional relation [3, 4]. Secondly, the computation of $\mathbf{t_{0ck}}$ could be carried out thanks pre-processing computations. Firstly, an electrokinetic's computation is provided. This could be achieved by imposing electric potential differences on a virtual cut in the multiply connected conductor and then compute the current density $\mathbf{j}_{0k}$. Secondly, computation of t_{0CK} can be achieved using the relation between $\mathbf{j}_{0k}$ and $\mathbf{t}_{0k}$ mentioned above. Once this information is known, the general solution can be carried out without further consideration of the cut. Interested readers are referred to more detailed literature in [4, 5]. We will here concentrate on the automatic generation and specific treatment of this cut.

This cut is an artificial handle in order to compute the current density $\mathbf{j}_{0C}$. In electromagnetic simulation, asking users to manually cut their multi-connected conductor during the modeling process is not consistent. Moreover, if users do not know how to create the right cuts, the obtained results could be completely incorrect. This could arise from complicated conductor's configurations with more than one hole for instance. Hence it is evident that we must automate the creation of these cuts and the linked current density $\mathbf{j}_{0C}$ computation.

3 Automatic Cut Algorithm – Application in NDT

The basic idea of creating the cut automatically is modeling the inflation of a virtual balloon in a hollow torus. The balloon is kept simply connected by increasing continuously its volume. This process is bounded because the torus volume is limited. This means at the final phase of the process, joining part of the balloon's boundary must form the required cuts [6].

Cuts are modeled by shell elements with double nodes having the same coordinates. Electric scalar potential are imposed on both sides of cuts to create the current distribution. Values of electric potential can be chosen arbitrarily and scaled to lead to a global unit current.

The algorithm is implemented and tested in a standard the finite elements code Flux [7] and lead to excellent results in some cases. In other applications, for example in NDT problems, this algorithm reveals a major drawback. In fact, NDT geometries contain one or more very thin cracks – crossing or not. When applying inflating balloon algorithm to these geometries, it can lead to "T-shaped" cuts (see Fig. 2).

This situation occurs because we did not implement any control on the inflating direction in the algorithm. Hence, balloon's boundaries developed themselves freely in all directions and they could meet each other. From a topological point of view, these crossing cuts cause any trouble because they effectively divide the multiply connected volume into simply connected one. All holes are linked to outward by these cuts.

But from a modeling point of view, these cuts are not straightforward to work with. In fact, on the junction of more than one branch of cuts, values of electric

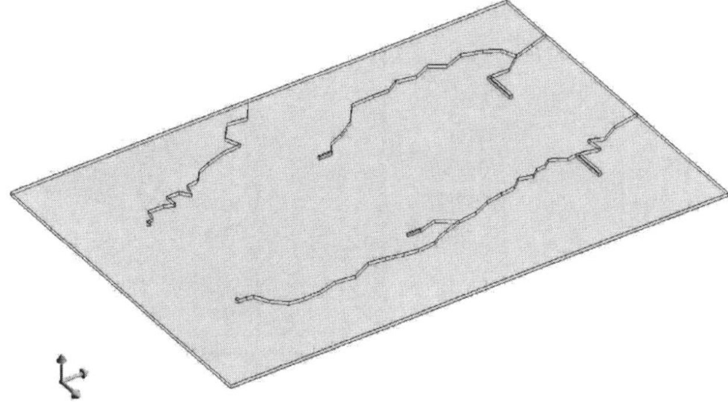

Fig. 2 Crossing automatic cuts

Fig. 3 "T-shaped" cut and protection wall around existing cut redirects future cut to exterior

potential can not be simply affected in existing formulations. Currently, these kinds of "T shaped" cuts leads to a failure $\mathbf{j}_{0C}$ computation.

This problem can be solved by modifying the numerical implementation of shell elements allowing the management of scalar potential value on nodes located in "T-shaped" area. Potential jumps must be attributed depending on number of branches at each junction. However, this solution is difficult to implement and we prefer to control the number of cuts created after inflating process and force the algorithm to create a separated cut for each hole.

This solution has been implemented with the following strategy. First of all, "T-shaped" cuts are considered and an extra branch is eliminated for each T to get simple cuts. The procedure is repeated until no more "T-shaped" will be found. At the end of this step, cuts are missing in comparison with the number of holes. In the following, 2D model and previous notations are used to facilitate the comprehension.

Inflating algorithm is the applied one more time but existing cuts have to be protected to prevent new cuts to join the old ones. If we call Ω the whole domain and C_K any existing cut, we form a special bounded sub-domain Ω_K for each C_K. If a future cut plugs into these Ω_K, they will be redirected to exterior or holes (see Fig. 3) hence preventing formation of crossing cuts. This is repeated until the number of cuts is equal to the number of holes.

Fig. 4 Formation mechanism of separated sub-domain

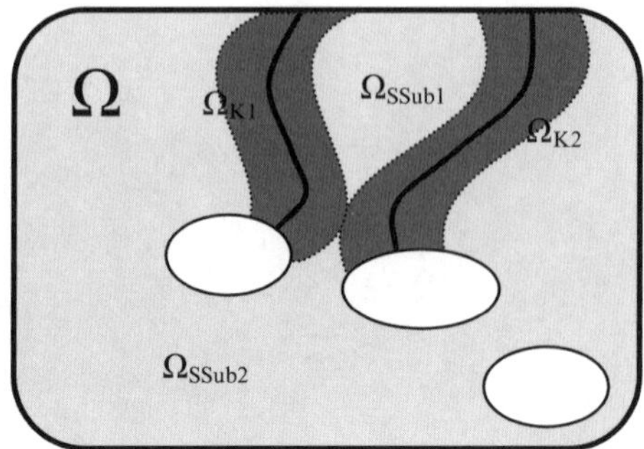

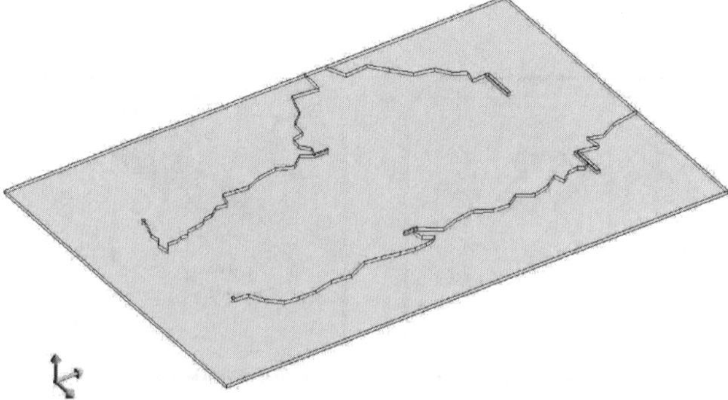

Fig. 5 Automatic cuts after treatment

One can observe that $\Sigma\Omega_K \neq \Omega$, or $\Omega/\Sigma\Omega_K \neq \emptyset$. This means in some cases, linked protected zones can divide the great domain into separated sub-domains Ω_{SSub}. Hence inflating algorithm will have some difficulties to generate correct number of cuts. This situation is occurring when dealing with low-meshed geometries in which sub-domains Ω_K have more chances to touch each other (see Fig. 4). To get rid of this problem, initialization phase of inflating algorithm must be carefully prepared. It must permit initial element having random walk onto Ω_{SSub} so that all positions could be scanned by the algorithm. This technique has been implemented in our work and gives interesting results (see Fig. 5).

We present here some computation results on a NDT test case with six crossing-cracks on a 1 mm thick plate (conductivity 59.6×10^6 siemens). The probe is an air-core type one and is located at 5 mm above the plate. It is fed with a current source at 100 kHz. To demonstrate the robustness of the automatic cut algorithm, we will compute the global inductance value in both cases of manual cut and the automatic one.

Figure 6 shows vector field of induced current in the conductive plate. Crossing cuts detected on Fig. 2 are replaced automatically with separated ones. All holes are linked to outside. Eddy current circulation is properly computed.

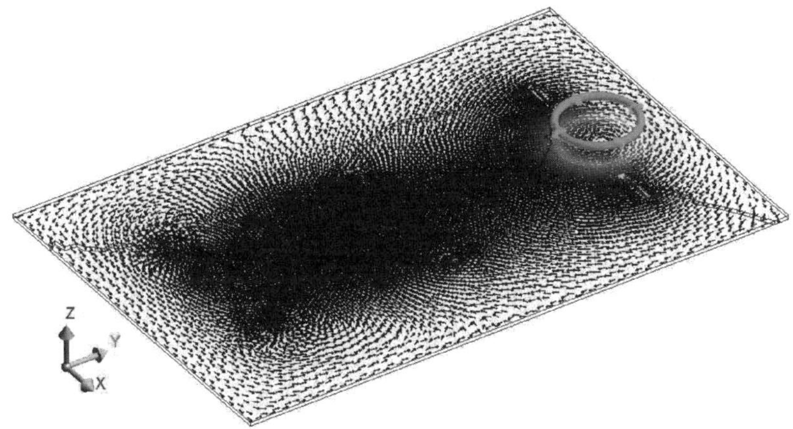

Fig. 6 Eddy current vector field on the conductive plate with automatic cuts

Table 1 Computation comparison

Result in case of	Manual cuts	Automatic cuts	Relative error
Losses in the plate	57.42 μW	55.62 μW	3.13%
Coil inductance	46.69 μH	45.58 μH	2.37%

In Table 1, we presents losses and coil inductance in both cases. A slight difference between two results can be observed. This must be principally due to the mesh variation between two cases.

4 Conclusion

The application of automatic cuts algorithm with some adapted treatments for EC-NDT geometries leads to an easier use of $T_0 - T - \Phi$ formulation. Computation result is comparable between manual and automatic cut. The creation of cuts is user-free and contributes to the enhancement of $T - T_0 - \Phi$.

References

1. Biro O., Preis K., Renhart W., Vrisk G., Richter K.R., Computation of 3-D current driven skin effect problems using a current vector potential, IEEE Transactions on Magnetics, 29(2), pp. 1325–1328, Mar. 1993.
2. Bouissou S., Piriou F., Kieny C., and Tanneau G., A numerical simulation of a power transformer using 3-D finite element method coupled to circuit equation, IEEE Transactions on Magnetics, 30(5), pp. 3224–3227, Sep. 1994.

3. Bedrosian G., Magnetostatic cancellation error revisited, IEEE Transactions on Magnetics, 27(5), pp. 4181–4184, Sep. 1991.
4. Meunier G., Luong H.T., and Marechal Y., Computation of coupled problem of 3D eddy current and electrical circuit by using $T0 - T - \varphi$ formulation, IEEE Transactions on Magnetics, 34(5), Part 1, pp. 3074–3077, Sept. 1998.
5. LeFloch Y., Meunier G., Guerin C., Labie P., Brunotte X., and Boudaud D., Coupled problem computation of 3-D multiply connected magnetic circuits and electrical circuits, IEEE Transactions on Magnetics, 39(3), Part 1, pp. 1725–1728, May 2003.
6. Phung A.T., Chadebec O., Labie P., Le Floch Y., and Meunier G., Automatic cuts for magnetic scalar potential formulations, IEEE Magnetics Transactions, 41(5), pp. 1668–1671, May 2005.
7. Flux Software (http://www.cedrat.com)

Refinement Strategy Using 3D FEM in Eddy Current Testing

Krebs Guillaume, Clenet Stéphane, and Abakar Ali

Abstract In this paper, a method is used for evaluate quickly and accurately default for the non-destructive control. The established approach is based on two finite element formulations. Two error estimators coupled with adaptive mesh software are then proposed. The efficiency of the procedure is showed on an example.

1 Introduction

Eddy current testing (ECT) is now often employed. To reduce cost experiments related to the ECT and to the default characterization (shapes, sizes, ...), analytical and numerical models are used. The finite element method (FEM) yields a precise description of the geometry problem but requires much attention to get a reliable result. In most case, the signal variations in ECT due to a default (rift in a steam pipe for example) are very low. In the FEM, the calculated signal due to the default is noised by the numerical error due to the discretisation. For calculating accurately the default perturbation, two calculations can be done considering or not default [1]. So, we show on Fig. 1 the two FE problems to be solved, the meshes used into the two cases are identical, only the material of the volume default is modified.

In the cases with and without default, one calculates the linkage fluxes and this generally for several positions of the sensor. The FE movement techniques are not discussed in this paper but some are described for example in [2]. Time harmonic approach is well adapted to this kind of problem, where several series of calculations

Krebs Guillaume and Clenet Stéphane
Laboratoire d'Electrotechnique et d'Electronique de puissance de Lille, L2EP-LAMEL, ENSAM, 8, Bd Louis XIV, 59046 Lille, France
Guillaume.KREBS@lille.ensam.fr

Abakar Ali
EDF R&D, LAMEL, 1 avenue du Général de Gaulle, 92141 Clamart Cedex, France
Ali.abakar@edf.fr

K. Guillaume et al.: *Refinement Strategy Using 3D FEM in Eddy Current Testing*, Studies in Computational Intelligence (SCI) **119**, 63–71 (2008)
www.springerlink.com

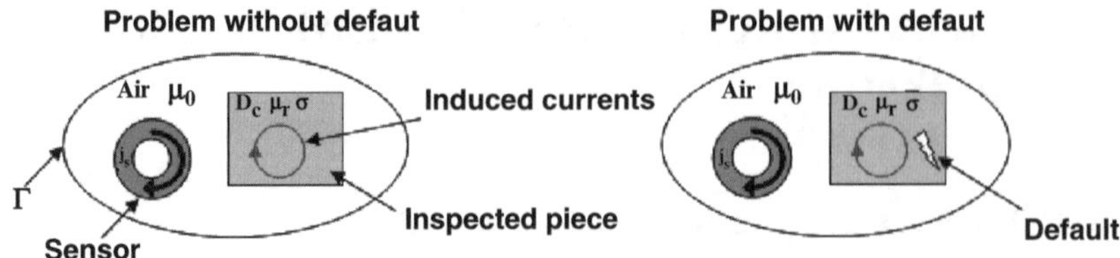

Fig. 1 Schematic diagrams of the modelled problem in the eddy current testing

are required (due to the sensor displacements and to the high working frequency). By using this method of resolution, the sensor variations dues to the default are then expressed as (for the real part of the linkages fluxes for example):

$$\Delta\Phi_r = (\Phi_{r\_nodef} \pm \Phi_{r\_mesh\_nodef}) - (\Phi_{r\_def} \pm \Phi_{r\_mesh\_def}). \tag{1}$$

With Φ_{r_nodef} and Φ_{r_def} that would be the results obtained with a "perfect" mesh by using a FEM formulation. The terms $\Phi_{r\_mesh\_nodef}$ and $\Phi_{r\_mesh\_def}$ are the perturbations introduced by the numerical errors. The weaker both last terms are, less $\Delta\Phi_r$ will be perturbed. However to obtain such accurate variations, mesh adaptation is needed, this in order to reduce the terms $\Phi_{r\_mesh\_nodef}$ and $\Phi_{r\_mesh\_def}$.

In this paper, we present a mesh adaptation procedure where the error distribution is calculated from the variation of the fields calculated using field distribution in the cases without and with defaults. The first part of this paper will deal with two FEM complementary formulations. From this, we will describe the local error estimators based on the non-verification of the behaviour laws [3, 4]. We will give also the procedures used with the mesh refinement software. In the last part, we will compare the results of the established methodologies on an example.

2 Formulations and Error Estimation

2.1 Used Finite Element Formulations

Magnetodynamic problems can be solved in frequency domain by two formulations. The first using the **A** vector and the φ scalar potentials, such as:

$$\begin{aligned} \mathrm{B} &= \mathbf{curl}\,\mathbf{A} \text{ with } \mathrm{A}\times\mathbf{n}|_{\Gamma_B} \\ \mathbf{E} &= -j\omega\mathbf{A} - \mathbf{grad}\,\varphi \text{ with } \varphi|_{\Gamma_H} = 0 \end{aligned}. \tag{2}$$

The system to solve can be expressed under the name, formulation $\mathrm{A}-\varphi$, by

$$\begin{aligned} \mathbf{curl}\frac{1}{\mu}\mathbf{curl}\,\mathbf{A} + \sigma(j\omega\mathbf{A} + \mathbf{grad}\,\varphi) &= \mathbf{J_S}. \\ \mathrm{div}(\sigma(j\omega\mathbf{A} + \mathbf{grad}\,\varphi)) &= 0 \end{aligned} \tag{3}$$

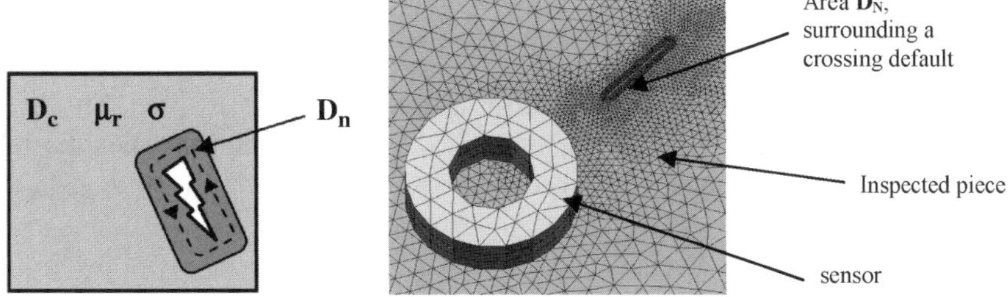

Fig. 2 Taking into account of crossing defaults, example of the team problem 8 from Compumag

The second formulation uses the **T** vector and the Ω scalar potentials, see (4), such as:

$$\begin{aligned} \mathbf{J_{ind}} &= \mathbf{curl\ T} \text{ with } \mathbf{T} \times \mathbf{n}|_{\Gamma_c} = 0 \\ \mathbf{H} &= \mathbf{H_s} + \mathbf{T} - \mathbf{grad}\,\Omega \text{ with } \Omega|_{\Gamma_H} = 0 \end{aligned} \quad (4)$$

So $T - \Omega$ formulation yields:

$$\begin{aligned} \mathbf{curl}\frac{1}{\sigma}\mathbf{curl\ T} + j\omega\mu(\mathbf{T} + \mathbf{grad}\,\Omega) &= -\mathbf{curl}\frac{1}{\mu}\mathbf{curl\ H_s} - j\omega\mathbf{H_s}. \\ \mathrm{div}(\sigma(j\omega\mathbf{A} + \mathbf{grad}\,\varphi)) &= 0 \end{aligned} \quad (5)$$

$\mathbf{H_s}$ is calculated from the relation $\mathbf{H_s} = \mathbf{K}i$ and from a turn vector density vector **Ns** such as:

$$\mathbf{curl\ H_s} = \mathbf{curl\ K_s}\, i = \mathbf{N_s} i. \quad (6)$$

With i the magnitude of the current in the sensor [1].

By using directly the $T - \Omega$ formulation, if a conductive region is not simply connected (holes), the Ampere's law would not be respected. Consequently, the adopted approach is to define inside an arbitrary area noted $\mathbf{D_N}$ surrounding the default [5], see Fig. 2.

An uniform current density $\mathbf{J_c}$ is supposed circulating in the area $\mathbf{D_n}$. In the same way as (6), vectors $\mathbf{N_c}$ and $\mathbf{K_c}$ can be defined. The area $\mathbf{D_n}$ is considered as short circuit, another electric equation can be deduced [6] and included in (5).

2.2 Flux Calculation

Considering the $A - \varphi$ and $T - \Omega$ formulation defined previously, the fluxes flowing through a coil "k" can be defined as [7]:

$$\begin{aligned} A - \varphi: \quad \Phi_k &= \int_D \mathbf{A}.\mathbf{N_{sk}}\, d\tau \\ T - \Omega: \quad \Phi_k &= \int_D \mathbf{B}.\mathbf{K_{sk}}\, d\tau \end{aligned} \quad (7)$$

With $\mathbf{N_{sk}}$ and $\mathbf{K_{sk}}$ represents the source fields due to the coil "k."

2.3 Error Estimators

From the previous formulations, two error estimators can be deduced. The $A-\varphi$ formulation leads to admissible solutions B_{adm} and E_{adm}, thus satisfies the relation (8). The $T-\Omega$ formulation leads to admissible solutions H_{adm} and J_{ind_adm} that satisfies (9)

$$\mathbf{curl\ E} = \frac{\partial B}{\partial t} \tag{8}$$

$$\mathbf{curl\ H} = \mathbf{J_{ind}} \tag{9}$$

Theses solutions $(B,\ E,\ H,\ J_{ind})$ satisfy both Maxwell equations but not the behaviour laws, if it was the case, it would be the exact solutions of the considered problem. Theses non-verifications are expressed by

$$\begin{cases} (\mathbf{B_{adm}} - \mu \mathbf{H_{adm}}) \neq 0 \\ (\mathbf{J_{ind\_adm}} - \sigma \mathbf{E_{adm}}) \neq 0. \end{cases} \tag{10}$$

One can give then two local error estimators, magnetic and electric. By using the energy normalisation the estimators are expressed by the relations (11). So, two global errors are given by the sum of the local error on each element [8].

$$\begin{aligned} \varepsilon_{elec\_1} &= \int_e (\mathbf{B_{adm}} - \mu \mathbf{H_{adm}}) \mu^{-1} (\mathbf{B_{adm}} - \mu \mathbf{H_{adm}}) dv \\ \varepsilon_{mag\_1} &= \int_e (\mathbf{J_{adm}} - \sigma \mathbf{E_{adm}}) \sigma^{-1} (\mathbf{J_{adm}} - \sigma \mathbf{E_{adm}})\, dv \end{aligned} . \tag{11}$$

3 Mesh Refinement Strategies

To obtain more accurate results, mesh adaptation can be employed, and for this, an error field is required. Usually the error field is deduced from the relations (11) coming from the default or non-default meshes, electrical and magnetic errors are then associated on each element. In theses cases, the refinement procedures deal usually with the high density error areas.

The problem is that the designed parts to refine are not necessarily interesting. Two identical meshes, one with and the other without defect are computed and the variation of the sensor is deduced from them. If the refinement is undertaken on the error distribution (for example, of the case with defect), the areas which will be re-meshed perhaps will induce disturbances on the results of the case without defect and thus on the sensor variations.

The global entity under interest is the variations due to default on the sensor linkage flux. Consequently, the area giving the higher local variations of fields (while comparing the cases with and without default) is the one around the default. The

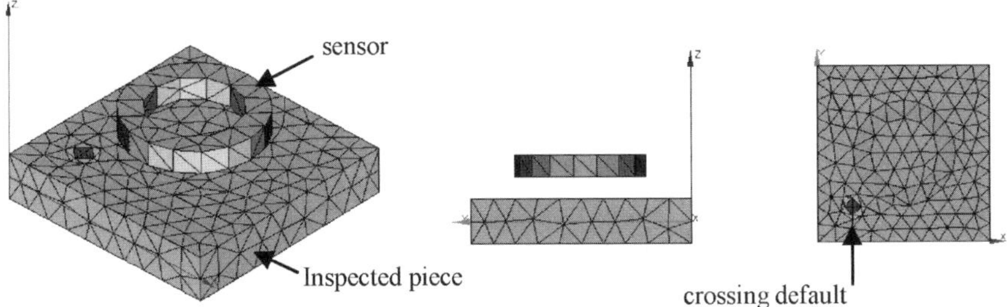

Fig. 3 Schemes of the traditional and proposed processes used for the refinement strategies

proposed solution is to replace in (11) the "local fields" by "the local variations of fields". So we define two new error estimators on each mesh elements by

$$\begin{cases} \varepsilon_{elec\_2} = \int_e [(\mathbf{B_{adm\_def}} - \mathbf{B_{adm\_nodef}}) - \mu(\mathbf{H_{adm\_def}} - \mathbf{H_{adm\_nodef}})]\,\mu^{-1} \\ \qquad \times [(\mathbf{B_{adm\_def}} - \mathbf{B_{adm\_nodef}}) - \mu(\mathbf{H_{adm\_def}} - \mathbf{H_{adm\_nodef}})]dv \\ \varepsilon_{mag\_2} = \int_e [(\mathbf{J_{ind\_adm\_def}} - \mathbf{J_{ind\_adm\_nodef}}) - \sigma(\mathbf{E_{adm\_def}} - \mathbf{E_{adm\_nodef}})]\,\sigma^{-1} \\ \qquad \times [(\mathbf{J_{ind\_adm\_def}} - \mathbf{J_{ind\_adm\_nodef}}) - \sigma(\mathbf{E_{adm\_def}} - \mathbf{E_{adm\_nodef}})]\,dv, \end{cases} \tag{12}$$

Where B_{adm_def} and H_{adm_def} are the fields computed on the mesh with default. The fields B_{adm_nodef} and H_{adm_nodef} are obtained in the case without default. Naturally for this strategy, four calculations are necessary for one position of the sensor. We show on Fig. 3, the scheme of the proposed strategy compared with the traditional one. The error estimator based on the non-verification of the electric behaviour law is not defined in the non-conductive region and also in the default.

4 Results and Comparison

Calculations were performed using the software CARMEL, with both FEM formulations. The proposed geometry and mesh were constructed by using the platform SALOME. The mesh refiner used is HOMARD from EDF R&D. The method is based on refinement and unrefinement of meshes using cutting of element. Various kinds of elements are concerned, for example triangles and quadrangles in 2D or tetrahedrons in 3D. HOMARD takes into account the sub-domain definitions with groups, attributes, boundary conditions, and periodicity.

The computed example (see Fig. 4) is a coil located above a non-magnetic but conductive metal piece (such as aluminium for example). The initial mesh is made of about 38,000 tetrahedrons for 7,500 nodes. The element size was chosen deliberately high in comparison with the desired frequency. A crossing default has been included in the metal sheet.

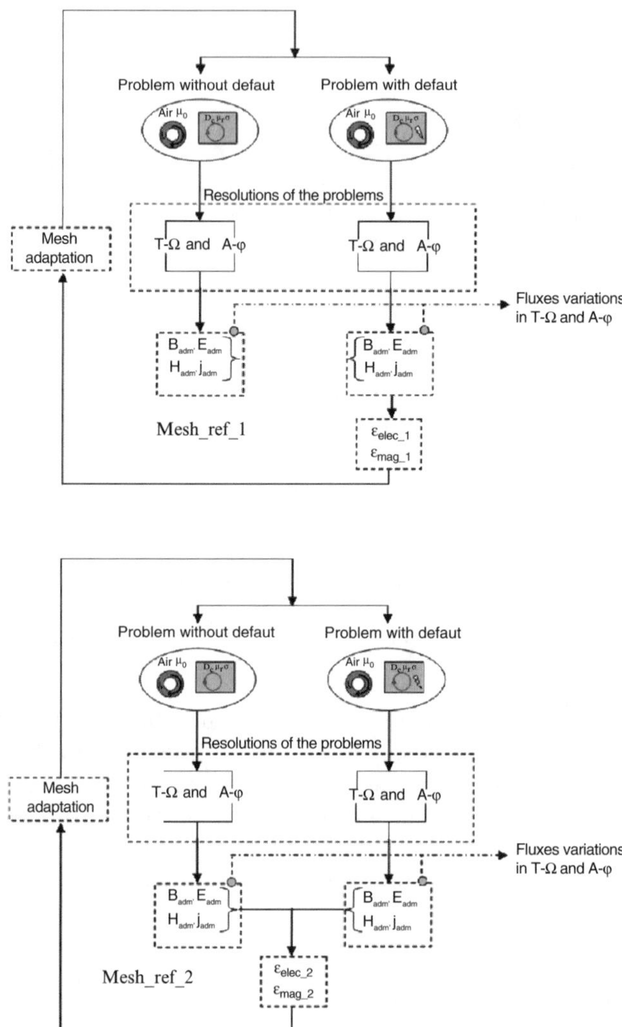

Fig. 4 The computed example used with the error estimators

The refinement conditions consist in taking at first the higher relative values (in percent) between the electrical and magnetic errors, and secondly to refine the elements on which the error was upper than 50%. For a frequency of 10 kHz (the skin effect is about 25% of the metal sheet), on Fig. 5 we compared the behaviours of both refinement schemes described by the flowcharts which corresponds to the application of the relationships (11) and (12) for error estimation. In the following, all results given with the classical scheme of refinement will be referenced as "Mesh_ref_1" otherwise the one based on the variation of the fields will be reference as "Mesh_ref_2". For the mesh refinement 1 (Fig. 3 on the left) and concerning the

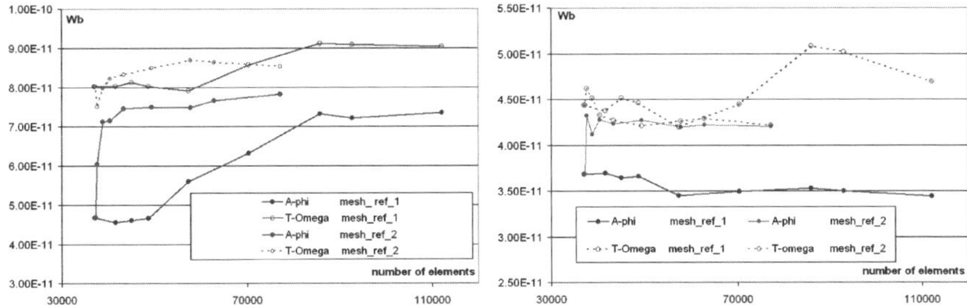

Fig. 5 Variations of the linkage flux for a frequency of 10 kHz, real and imaginary parts

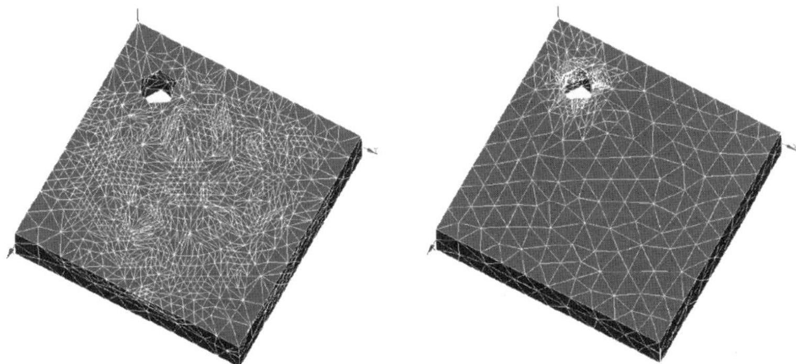

Fig. 6 Metal piece of the computed problem, mesh refinement 1 and 2

fluxes variations, only the fields computed from the case with default are used for error estimation. Nethertheless, we noticed that this error estimation was better than the one without default.

The mesh refinement 2 (Fig. 3 on the right) is better than the other one. For this example, a solution for the real part of the flux is enclosed in three refinement steps. Concerning the imaginary part, only two steps are almost needed. For the imaginary part, with the mesh refinement 1, it seems to have difficulties to converge for the given number of elements. The curves have non-linearities due to the fact that in certain cases, the refinement is effected either from the electric or magnetic point of view. We present on Fig. 6, the metal piece for the two last steps (about 100,000 elements) of the refinement methods, the default area is only concerned by our process.

Instead of looking at the variations, we took into account the flux shapes themselves and this according to the number of elements. We can see on Fig. 7 that for the classical scheme there is a convergence between both fluxes calculated using both formulation. There are few modifications of the flux values when the refinement on the variation is employed.

For a frequency of 100 kHz, the skin effect is about 8% of the metal piece thickness. We show on Fig. 8 the variation of the linkage flux. One needs a rather high number of elements, but either for real or imaginary part, solutions are approached.

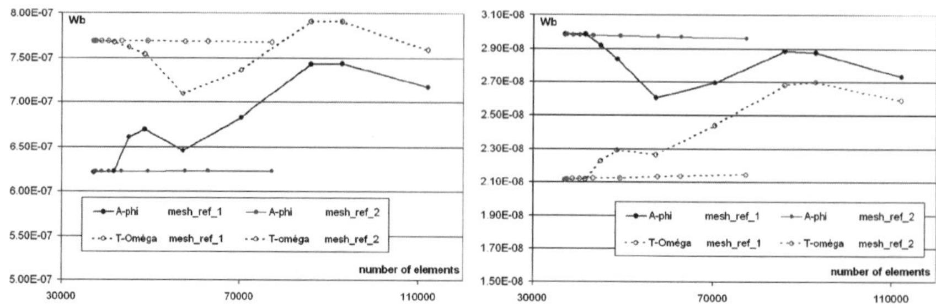

Fig. 7 Evolution of the linkage flux, function of the number of elements

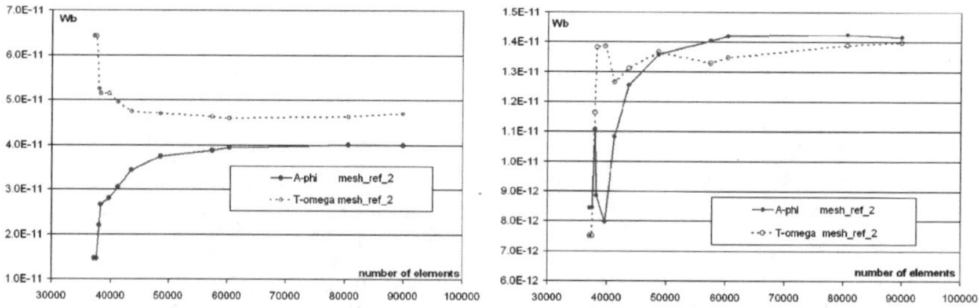

Fig. 8 Variation of the linkage flux for a frequency of 100 kHz, real and imaginary parts

The taking into account of the skin effect needs generally three layers of elements in the thickness for correctly modelling it. With the refinement on the variation, it's not necessary to have this condition; the initial mesh can be very small, even badly constructed. In the ECT modelling case, the sensor(s) move along the default, so the following approach can be used with the refinement on the variation. The initial mesh is chosen deliberately small. A first set of refinement is performed on the initial position until obtain correct value on the flux(es) variations. After that, we take back the initial mesh and a movement method is employed (such as "locked step", "3D interpolation",...). Then, a new set of refinement can be applied.

5 Conclusion

In this paper, we presented a refinement strategy using 3D FEM and applied it to the eddy current testing. In a first part, we presented two complementary FEM formulations. Error estimators based on both formulations were introduced. We proposed a refinement strategy, based on the error difference between the case with and without default. In the last part, we compared the estimators on a simple problem. We show that the studied strategy leaded more quickly and accurately to correct values of the coil parameters variations. Although this strategy needs more calculations at each refinement steps, the mesh should not be very fine, because only the default area is refined.

Acknowledgements Theses works are supported by the French RNTL program PLAYA (promotion of the eddy current simulation).

References

1. T. Henneron, Y. Le Menach, J.-P. Ducreux, O. Moreau, J.-C. Verite, S. Clenet, and F. Piriou, *Source field computation in NDT applications*, Electromagnetic Field Computation, 2006 12th Biennial IEEE Conference, pp. 428–428.
2. M. Rachek and M. Feliachi, *3-D movement simulation techniques using FE methods: application to eddy current non-destructive testing*, NDT & E International, 40(1), January 2007, pp. 35–42.
3. C. Li, Z. Ren, and A. Razek, *An approach to adaptive mesh refinement for three-dimensional eddy-current computations*, IEEE Transactions on Magnetics, 30(1), January 1994.
4. C. Li, Z. Ren, and A. Razek, *Application of complementary formulations and adaptive mesh refinements to non-linear magnetostatics problems*, IEEE Transactions on Magnetics, 30(1), January 1994.
5. Z. Ren, *T-formulation for eddy-current problems in multiply connected regions*, IEEE Transactions on Magnetics, 38(2), March 2002.
6. T. Henneron, S. Clénet, and F. Piriou, *Calculation of extra copper losses with imposed current magnetodynamic formulations*, IEEE Transactions on Magnetics, 42(4), April 2006.
7. Y. Le Menach, S. Clénet, and F. Piriou, *Numerical model to discretize source fields in the 3D finite element method*, IEEE Transactions on Magnetics, 36(4), July 2000.
8. M. Bensetti, L. Santandrea, Y. Choua, Y. Le Bihan, and C. Marchand, *Adaptive mesh refinement and probe signal calculation in eddy current NDT by complementary formulations*, COMPUMAG Conference, Aachen, June 2007.

The Use of L-Curve and U-Curve in Inverse Electromagnetic Modelling

Dorota Krawczyk-Stańdo and Marek Rudnicki

Abstract Regularization methods are used for computing stable solutions to the ill-posed problems. The well-known form of regularization is that of Tikhonov in which the regularized solution is searched as a minimize of the weighted combination of the residual norm and a side constraint-controlled by the regularization parameter. For the practical choice of regularization parameter α we can use the well-known L-curve criterion, or introduced by us U-curve criterion. The efficiency of the approach is demonstrated on examples of synthesis of magnetic field. The paper is finished with the comparison of the two mentioned above methods made on numerical examples.

1 Introduction

The Fredholm integral equations of the first kind are used to model a variety of real applications, such as: medical, imaging, geophysical prospecting, image deblurring, and deconvolution of a measurement instrument's response and of course electromagnetism. The Fredholm integral equation of the first kind takes the generic form:

$$\int_0^1 K(s,t) f(t)\, dt = g(s), \quad 0 \le s \le 1. \tag{1}$$

Dorota Krawczyk-Stańdo
Technical University of Lodz, Center of Mathematics and Physics, 90-924 Łódź, ul. Al. Politechniki 11, Poland
krawczyk@p.lodz.pl

Marek Rudnicki
Technical University of Lodz, Institute of Computer Science, 90-924 Łódź, ul. Wólczańska 215, Poland
rudnicki@ics.p.lodz.pl

D. Krawczyk-Stańdo and M. Rudnicki: *The Use of L-Curve and U-Curve in Inverse Electromagnetic Modelling*, Studies in Computational Intelligence (SCI) **119**, 73–82 (2008)
www.springerlink.com

This equation establishes a linear relationship between the two functions f and g, and the kernel K describes the precise relationship between the two quantities. The inverse problem consists of computing f given the right-hand side and the kernel. Inverse problems, belong to the class of ill-posed problems. The linear problem is well-posed if it satisfies the following three requirements: existence, uniqueness, stability. If the problem violates one or more of these requirements, it is said to be ill-posed. We can only hope to compute useful solution to these problems if we fully understand their inherent difficulties. When the discretization of the problem is carried out we receive a matrix equation in C^m,

$$Ku = f, \tag{2}$$

where K is an $m \times n$ matrix with a large condition number, $m \geq n$. A linear last squares solution of the system (1) is a solution to the problem

$$\min_{u \in c^n} \|Ku - f\|^2, \tag{3}$$

where the Euclidian vector norm in C^m is used. For our least squares problems we use the singular value decomposition (SVD) of the matrix K [1].

The least squares solution of the minimal norm is minimal norm solution of the normal equation $K^* \, Ku = K^* \, f$, and thus if

$$f = \sum_{i=1}^{m} f_i u_i, \text{ where } f_i = u_i^* f, \; i = 1, \ldots, m \tag{4}$$

then

$$u = \sum_{i=1}^{r} \frac{f_i}{\sigma_i} v_i, \tag{5}$$

where the σ_i are called the singular values of K. The numerical methods for solving ill-posed problems are based on so-called regularization methods. The main objective of regularization is to incorporate more information about the desired solution in order to stabilize the problem and find a useful and stable solution. The well-known form of regularization is that of Tikhonov [2]. It consists of substituting the least squares problem (3) with the problem of suitably chosen Tikhonov functional. The most basic version of this method can be presented on

$$\min_{u \in C^n} \left\{ \|Ku - f\|^2 + \alpha^2 \|u\|^2 \right\}, \tag{6}$$

where $\alpha \in R$ is called the regularization parameter. The Tikhonov regularization is the method in which the regularized solution is searched as a minimizer of the weighted combination of the residual norm and a side constraint. The quality of the regularized solution is controlled by the regularization parameter.

The regularized solution of (6) satisfies the normal equation $K^* \, Ku + \alpha^2 u = K^* \, f$ and takes the form

$$u_\alpha = \sum_{i=1}^{r} \frac{\sigma_i f_i}{\sigma_i^2 + \alpha^2} v_i. \tag{7}$$

Because of $\alpha > 0$, the problem of computing u_α becomes less ill-conditioned than that of computing u.

It is easily found [3, 4] that

$$x(\alpha) = \|Ku_\alpha - f\|^2 = \sum_{i=1}^{r} \frac{\alpha^4 f_i^2}{\left(\sigma_i^2 + \alpha^2\right)^2} + \|f_\perp\|^2, \text{ where } f_\perp = \sum_{i=r+1}^{m} f_i u_i, \quad (8)$$

$$y(\alpha) = \|u_\alpha\|^2 = \sum_{i=1}^{r} \frac{\sigma_i^2 f_i^2}{\left(\sigma_i^2 + \alpha^2\right)^2}. \quad (9)$$

The main thing that we are still missing is a reliable technique for choosing the regularization parameter. Unfortunately, such a method has yet to be found At our disposal we have a collection of methods (the well known: discrepancy principle, generalized cross validation, L-curve criterion and introduced by us [5] U-curve criterion), which – under certain assumptions – tend to work well, but all of them can – and will – occasionally fail to produce good results.

In Sect. 2, we recall the L-curve criterion and U-curve criterion for choosing the regularization parameter. Section 3 presents numerical results obtained using the new U-curve criterion and compared them with those resulting from the application of the L-curve criterion.

2 L-Curve Criterion and U-Curve Criterion for Choosing Regularization Parameter

The more recent method of choosing regularization parameter is to base regularization parameter on the so-called L-curve, see [3, 6]. For Tikhonov regularization L-curve is a parametric plot of $(x(\alpha), y(\alpha))$, where $x(\alpha)$ and $y(\alpha)$ given by (8), (9) for all $\alpha > 0$ [3, 4, 6] contains many properties of the L-curve for Tikhonov regularization. The L-curve has two characteristic parts: the more horizontal where the solution is dominated by the regularization errors, the vertical part where the solution is dominated by the right-hand errors. The solutions are over- and under-smoothed, respectively. The corner of the L-curve corresponds to a good balance between minimization of the sizes, and the corresponding regularization parameter α is a good one. There are two meanings of the "corner" [3]: the point where the curve closest to the origin, the point where the curvature is a maximum. Figure 1 shows an example of a typical L-curve.

The new, introduced by us, method of choosing regularization parameter is to base regularization parameter on the so-called U-curve, see [5]. Let us take the following function into account

$$U(\alpha) = \frac{1}{x(\alpha)} + \frac{1}{y(\alpha)}, \text{ for } \alpha > 0, \quad (10)$$

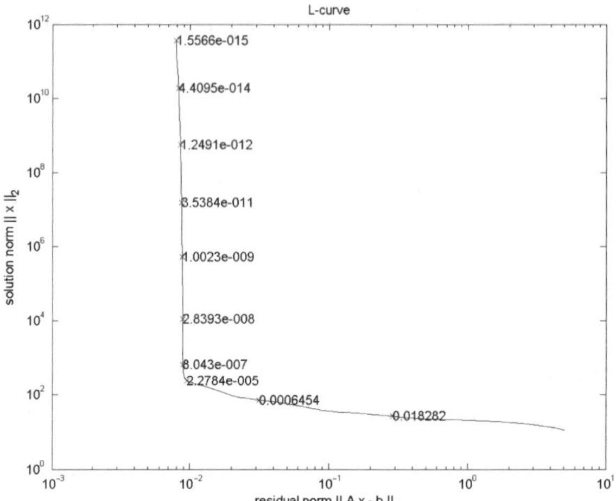

Fig. 1 The plot of typical L-curve

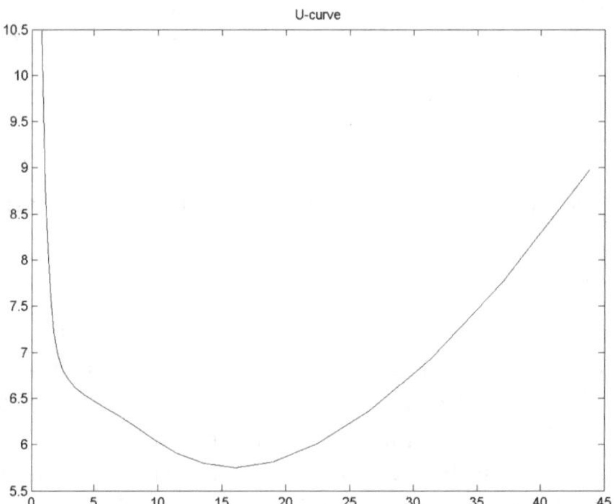

Fig. 2 The plot of typical U-curve

where $x(\alpha)$ and $y(\alpha)$ are defined by (8), (9) and measure the size of the regularized solution and the corresponding residual, for all $\alpha > 0$. By U-curve for Tikhonov regularization we understand the plot of $U(\alpha)$, i.e. the plot of the sum of the reciprocals of the regularized solution norm and the corresponding residual norm, for $\alpha > 0$. The U-curve consists of three characteristic parts, namely: on the left and right side – almost "vertical" parts, in the middle – almost "horizontal" part. The vertical parts correspond to the regularization parameter, for which the solution norm and the residual norm are dominated by each other, respectively. The more horizontal part corresponds to the regularization parameter, for which the solution norm and the residual norm are close to each other. Figure 2 shows an example of a typical U-curve. The objective of the U-curve criterion for selecting the regularization parameter is to choose a parameter which corresponds to the corner between left vertical part and horizontal part of U-curve. It is located exactly where the $x(\alpha)$ and $y(\alpha)$ gradually become balanced (stop to be dominated by each other) and begin to be close to each other). That is why it corner of U-curve corresponds to a good balance between minimization of the sizes $x(\alpha)$ and $y(\alpha)$, and the corresponding regularization parameter α is the good one. This corner we can selected as the point where the curve closest to origin or as the point where the curvature is maximum.

The regularization parameter appropriate for U-curve criterion is calculated numerically by applying routines, available in the package Regularization Tools of Mat lab – appropriately modified by us.

3 Numerical Results

Example 1. Let us consider the following Fredholm integral equation of the first kind in one dimension

$$\int_0^1 e^{xy} f(y)\, dy = \frac{1}{x}\left(e^x - 1\right), \; 0 \leq x \leq 1$$

Its unique solution is given by $f = 1$, see [7].

The numerical results corresponding to noise of order of 10^{-1} are shown below (see Figs. 3 and 4). The exemplary solution is presented in Fig. 5.

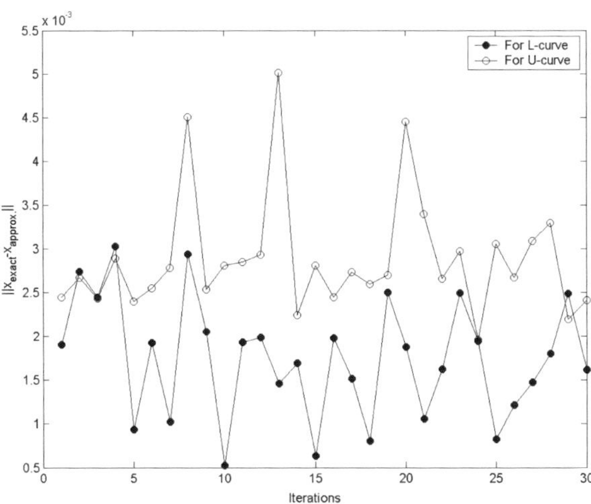

Fig. 3 The plots of error norm

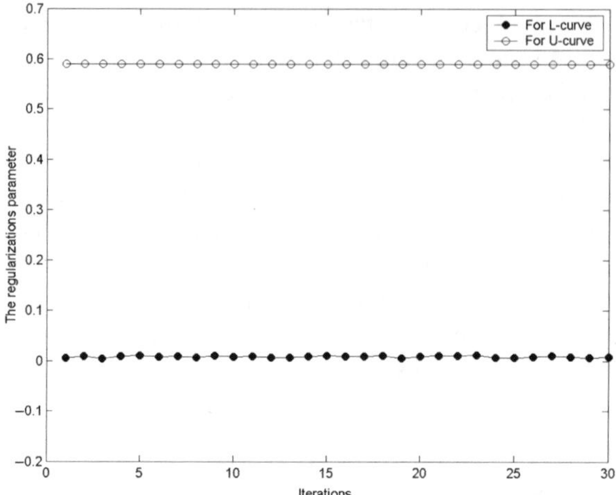

Fig. 4 The plots of regularization parameter

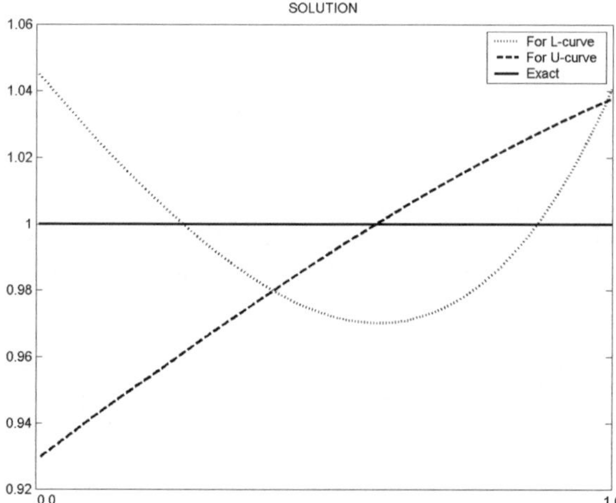

Fig. 5 The plots of solution

The numerical results corresponding to noise of order of 10^{-2} are shown below (see Figs. 6 and 7). The exemplary solution is presented in Fig. 8.

Example 2. Let us consider as an example the test problem *"philips"* [8], with given exact solution.

The numerical results (for noise of order 10^{-1} are shown below (see Figs. 9 and 10). The exemplary solution is presented in Fig. 11. The numerical results for noise of order of 10^{-2} are shown below (see Figs. 12 and 13). The exemplary solution is presented in Fig. 14.

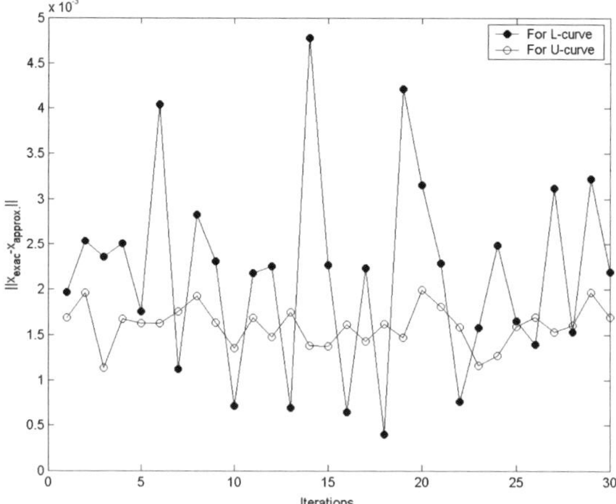

Fig. 6 The plots of error norm

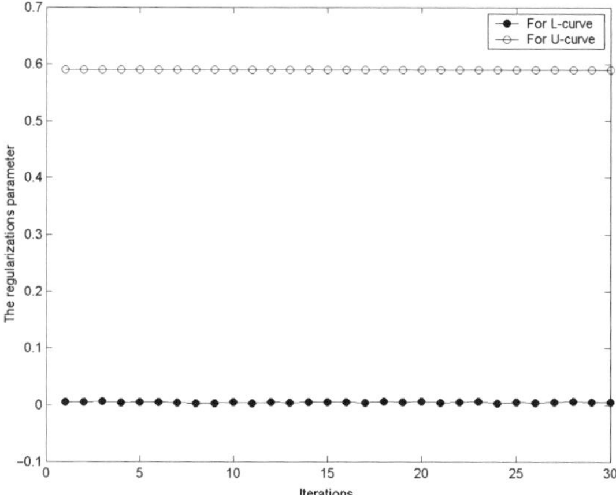

Fig. 7 The plots of regularization parameter

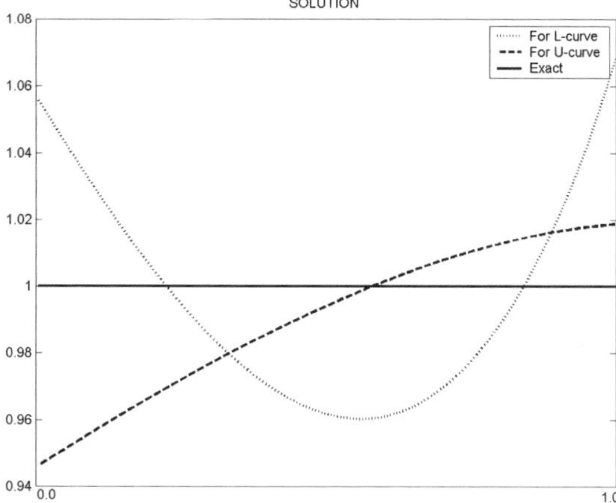

Fig. 8 The plots of solution

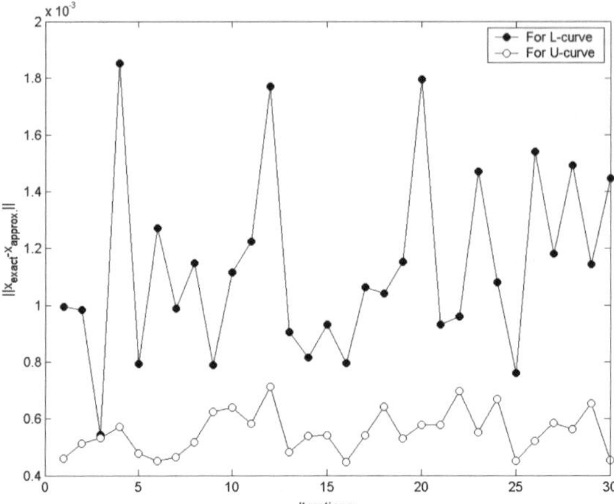

Fig. 9 The plots of error norm

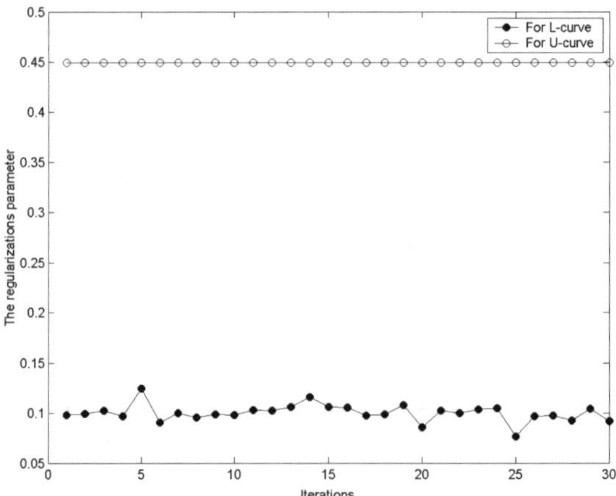

Fig. 10 The plots of regularization parameter

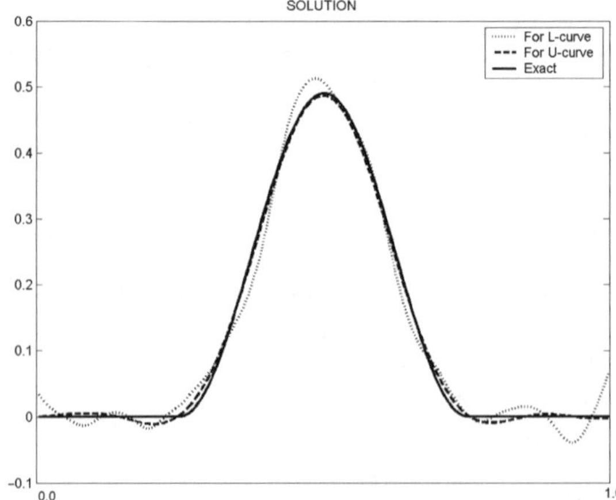

Fig. 11 The plots of solution

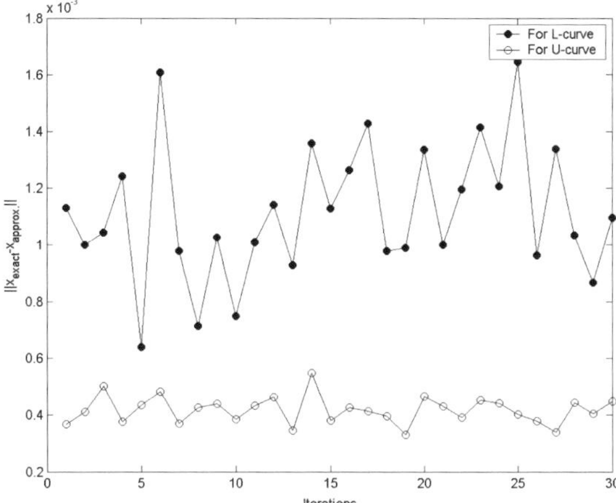

Fig. 12 The plots of error norm

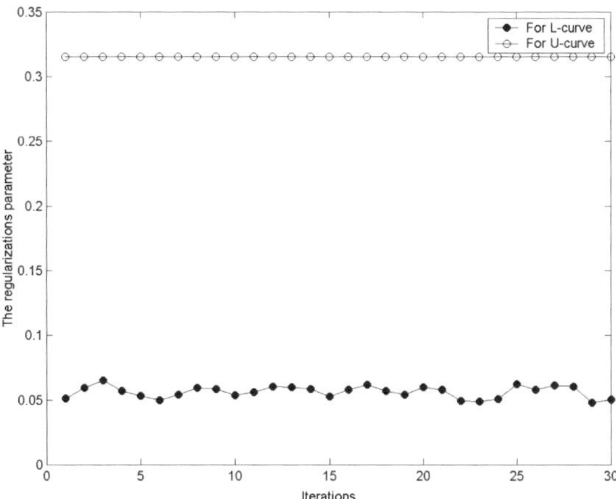

Fig. 13 The plots of regularization parameter

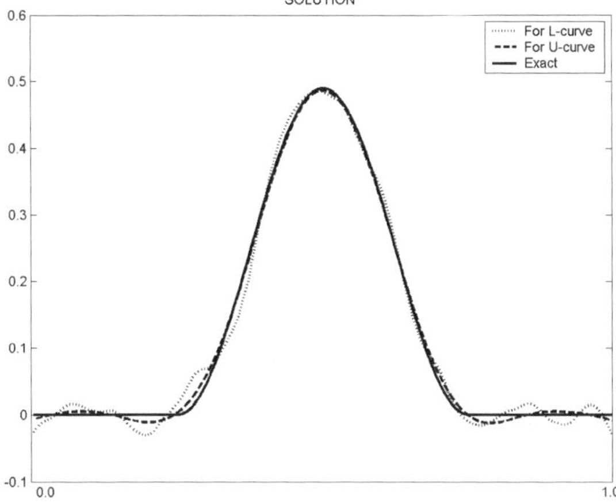

Fig. 14 The plots of solution

4 Conclusions

The results which we have found from both L-curve and U-curve methods are comparable and are close to the results which we published in [7, 9]. Moreover, we can observe the following:

1. For perturbation of order of 10^{-1} the error norm $||x_{exact} - x_{approx^*}||$ appears to be worse compared to the one found by the U-curve versus the L-curve.
2. The U-curve method is more stable against the L-curve method (see Figs. 10 and 13).

Open problem. Searching for the dependence between regularization parameters that can be found by considered methods: L-curve and U-curve.

References

1. G. Wahba, Practical approximate solutions to linear operator equations when data are noisy, SIAM Journal of Numerical Analysis, Vol. 14, No. 4, pp. 651–667, 1977
2. N. Groetsch, The theory of Tikhonov regularization for Fredholm integral equations of the first kind. Pitman, London, 1984
3. P.C. Hansen and D.P. O'Leary, The use of L-curve in the regularization of discrete ill-posed problems, SIAM Journal of Scientific Computing, Vol. 14, pp. 1487–1503, 1993
4. T. Regińska, A regularization parameter in discrete ill-posed problems, SIAM Journal Scientific Computing, Vol. 17, No. 3, pp. 740–749, 1996
5. D. Krawczyk-Stańdo and M. Rudnicki, Regularization parameter in discrete ill-posed problems-the use of the U-curve, International Journal of Applied Mathematics and Computer Science, Vol. 17, No. 2, pp. 101–108, 2007
6. P.C. Hansen, Analysis of discrete ill-posed problems by means of the L-curve, SIAM Review, Vol. 34, No. 4, pp. 561–580, 1992
7. P. Neittaanmaki, M. Rudnicki, and A. Savini, Inverse Problems and Optimal Design in Electricity and Magnetism, Oxford, Clarendon Press, 1996
8. P.C. Hansen, Regularization tools, a Mat Lab Package for Analysis and Solution of Discrete Ill-Posed Problems – Report UNIC-92-03, 1993
9. D. Krawczyk-Stando and M. Rudnicki, Regularised synthesis of the magnetic field using the L-curve approach, International Journal of Applied Electromagnetics and Mechanics, Vol. 22, No. 3,4, pp. 233–242, 2005

Optimization of the HTSC-PM Interaction in Magnetic Bearings by a Multiobjective Design

Paolo Di Barba and Ryszard Palka

Abstract The paper deals with the shape design of the excitation system of a high-temperature-superconductor (HTSC) magnetic bearing by means of a procedure of multiobjective design. As a result of the proposed method, a set of optimal geometries of the basic excitation structure was extracted from all possible solutions in the objective space. Finally, the sensitivity of the levitation force acting on the HTSC with respect to geometry fluctuations was estimated; this way, the robustness of optimal solutions can be controlled.

1 Introduction

Superconducting magnetic bearings (SMB) are one of the most promising applications of bulk high-temperature superconductors. The special advantages of this kind of bearings are: contact-free operation and thus no wear-out, no need of control- and sensor-units, high reliability and no EMC-problems [1–3]. Of outstanding interest is their use for high speed rotating machines, e.g. turbo machinery. The forces of SMB are caused by the interaction between superconducting currents induced in the HTSC and the external magnetic field generated by the excitation unit consisting of permanent magnets (PM). Figure 1 shows the bulk superconductor in two different positions within the magnetic field of the main configuration used for SMB.

In HTSC systems, the magnetic field experienced in the activation position (g_{act}) remains unaltered after movement to the operational position (g_{op}). The calculation method in this case (trapped field model, Fig. 1) consists of four steps [2, 3]. Two

Paolo Di Barba
University of Pavia, 27100 Pavia, Italy
paolo.dibarba@unipv.it

Ryszard Palka
Szczecin University of Technology, 70-313 Szczecin, Poland
rpalka@ps.pl

P.D. Barba and R. Palka: *Optimization of the HTSC-PM Interaction in Magnetic Bearings by a Multiobjective Design*, Studies in Computational Intelligence (SCI) **119**, 83–90 (2008)
www.springerlink.com

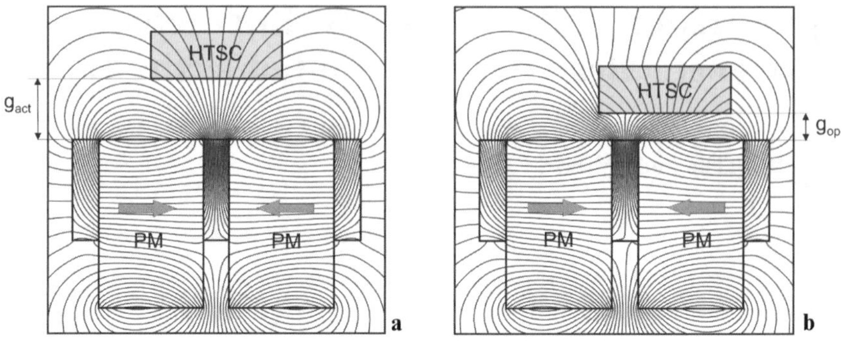

Fig. 1 High-temperature superconductor bulk acting with the flux concentrating arrangement used for superconducting bearings. Initial position g_{act} **a** and final position g_{op} **b** – the field distribution within the HTSC does not change (*trapped field model*)

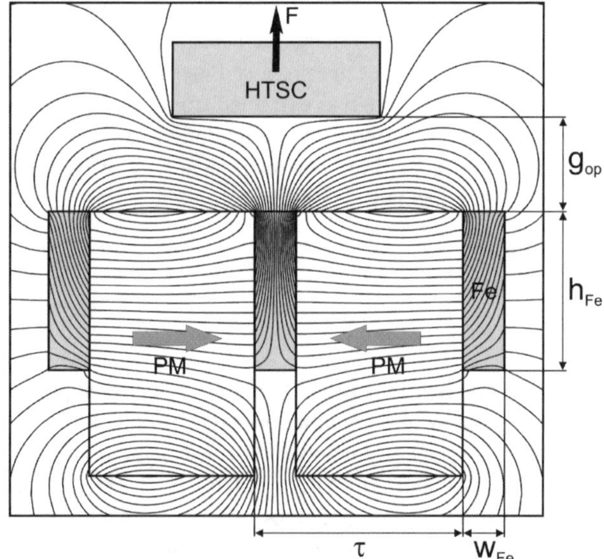

Fig. 2 High-temperature bulk superconductor in the Meissner state acting with the flux concentrating arrangement used for superconducting bearings

of them require that the Meissner effect within the HTSC in the initial and the final position has to be evaluated. The Meissner state is typical for type-I superconductors, where reversible surface currents are induced in the superconducting body, perfectly screening the applied magnetic field. In HTSC such behaviour can be observed as long as the external field is lower than the first critical magnetic field. However, this perfect diamagnetic model should be used as a theoretical limit to maximum force in view of the design of repulsive HTSC-PM configurations. Additionally, the Meissner state within the HTSC is equivalent to the zero field cooling activation mode of SMB. All above reasons ask for geometry optimization of the flux concentrating arrangement shown in Fig. 1, assuming the Meissner state in the HTSC (Fig. 2).

The optimization of the basic structure in Fig. 2 is essential for any practical application of SMB.

2 Multiobjective Optimization of the PM-HTSC Interaction

In order to optimize the structure shown in Fig. 2, the vector $g = (h_{Fe}, w_{Fe}, \tau)$ of three design variables was considered. Preliminarily, two design criteria were defined (see Fig. 2):

1. The levitation force density $f_1(g)$ in the vertical direction, for a given activation position of the superconductor, to be maximized;
2. The cross-sectional area f_2 of the HTSC, to be prescribed.

   The optimization target can be stated as follows: *find the maximum levitation force density that can be obtained for a given HTSC area; then, identify the geometry of the relevant excitation system.* The trade-off solutions are given by the non-dominated solutions in the (f_1, f_2) space [4]: they are those for which the decrease of a criterion is not possible without the simultaneous increase of at least one of the other criteria.

   For the sake of a realistic design, however, the sensitivity of the levitation force with respect to random fluctuations of the geometry should be taken into account. Consequently, a third design criteria, i.e.
3. The sensitivity s(g) of a feasible solution g, to be controlled, was considered.

3 Numerical Model of Design Sensitivity

It is well known that optimal solutions which are remarkably perturbed by small deviations of the design variables around their nominal value are not desirable; in fact, they would determine strong tolerance specifications in the fabrication process. Therefore, low-sensitivity – or robust – solutions are preferred.

An effective model to estimate the sensitivity of a design point is based on the definition of cluster. In this respect, given a set Ω of $n_p \gg 1$ points sampling the n_v-dimensional design space (here, $n_v = 3$), the distance

$$d(g_i, g_j) = \left[\sum_{k=1}^{n_v} \left[g_i(k) - g_j(k)\right]^2\right]^{\frac{1}{2}}, \; i = 1, \; n_p - 1, \; j = 2, n_p, j > i, \qquad (1)$$

between *i*th and *j*th point is evaluated for all possible pairs of points in Ω. Of course, a point in the design space is represented by a n_v-dimensional vector g. Given a point $\tilde{g} \in \Omega$, the associated cluster ω is formed by points $g \in \Omega$ such that $d(g, \tilde{g})$ is lower than a known threshold δ; the latter depends on set Ω and is defined as

$$\delta = (n_v)^{\frac{1}{2}} \left\{ (n_p)^{-1} \prod_{k=1}^{n_v} \left[\sup_{\Omega} g(k) - \inf_{\Omega} g(k) \right] \right\}^{\frac{1}{n_v}}. \quad (2)$$

Geometrically, $\prod_{k=1}^{n_v} \left[\sup_{\Omega} g(k) - \inf_{\Omega} g(k) \right]$ is the volume of the hypercube incorporating set Ω; therefore, δ can be viewed as the elementary hyper-diagonal associated to Ω. Then, the sensitivity $s(\tilde{g})$ can be evaluated, e.g. according to the ∞-norm within the cluster ω associated to $\tilde{g}$:

$$s(\tilde{g}) = [f_1(\tilde{g})]^{-1} \left[\sup_{g} f_1(g) - \inf_{g} f_1(g) \right] \equiv \frac{\Delta f_1}{f_1},\ f_1(\tilde{g}) \neq 0, \quad (3)$$

where f_1 is the levitation force density, for all $g \in \Omega$ such that $d(g, \tilde{g}) < \delta$. Indeed, the bigger the number of cluster elements, the more accurate the sensitivity evaluation.

According to an enumerative search, a discrete representation of the three-dimensional design space was obtained by randomly sampling the design space with about 2,000 points fulfilling the problem constraints; for each point, the magnetic field was simulated and the sensitivity of criterion f_1 was then estimated. In Fig. 3 the sensitivity of all points g in Ω is shown in terms of the following variation

$$\left[\sum_{k=1}^{n_v} [g(k)]^2 \right]^{-\frac{1}{2}} \left[\sum_{k=1}^{n_v} \left[\sup_{\omega} g(k) - \inf_{\omega} g(k) \right]^2 \right]^{\frac{1}{2}} \equiv \frac{\Delta g}{g},\ g \in \omega,\ g \neq \{0\}. \quad (4)$$

It can be remarked that the random distribution of points concentrates in a region of low/moderate variation of design point g.

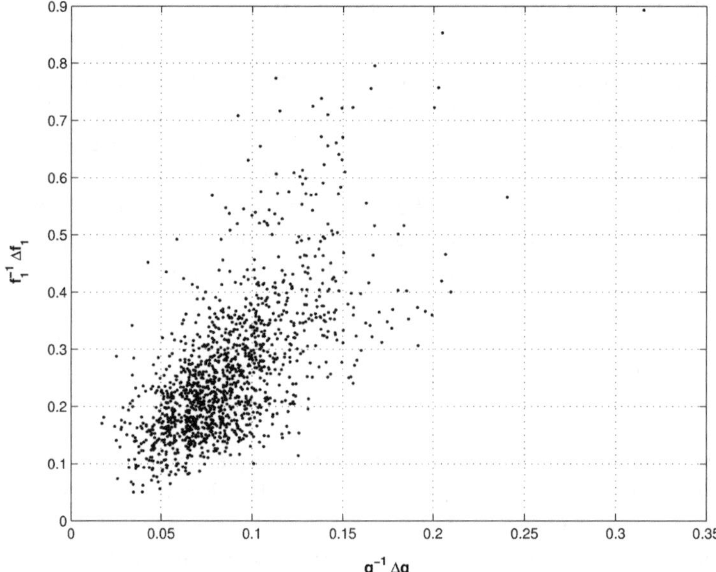

Fig. 3 Variation of design criterion f_1 vs. variation of design point g (1,915 field solutions)

4 Design Problem: Performance Vs. Robustness

A discrete representation of the objective space (f_1, s) is shown in Fig. 4, where the upper limit of the distribution is the front of worst-case solutions, i.e. the set of solutions exhibiting the highest sensitivity with respect to geometry fluctuations.

The field distributions for two worst case solutions (i.e. front-centre and one out of two front-ends as in Fig. 4) are shown in Fig. 5.

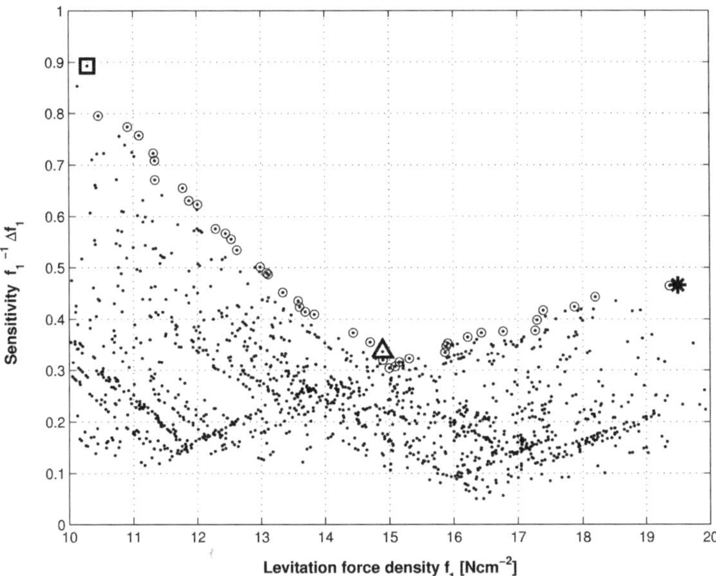

Fig. 4 Sensitivity vs. levitation force density: sampled objective space (1,915 field solutions, dot) with 43 worst-case solutions (*circle*) highlighted. Front-end (*square, asterisk*) and front-centre (*upward triangle*) solutions are also outlined

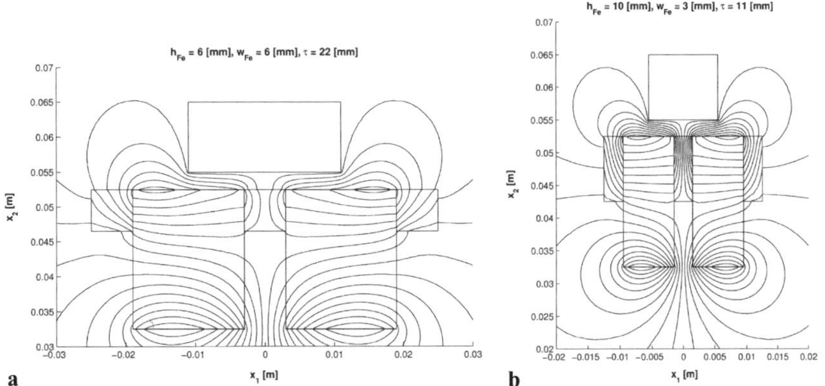

Fig. 5 Geometry and flux lines (corresponding to the solutions marked in Fig. 4) **a** by an upward triangle: $B_{PM} = 332\,\text{mT}$, $f_1 = 14.893\,\text{N cm}^{-2}$, $f_2 = 2.2\,\text{cm}^2$, $f_1^{-1}\Delta f_1 = 0.337$, **b** by an asterisk: $B_{PM} = 192\,\text{mT}$, $f_1 = 19.491\,\text{N cm}^{-2}$, $f_2 = 1.1\,\text{cm}^2$, $f_1^{-1}\Delta f_1 = 0.466$. B_{PM} is the induction at the centre of each permanent magnet

Comparing Fig. 4 and Fig. 5, it can be noted that front-ends and front-centre differ in both performance (levitation force with relevant sensitivity) and shape (size of field excitation arrangement). In view of an automated procedure of robust optimal design, a set $\{g_\ell\}$ of n_s solutions, non-dominated in the (f_1, f_2) space and such that $0 < s(g_\ell) \leq s^*$, $\forall \ell = 1, n_s$ where $s^* \in]0, 1]$ is a prescribed threshold of acceptable sensitivity, should be identified.

5 Results

In Fig. 6 the (f_1, f_2) space corresponding to two thresholds s^* of sensitivity is represented. In particular, $s^* < 0.15$ holds in the left picture, with 363 out of 1,915 solutions sampling the whole design space, while $s^* < 0.1$ holds in the right picture, capturing 97 out of 1,915 solutions. In both cases, non-dominated solutions (upper triangle) with relevant front-end solutions (asterisk, circle) are highlighted.

The solutions which are non-dominated solve the following problem:

given $\sigma > 0$ and $0 < s^* < 1$, find sup $f_1(g)$ subject to $f_2(\tau) = \sigma$ and $0 < s \leq s^*$, where σ and s^* are prescribed values of superconductor area and sensitivity,

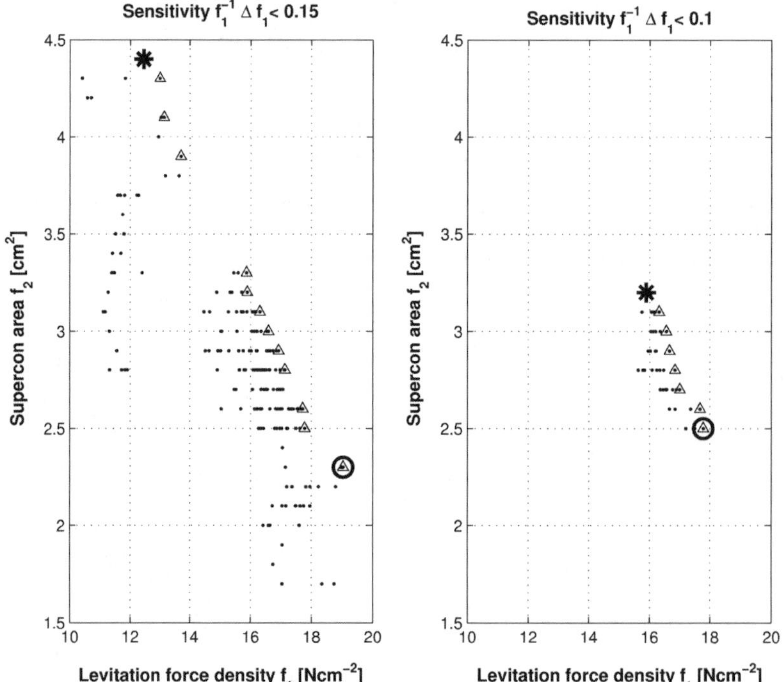

Fig. 6 Superconductor area vs. levitation force density

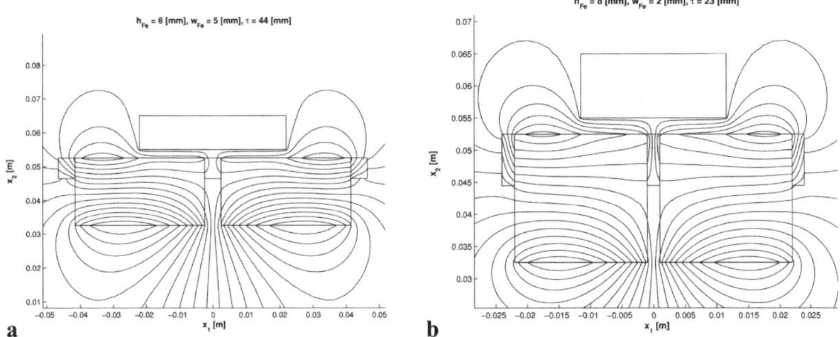

Fig. 7 Geometry and flux lines (corresponding to the solution marked in Fig. 6 left) **a** by an asterisk: $B_{PM} = 581\,mT$, $f_1 = 12.442\,N\,cm^{-2}$, $f_2 = 4.4\,cm^2$, $f_1^{-1}\Delta f_1 = 0.1485$, **b** by a circle: $B_{PM} = 384\,mT$, $f_1 = 19.03\,Ncm^{-2}$, $f_2 = 2.3\,cm^2$, $f_1^{-1}\Delta f_1 = 0.1396$

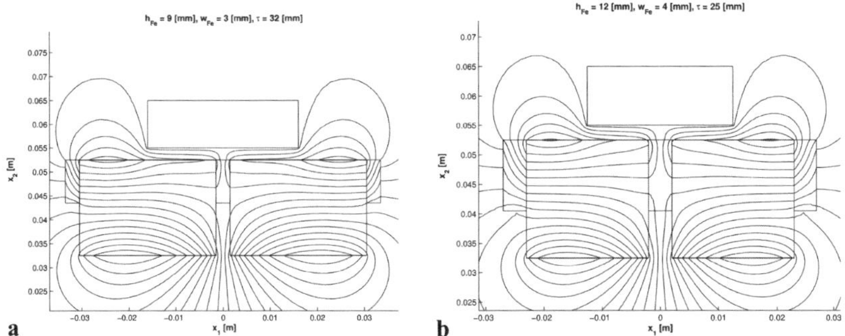

Fig. 8 Geometry and flux lines (corresponding to the solution marked in Fig. 6 right) **a** by an asterisk: $B_{PM} = 494\,mT$, $f_1 = 15.887\,N\,cm^{-2}$, $f_2 = 3.2\,cm^2$, $f_1^{-1}\Delta f_1 = 0.0958$, **b** by a circle: $B_{PM} = 405\,mT$, $f_1 = 17.778\,N\,cm^{-2}$, $f_2 = 2.5\,cm^2$, $f_1^{-1}\Delta f_1 = 0.0923$

respectively. In general, it can be stated that the set of non-dominated solutions in the (f_1, f_2) space changes according to the value of s^*. In Figs. 7 and 8 the two pairs of geometries corresponding to the front-end solutions are represented with reference to Fig. 6.

Again, it can be remarked that the front-ends differ in both performance and shape.

The comparison between some selected SMB basic structures are shown in Table 1. The exploitation of the HTSC-mass has been enhanced by reducing the HTSC-thickness. Further improvements have been achieved by reduction of the height of the flux-collecting iron core. The suspension stiffness of the SMB structure has been evaluated using the formula: $S_s = -\partial f_1/\partial x_2$.

Table 1 Partial optimization of the SMB basic structure

	a	b	c
Remarks	Basic configuration (%)	Reduced HTSC height (%)	Reduced iron height (%)
HTSC volume	100	70	70
Repulsive force	100	100	110
Stiffness (suspension)	100	100	107

6 Conclusion

A numerical procedure for the optimization of the superconducting bearing has been presented. The definition of non-dominated solution, along with the adoption of a technique of enumerative search, made it possible to investigate the objective space of levitation force density vs. superconductor area. During the search, an upper bound to the design sensitivity was considered. The proposed method puts the ground for the robust shape design of a class of superconducting magnetic bearings.

References

1. Krabbes G., Fuchs G., Canders W.-R., May H., Palka R., High Temperature Superconductor Bulk Materials, Wiley, New York, 2006
2. May H., Palka R., Portabella E., Canders W.-R., Evaluation of the Magnetic Field-High Temperature Superconductor Interactions, COMPEL 23, No. 1, 2004, pp. 286–304
3. Palka R., Modelling of High Temperature Superconductors and Their Practical Applications, International Compumag Society Newsletter, Vol. 12, No. 3, Nov. 2005, pp. 3–12
4. Di Barba P., Multiobjective Design Optimization: a Microeconomics-Inspired Strategy Applied to Electromagnetics, IJAEM 21, 2005, pp. 101–117

Part II
Computer Methods and Engineering Software

Optimization of an Electromagnetic Linear Actuator Operating in Error Actuated Control System

Lech Nowak, Kazimierz Radziuk, and Krzysztof Kowalski

Abstract In the paper an algorithm and computer software for the optimisation of error actuated control systems containing of an electromagnetic linear actuator is presented. The software consists of two main parts: (a) optimisation solver and (b) numerical model of the actuator. For the optimisation, the genetic algorithm (GA) has been used. The field discrete mathematical description is employed. The field-circuit model of the electromagnetic actuator includes: (a) the equation of a transient electromagnetic field, (b) equations which describe electric circuits of the system, including feedback, (c) the equation of mechanical motion and (d) the equation describing and the proportional-plus-integral-plus-derivative (PID) controller.

1 Introduction

Linear movement electromagnetic actuators constitute a significant part of electromechanical energy converters. Because of elimination of the transmission and coupling mechanisms such linear movement driving devices have many advantages as compared to classical electric drives with rotational machines. Very often, e.g. in magnetic levitation and magnetic bearing systems, a linear control characteristic, (i.e. proportionality between electric input signal and output mechanical quantity) is required. In such case the closed loop error actuated control system is applied [1,2].

In the paper an algorithm and computer software for the optimisation of the error actuated control systems with electromagnetic actuator is presented. The actuator dimensions and the settings of proportional-plus-integral-plus-derivative (PID) controller settings are assumed as a design variables.

Lech Nowak, Kazimierz Radziuk, and Krzysztof Kowalski
Poznań University of Technology, Institute of Industrial Electrical Engineering,
ul. Piotrowo 3a, PL 60-965 Poznań, Poland
lech.nowak@put.poznan.pl, kazimierz.radziuk@put.poznan.pl, krzysztof.kowalski@put.poznan.pl

L. Nowak et al.: *Optimization of an Electromagnetic Linear Actuator Operating in Error Actuated Control System*, Studies in Computational Intelligence (SCI) **119**, 93–100 (2008)
www.springerlink.com

The software consists of two main parts: (a) optimisation solver and (b) numerical model of the actuator. The genetic algorithm has been used for the optimisation. The field-circuit discrete mathematical description for the dynamics of an electromagnetic actuator is employed. The model includes coupled equation describing different phenomena. It consist of: (a) the equation of a transient electromagnetic field in a non-linear conducting and moving medium, (b) equations which describe electric circuits of the converter and the supply system, (c) the equation of mechanical motion and (d) the equations which describe closed-loop control. Models of a sensor and a PID controller which form the feedback loop are also included. Numerical implementation is based on the finite elements, Newton–Raphson procedure and the step-by-step algorithm.

2 Optimisation Procedure

For the optimisation the genetic algorithm has been adopted. For the crossover and mutation the binary (32-bit) representation of the design variables (chromosomes) is applied. Their dimensionless decimal counterparts x_i belongs to the range $< 0,\ 1 >$, while the number 1 corresponds to 2^{32}.

The optimisation solver has been elaborated regardless of the software for field circuit simulation of the actuator and feedback. Both units are connected through the transformation of the real design variables s_i uses in the mathematical model of the actuator and their dimensionless counterparts x_i used in the optimisation procedure. The set of the data must include the lower s_{imin} and upper s_{imax} limits (expected values) of each variable s_i. The variable transformations are given by the relations:

$$x_i = (s_i - s_{imin})/(s_{imax} - s_{imin}), \text{ and } s_i = s_{imin} + (s_{imax} - s_{imin})x_i \tag{1}$$

If $s_i \in \langle s_{imin}, s_{imax} \rangle$ then $x_i \in \langle 0, 1 \rangle$.

In the elaborated genetic procedure objective function $f(\mathbf{X})$ (individuals adaptation) is maximised. But the natural criterion of the optimisation of the dynamic system under the consideration is to minimize the set-up time T_{set} of the error actuated system. Therefore, the criterion has to be modified. Assuming the time of reference T_{ref} the objective function is constructed as follows:

$$f(\mathbf{X}) = 1 - T_{set}/T_{ref} \tag{2}$$

Value of T_{ref} should be of the order $2\,(T_{set})_{opt}$, where $(T_{set})_{opt}$ is the expected value of the setting time for optimal individual. In such case, the conversion (4) practically doesn't change the measure of the differences between individuals, what ensures unchanged probability of individuals reproduction.

3 Numerical Implementation of Field Circuit Model of the Dynamics of Actuator Operating in Error Actuated System

The most common design of electromagnetic actuators is known as the plunger-type structure – Fig. 1 [3, 4]. A roller-shaped mover plunges into a cylindrical coil positioned inside the ferromagnetic core. Such actuators are widely used as servos in automation systems or as driving devices of hydraulic and pneumatic electro-valves. The reverse movement is caused by a return spring.

The complete model of electromagnetic and mechanical phenomena comprises: the equation describing a transient electromagnetic field, the Kirchhoff's equation describing electric circuits (including feedback with PID controller) and the equation of mechanical motion [2].

Numerical implementation is based on the finite element method (FEM) and the Newton–Raphson process. The Crack–Nicholson difference scheme for time-stepping has been employed. As a result, for the nth instant, the following non-linear set of equations is obtained [4]:

$$\mathbf{M}_n(\Phi_n,\xi)\cdot\Phi_n = \mathbf{N}i_n + 2\mathbf{G}(\Delta t)^{-1}\Phi_{n-1} + \mathbf{G}\left.\frac{\mathrm{d}\Phi}{\mathrm{d}t}\right|_{n-1}, \tag{3}$$

where: $\mathbf{M}_n(\Phi_n,\xi) = \mathbf{S}_n(\Phi_n,\xi) + 2(\Delta t)^{-1}\mathbf{G}$.

Here, $\mathbf{S}$ is the stiffness matrix, Φ is the vector of nodal potentials, $\mathbf{G}$ is the conductance matrix, $\mathbf{N}$ is the matrix of turn numbers, i is the exciting current. If current $i(t)$ is enforced then (3) may be solved directly at each time-step. Otherwise, the

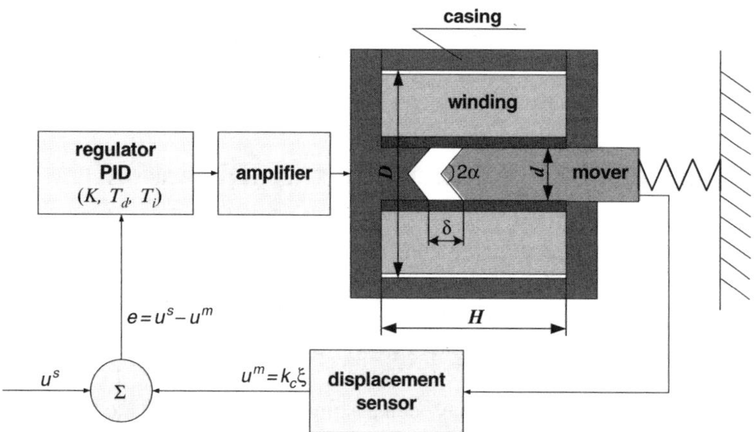

Fig. 1 The plunger type actuator operating in error actuated control system

circuit equations must be taken into account [2,4]. According to the applied difference scheme, it assumes the following discrete form:

$$2(\Delta t)^{-1}\mathbf{N}^{\mathrm{T}}\Phi_n + R\,i_n = u_n + \mathbf{N}^{\mathrm{T}}\left\{2(\Delta t)^{-1}\Phi_{n-1} + \left.\frac{\mathrm{d}\Phi}{\mathrm{d}t}\right|_{n-1}\right\}, \tag{4}$$

where: u_n is the supply voltage value at $t = t_n$, R is the winding resistance.

The formula for the mover displacement ξ_n has been derived from the equation of motion [2]:

$$\xi_n = \xi_{n-1} + \frac{\left[m + k_f 0.25\Delta t\right]\mathrm{v}_{n-1}}{m + 0.5k_f\Delta t}\Delta t + 0.25\frac{(F_{en} - F_{lon}) + m\left.\frac{\partial \mathrm{v}}{\partial t}\right|_{n-1}}{m + 0.5k_f\Delta t}(\Delta t)^2, \tag{5}$$

where: m is the mass of moving parts, v is the velocity, k_f is the friction coefficient, F_e, F_{lo} are the electromagnetic and loading forces, respectively.

The position ξ_n at the nth time-step is not known in advance, hence the forces $F_{en} = F_e(\Phi_n, \xi_n)$ and $F_{lon} = F_{lo}(\xi_n)$ in the right-hand side of (5) are not known explicitly. Therefore, in order to determine ξ_n, iterative calculations are required [2,4].

Finally, the equation describing the PID controller was included. For the nth instant, the difference scheme gives the following discrete formula [1,2]:

$$u_n = Ke_n + KT_d\left\{\frac{e_n - e_{n-1}}{0.5\Delta t} - \left.\frac{\mathrm{d}e}{\mathrm{d}t}\right|_{n-1}\right\}\left.\frac{\mathrm{d}e}{\mathrm{d}t}\right|_n + \frac{K}{T_i}\left\{J_{n-1} + 0.5\left[e_n + e_{n-1}\right]\Delta t\right\}. \tag{6}$$

Here: u_n is the controller output signal; $e_n = u^s - k_c\xi_n$ is the error equal to the difference between the set value u^s proportional to the set position ξ^s and the measured sensor output voltage proportional to the actual position ξ_n; K is the controller amplification; T_d, T_i are the controller differentiation and integration constants, J_n is the value of the integral at the previous instant. The displacement ξ_n is determined iteratively and so is the error e_n [2].

4 Optimisation of the Actuator Structure (Dimensions)

Dynamic operation of the actuator with feedback, after the application of the DC supply voltage $U = 60\,\mathrm{V}$, has been considered. The magnetic air gap $\delta(t)$ has changed during the movement from the initial stable value $\delta_b = 8\,\mathrm{mm}$ to the final stable value $\delta_f = 2\,\mathrm{mm}$. The displacement $\xi(t) = \delta_b - \delta(t)$ has changed in interval 0–6 mm and the relative value $x(t) = \xi(t)/(\delta_b - \delta_f)$ – in interval $(0,\ 1)$. The set value $\xi^s = 3\,\mathrm{mm}$ ($\delta^s = 5\,\mathrm{mm}$, $x^s = 0.5$) has been assumed. It corresponds to the set voltage signal $u^s = k_c\xi^s = 0.3\,\mathrm{V}$. It has been assumed, that mover position is set up when the relative error decreases below 5%. The set-up time T_{set} was minimised. The following regulator settings have been assumed: $K = 100$, $T_d = 12\,\mathrm{ms}$, $T_i = 14\,\mathrm{ms}$.

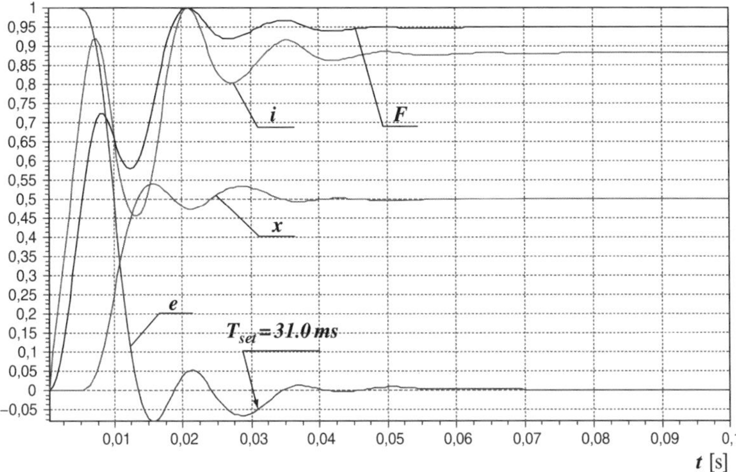

Fig. 2 Dynamic response after structure optimisation

First, the relatively simple problem was being solved. Initially, two design variables: relative angle of the tapered plunger $s_1 = \alpha/90°$ and its relative diameter $s_2 = d/\delta_b$ (see Fig. 1) have been assumed. These parameters have the crucial influence on the winding inductance, the electromagnetic force and mass of the moving part. Therefore, they have significant influence on dynamic operation of the actuator. The outer dimensions (Fig. 1) had the constant values $D = 26\,\text{mm}$, $H = 22\,\text{mm}$ during the optimisation process. Number of individuals $n_{inv} = 200$ has been assumed. After 28 generations the optimal design variables $\alpha = 30.3°$, $d = 11.01\,\text{mm}$ and optimal set-up time $T_{set} = 31\,\text{ms}$ have been obtained, what is by 42 % less than for best variant drawn during the initiation stage. The dynamic response of optimal variant is presented in Fig. 2. The time-variations of the relative values of the current $i(t)$, the force $F_e(t)$, the displacement $x(t)$ and the error $e(t)$ are shown.

5 Optimisation of the PID Regulator Settings

This time, as the design variables the regulator three settings: $s_1 = K/100$, $s_2 = T_d/T_{cr}$, $s_3 = T_i/T_{cr}$, have been assumed. Here $T_{cr} = 0.025\,s$ is so called critical (reference) value. The set-up time T_{set} was minimised. The actuator structure optimal parameters $\alpha = 30.3°$, $d = 11.01\,\text{mm}$ calculated in previous section are taken into computation. As a result of optimisation the optimal PID regulator setting have been obtained $K = 127$, $T_d = 7.5\,\text{ms}$, $T_i = 9.275\,\text{ms}$. For these optimal settings, the set-up time T_{set} decreased to the value 27.5 ms. The dynamic response for this case is shown in Fig. 3.

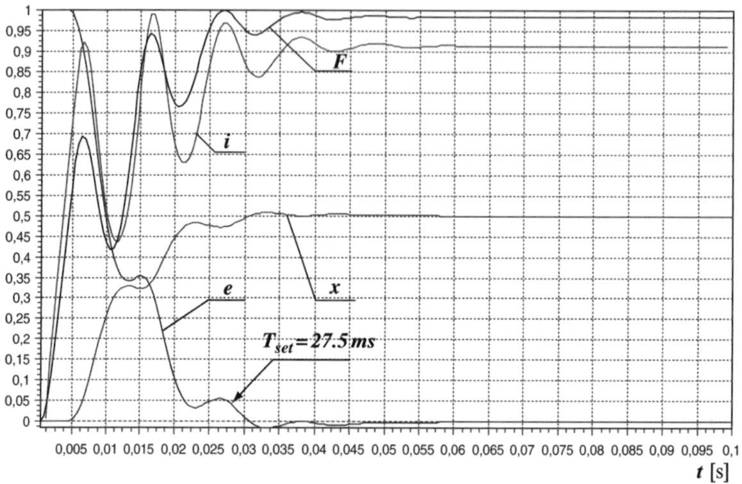

Fig. 3 Dynamic response after optimisation of the regulator settings

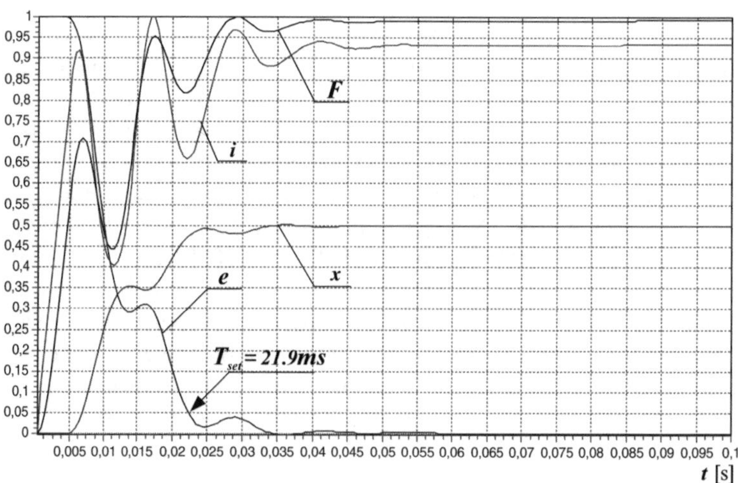

Fig. 4 Dynamic response after second optimization of the structure for optimal regulator settings

Such iterative procedure, i.e. alternately optimisation of the actuator structure and regulator settings may lead to the optimal comprehensive variant of the whole system, i.e. system optimal in relation to the both group of variables. As an example, the results obtained after renewed optimisation of the actuator structure (assuming obtained optimal regulator settings) are shown in Fig. 4. Due to this additional step the set-up time T_{set} decreased by 25.5 % – from 27.5 to 21.9 ms.

However, such iterative procedure is very time consuming. Much more effective is comprehensive optimisation of the system, assuming five design variables. This is because the computation time in genetic algorithm doesn't depend on the number of variables as strongly as in case of deterministic methods.

6 Complex Optimisation of the Actuator Structure and PID Regulator Settings

In this section, the results of simultaneous optimisation of both: the actuator parameters and regulator settings are presented. First, the simpler problem has been solved. *Five* design variables: $s_1 = \alpha/90°$, $s_2 = d/\delta_b$, $s_3 = K/100$, $s_4 = T_d/T_{cr}$, $s_5 = T_i/T_{cr}$ have been assumed. The following optimal values of these variables have been obtained: $\alpha = 28.33°$, $d = 10.17\,\text{mm}$, $K = 140.80$, $T_d = 7.47\,\text{m}s$, $T_i = 5.58\,\text{ms}$. Substantial improvement of the dynamic properties has been achieved. The set-up time decreased to 19.6 ms. The response for this optimal variant is presented in Fig. 5.

Next, the outside dimensions have been also included into the optimisation process. The extended complex problem with *seven* variables has been solved. The variables are assumed as follows: $s_1 = \alpha/90°$, $s_2 = d/\delta_b$, $s_3 = D/d$, $s_4 = H/\delta_b$ $s_5 = K/100$, $s_6 = T_d/T_{cr}$, $s_7 = T_i/T_{cr}$ – see Fig. 1. The following optimal values of actuator and regulator parameters has been obtained: $\alpha = 35.23°$, $d = 10.52\,\text{mm}$, $D = 15.15\,\text{mm}$, $H = 38.80\,\text{mm}$, $K = 154.9$, $T_d = 6.28\,\text{m}s$, $T_i = 6.67\,\text{m}s$. As a result of such comprehensive optimisation, the lowest time $T_{set} = 10.78\,\text{ms}$ has been attained. Results are presented in Fig. 6. The comprehensive, simultaneous optimisation of all main actuator dimensions and regulator settings enabled to reduce the set-up time almost by 65% in comparison to previous optimal system, presented in Fig. 5.

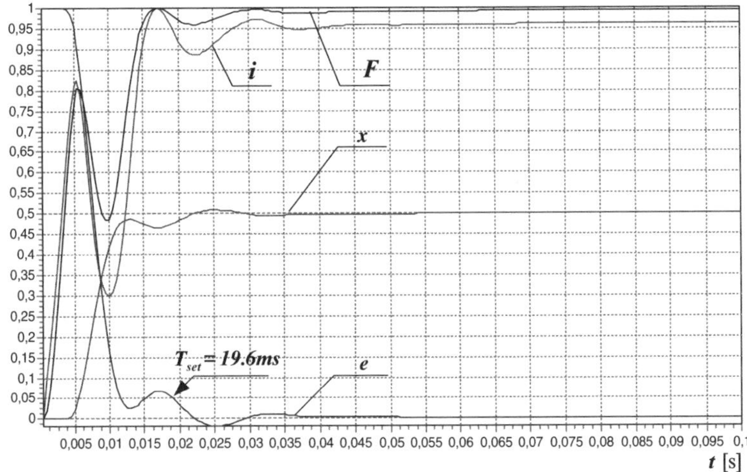

Fig. 5 Dynamic response after comprehensive optimisation (five variables)

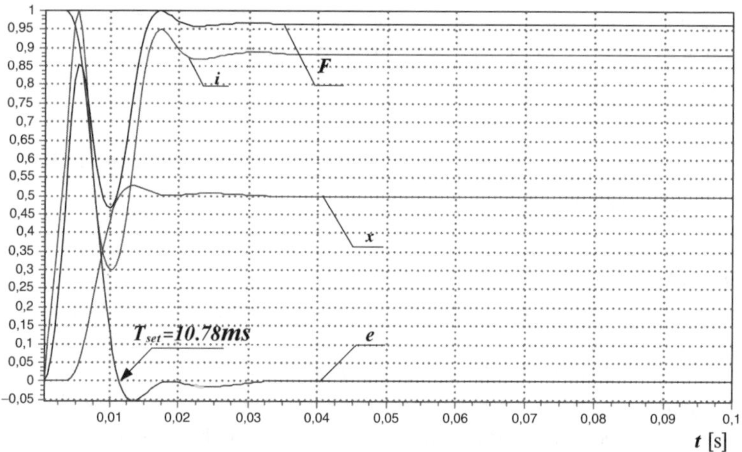

Fig. 6 Dynamic response after comprehensive optimisation (seven variables)

7 Conclusions

Error actuated control system for the linear displacement control is considered. Algorithm and computer software for the optimisation of the actuator structure and regulator settings has been presented. Two different approaches have been proposed. In the first procedure, the actuator structure and regulator settings are optimised in turns. In the second approach, which is more effective, all parameters of the system are optimised simultaneously. It has been proved that number of free design variables has significant effect on the dynamic properties of the system.

References

1. Kiam H.A., Chong G., Yun L., PID control system analysis, design, and technology, IEEE Transactions on Control System Technology,Vol. 13, 2005, pp. 559–576
2. Nowak L., Kowalski K., Radziuk K., Dynamics of an electromagnetic linear actuator operating in error actuated control system, Proceedings of 19th Symposium Electromagnetic Phenomena in Nonlinear Circuits-Science, Maribor, Slovenia, June, 28–30, 2006
3. Ho S.L., Li Y., Lin X., Xu J.Y., Lo Wc., Wong H.C., Design and dynamic analyses of permanent magnetic actuator for vacuum circuit breaker, 14th Conference on the Computation of Electromagnetic Fields, 2003, Saratoga Springs, New York, USA
4. Nowak L., Radziuk K., Transient Analysis of PWM-Excited Electromagnetic Actuators, IEE Proceedings-Science, Measurement and Technology, Vol. 149, 2002, pp. 199–202

LMAT_SIMEL – The Electric Field Numerical Calculator of the Package LMAT_SIMX for Very High Voltage Power Lines

Carlos Lemos Antunes, José Cecílio, and Hugo Valente

Abstract In this paper it is presented the software tool LMAT_SIMEL, the Electric Field numerical calculator for the package LMAT_SIMEX that allows the calculation and simulation of ELF electric and magnetic fields emanated from very high voltage power lines.

The electric field is calculated using a 3D integral numerical approach and makes use of the image method. The catenary of the Line(s) approached by straight lineal segments, where each segment makes a contribution to the electric field at point of analysis and each segment is considered as a filamentary wire. The Lines in analysis can have different configurations and orientations and the field solution can be expressed in terms of the resultant components in the x, y, z direction due to each separate line or all the Lines. The influence of vegetation and terrain elevations is not taken into consideration.

This software was developed in MATLAB environment is very easy to use and has a user friendly interface.

1 Introduction

The LMAT_SIMEL is the software tool developed in MATLAB environment [1, 2], used for the calculation and simulation of ELF electric fields emanated from very high voltage power lines. In this software tool, the information about

Carlos Lemos Antunes
Laboratory of CAD/CAE, Electrical Engineering Department, University of Coimbra, Pólo II, 3030 – 290 Coimbra, Portugal
carlos.antunes@enaco.pt, clemos.antunes@apdee.org

Carlos Lemos Antunes and José Cecílio
APDEE – Assoc. Port. Prom. Desenv. Eng. Electrotécnica, Rua Eládio Alvarez, Ap. 4102, 3030 – 281 Coimbra, Portugal

Hugo Valente
REN – Rede Eléctrica Nacional, Av. Estados Unidos da América 55, 1749 – 061 Lisboa, Portugal

C.L. Antunes et al.: *LMAT_SIMEL – The Electric Field Numerical Calculator of the Package LMAT_SIMX for Very High Voltage Power Lines*, Studies in Computational Intelligence (SCI) **119**, 101–109 (2008)
www.springerlink.com

Line(s) is loaded from database generated by the module LMAT_GEOMODEL [3]. This module also allows the user to choose the points or the plane where the electric field is to be calculated.

The LMAT_SIMEL calculates the electric field resulting from general 3D line(s) configurations. The conductors are considered filamentary wires of arbitrary geometric configuration with known imposed voltages: phase-earth or zero if it corresponds to the guard conductor and the catenary is approximated by straight lineal segments. The electric field can be calculated along any path or on any plane. The earth is considered as a perfect conductor at zero voltage reference value and its influence is taken into account using the method of images. The influence of vegetation and terrain elevations is not taken into consideration. The modules related in this paper are part of the package LMAT_SIMX that allows the calculation and simulation of ELF electric and magnetic fields emanated from very high voltage power lines.

This software tool is very easy to use and has a user friendly interface.

2 Formulation

The method of image charges (also known as the method of images and method of mirror charges) is a known 2D problem-solving tool in electrostatics. The name originates from the replacement of certain elements in the original layout with imaginary charges, which replicates the boundary conditions of the problem.

For 3D situations one has to generalize the 2D case and so image Line(s) have to be considered.

The catenary of each Line(s) and the corresponding image is approximated by straight N_S lineal segments, each with length L_i.

The phasor electric field $\bar{\hat{E}}$ at any point P(x, y, z) due to a Line, is calculated by

$$\bar{\hat{E}} = \frac{1}{4\pi\varepsilon_0} \cdot \sum_{i=1}^{N_S} L_i \cdot \int_0^1 \frac{\hat{\lambda}_i(s)}{|\bar{r}-\bar{r}'|^2} \cdot ds \cdot \hat{a}, \tag{1}$$

where the point P(x, y, z) is defined by $\bar{r}$ and the phasor charge density in the segment i is located at $\bar{r}'$, with $\hat{a}$ as the unit vector in direction $(\bar{r}-\bar{r}')$.

The expression (1) yields:

$$\bar{\hat{E}} = \frac{1}{4\pi\varepsilon_0} \cdot \sum_{i=1}^{N_S} L_i \cdot \int_0^1 \frac{\hat{\lambda}_i(s)}{|\bar{r}-\bar{r}'|^3} \cdot ds \cdot (\bar{r}-\bar{r}'). \tag{2}$$

This expression has to be applied for all Line(s) and Line(s) images to obtain the resultant electric field.

It is seen that the phasor linear charge density has to be previously calculated for all the Line(s) and their images. For each line segment the charge distribution is approached by a cubic spline polynomial as (3).

$$\hat{\lambda}(s) = c_0 + c_1 s + c_2 s^2 + c_3 s^3, \tag{3}$$

where $c_0 = \left(1 - \frac{3s^2}{L_1^2} + \frac{2s^3}{L_1^3}\right)$, $c_1 = \left(s - \frac{2s^2}{L_1} + \frac{s^3}{L_1^2}\right)$, $c_2 = \left(\frac{3s^2}{L_1^2} - \frac{2s^3}{L_1^3}\right)$ *and* $c_3 = \left(-\frac{s^2}{L_1} + \frac{s^3}{L_1^2}\right)$. s is an adimensional parameter ($s = 0$ at the beginning of the segment and $s = 1$ at the end of that segment).

For each segment node the charge density $\hat{\lambda}$ and the corresponding derivative $\hat{\lambda}'$ have to be calculated. Continuity conditions (of level 2) at interconnection segment nodes and relaxed natural boundary conditions at extreme points defining the Line are applied, in order to make the systems of equations possible to solve.

For the example, with two segments (4) is obtained:

$$\begin{bmatrix} \frac{2}{L_1} & \frac{1}{L_1} & 0 \\ -\frac{4}{L_1} + \frac{6}{L_1^2} & -\frac{2}{L_1} + \frac{6}{L_1^2} + \frac{2}{L_2} & \frac{2}{L_2} \\ 0 & -\frac{2}{L_2} + \frac{3}{L_2^2} & -\frac{1}{L_2} + \frac{3}{L_2^2} \end{bmatrix} \cdot \begin{bmatrix} \hat{\lambda}_0' \\ \hat{\lambda}_1' \\ \hat{\lambda}_2' \end{bmatrix}$$

$$= \begin{bmatrix} -\frac{3}{L_1^2} & \frac{3}{L_1^2} & 0 \\ \frac{6}{L_1^2} - \frac{12}{L_1^3} & -\frac{6}{L_1^2} + \frac{12}{L_1^3} - \frac{6}{L_2^2} & \frac{6}{L_2^2} \\ 0 & \frac{3}{L_2^2} - \frac{6}{L_2^3} & \frac{6}{L_2^3} - \frac{3}{L_2^3} \end{bmatrix} \cdot \begin{bmatrix} \hat{\lambda}_0 \\ \hat{\lambda}_1 \\ \hat{\lambda}_2 \end{bmatrix}, \tag{4}$$

where L_1 is the length of the first segment and L_2 of the second.

For all segment nodes it is obtained the relation between $\left[\hat{\lambda}\right]$ and $\left[\hat{\lambda}'\right]$, as:

$$[C] \cdot \left[\hat{\lambda}'\right] = [D] \cdot \left[\hat{\lambda}\right]. \tag{5}$$

The phasor electric potential $\hat{V}_P$ at a point P(x, y, z) due to a line with N_S segments is given by

$$\hat{V}_P = \hat{V}(\bar{r}) = \frac{1}{4\pi\varepsilon_0} \cdot \sum_{i=1}^{N_S} L_i \cdot \int_{s=0}^{s=1} \frac{\hat{\lambda}_i(s)}{|\bar{r} - \bar{r}'|} \cdot ds, \tag{6}$$

or after some manipulation

$$\hat{V}_P = \frac{t}{4\pi\varepsilon_0} \int_0^1 \frac{\left[\hat{\lambda}_0 \; \hat{\lambda}_1\right] \cdot \begin{bmatrix} k_o(s) \\ k_1(S) \end{bmatrix}}{|\bar{r} - \bar{r}'|} ds + \frac{t}{4\pi\varepsilon_0} \int_0^1 \frac{\left[\hat{\lambda}_0' \; \hat{\lambda}_1'\right] \cdot \begin{bmatrix} k_0'(s) \\ k_1'(S) \end{bmatrix}}{|\bar{r} - \bar{r}'|} ds, \tag{7}$$

where $k(s)$ is a constant term associated with $\hat{\lambda}$, and $k'(s)$ is associated with $\hat{\lambda}'$.

If the points are chosen as the segment nodes of the lines, the corresponding voltage values are the voltage reference values of the Lines and the following equation holds:

$$\left[\hat{V}\right] = [A] \cdot \left[\hat{\lambda}\right] + [B] \cdot \left[\hat{\lambda}'\right], \tag{8}$$

as

$$\left[\hat{\lambda}'\right] = [C]^{-1} \cdot [D] \cdot \left[\hat{\lambda}\right] \tag{9}$$

The charge density at each node segment is given by

$$\left[\hat{\lambda}\right] = \left([A] + [B] \cdot [C]^{-1} \cdot [D]\right)^{-1} \cdot \left[\hat{V}\right]. \tag{10}$$

For each segment of a line, say *i*, the components E_{xi}, E_{yi} and E_{zi} of the electric field at a given point are then calculated, and the procedure repeated for all the segments of that Line and for other Lines (including the image lines).

The value of the resultant components E_{xR}, E_{yR} and E_{zR} will be the arithmetic sum of the contributions of each segment.

The electric field is a vector where components in the x, y, z directions are also phasors, thus varying in time is a simusoidal way. For electric field exposure assessments we are interested in the effective value E_{ef}.

In this case we define E_{ef} as:

$$E_{ef} = \sqrt{\frac{\bar{\hat{E}} \cdot \bar{\hat{E}}^*}{2}} = \sqrt{\frac{\left(E_{xReal}^2 + E_{xImag}^2\right) + \left(E_{yReal}^2 + E_{yImag}^2\right) + \left(E_{zReal}^2 + E_{zImag}^2\right)}{2}}, \tag{11}$$

where $\bar{\hat{E}}$ is the electric field value, which is a complex value.

The electric field value has three components $\hat{E}_x$, $\hat{E}_y$ and $\hat{E}_z$, where each component is a complex number with a real and imaginary part, i.e., for example, $\hat{E}_x = E_{x\mathrm{Real}} + jE_{x\mathrm{Imag}}$.

$\bar{\hat{E}}^*$ is the complex conjugate of $\bar{\hat{E}}$.

3 User Interface

When LMAT_SIMEL is initialized the window (Fig. 1) appears on the screen. In this window, the user selects the Line$^{(a)}$ and introduces the corresponding state, i.e., the rms line voltage and electrical current in amplitude and phase$^{(b)}$. After, the user chooses one of the two different formats for introducing the points where the electric field is to be calculated$^{(c)}$. In this section, the user introduces the points in a discrete way, one by one or introduces the *z* coordinate of plane and decides the grid of points in that plane, by indicating the range in the *x* and *y* direction and the increment$^{(d)}$. In the window (Fig. 1) it is also shown the geometric configuration of the Line(s)$^{(e)}$.

When the user introduces all information, and if interested to view the geometric configuration of the Line(s) and the points where the electric field is to be calculated it is necessary to click on "pre-view configuration"$^{(f)}$ and a sketch appears in window$^{(e)}$.

When the user completes all input information it is necessary to click on "solve"$^{(g)}$ to start the calculation process. This calculation process may take considerable computation time depending on the level of refinement of the grid. Once the calculation is completed, a sketch of electric field distribution is shown$^{(h)}$.

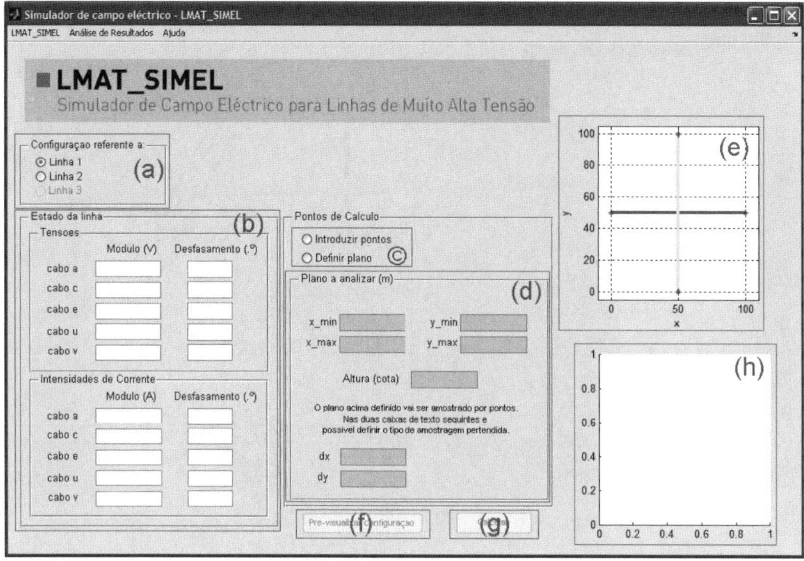

Fig. 1 Interface

In this interface some other menus also exist that allow the saving of the current configurations, the load of a configuration from file and call the module LMAT_VISUAL.ELF.

4 Results

It is presented as an illustration example the electric field emanated from two high voltage power lines (220 kV) carrying each 1,140 A per phase conductor. The lines are placed orthogonally to each other with 100 m of the length (Fig. 2), the catenary parameter for each Line is 1,460 with different heights of 10 and 20 m. In this example the solution plane is defined by a span of all lines (Fig. 3). The representative grid of the plane is shown with cyan colour.

The data referred to the lines and their corresponding electrical state is defined by the module LMAT_GEOMODEL [3] of the package LMAT_SIMX. The distribution of resultant electric field on the plane of analysis (ground level) is shown in Fig. 4.

To get a deeper analysis of the results it is necessary to use the LMAT_VISUAL. ELF [4] module. This allows the visualization and manipulation of ELF magnetic/electric fields.

In Fig. 5 it is shown the 3D distribution of the electric field at ground level graded by a scale colours pre-defined by the user. When it is important to have a smoother field distribution in the plane of analysis and the number of solution points is insufficient (i.e. existing a shaped or faceted field distribution) the LMAT_VISUAL.ELF

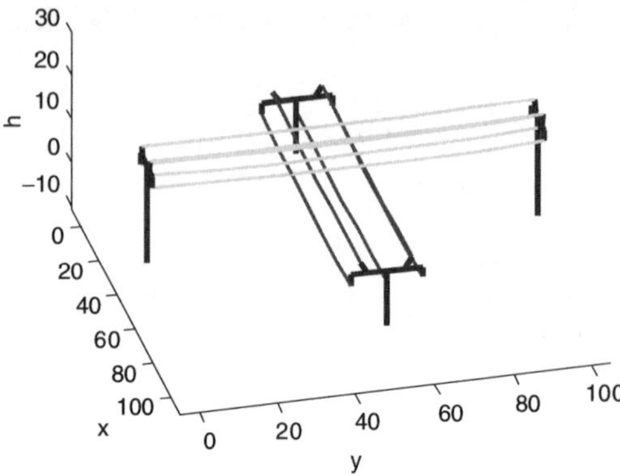

Fig. 2 Geometric configuration of the lines

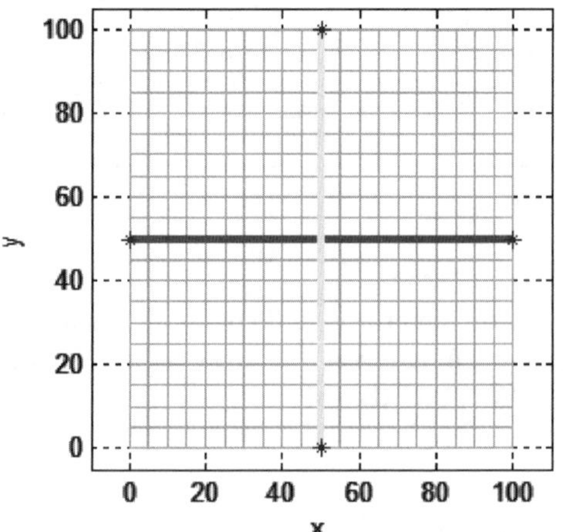

Fig. 3 Geometric configuration of the plane

allows the user to opt making an interpolation of the results (Fig. 5), to obtain a more refined grid. This interpolation is a *spline interpolation* based in function *interp2* from Matlab.

In the example chosen the maximum and minimum electric field values are $4,24\,\mathrm{kV\,m^{-1}}$ and $38,7\,\mathrm{V\,m^{-1}}$, respectively, where the maximum electric field value is found at point $(x = 60,\ y = 60,\ z = 0)$.

To obtain a 2D distribution, i.e., view a 2D section of the 3D solution profile, it is necessary to specify the plane (x or y), as well the field components to be represented. For the example in analysis, for plane $y = 60$ which crosses the maximum value of the electric field the 2D distribution is shown in Fig. 6.

In this example the components in the x and y directions of the resultant electric field and the contribution of each Line are equal zero, and the component in the z direction is equal of resultant field value.

In Fig. 7 it is shown the visualization of a 2D projection of the field distribution, contour lines corresponding to isovalues of the field using the representation of coloured shaded plots.

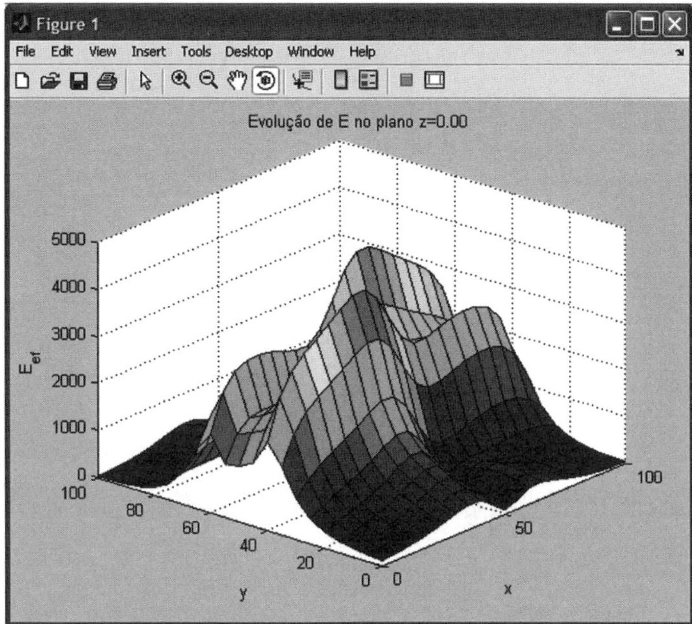

Fig. 4 Distribution of electric field in the plane (ground level)

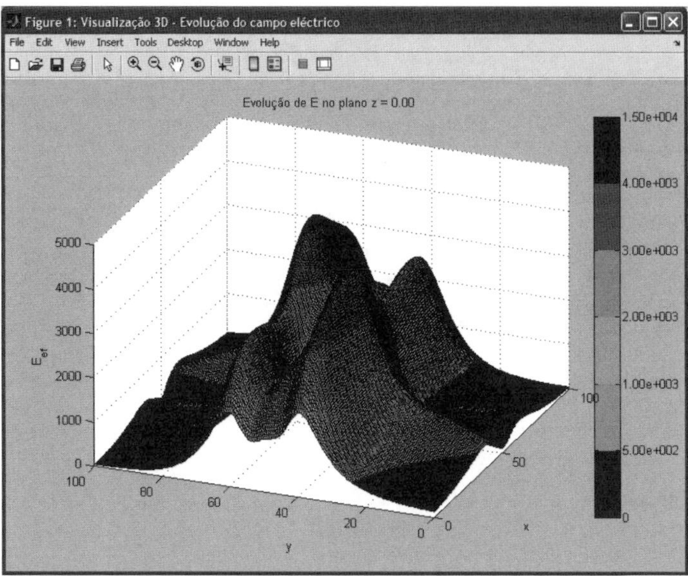

Fig. 5 Smoother distribution of electric field on the plane with attributed scale colours

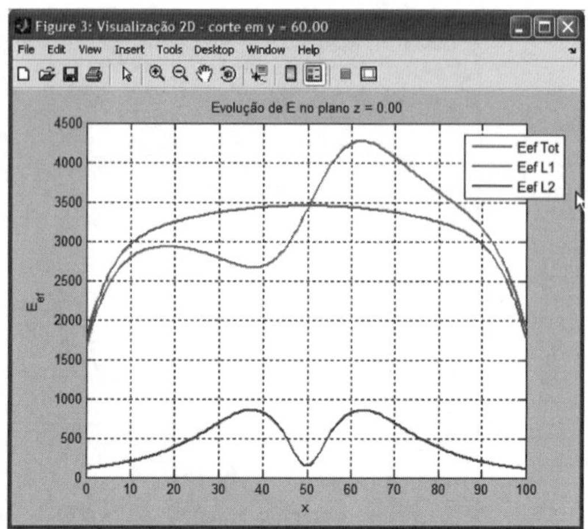

Fig. 6 2D field distribution in a plane

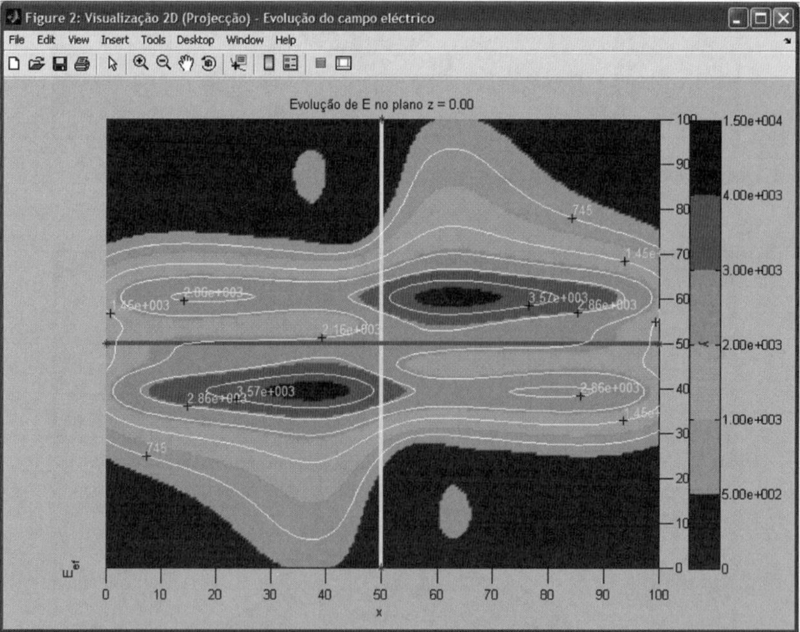

Fig. 7 2D projection of the field

5 Conclusions

It was presented the major useful facilities implemented in this powerful design tool. This software is part of a more complete package LMAT_SIMX that allows the analysis and simulation of electrical and magnetic fields emanated from very high voltage power lines, developed by the authors.

Acknowledgements The authors gratefully acknowledge REN-Redes Energéticas Nacionais SGPS, SA for the financial support received under Project COIMBRA_EMF.ELF.

References

1. Mathworks, "Creating Graphical User Interfaces", September 2006, Revised for MATLAB 7.3
2. http://www.mathworks.com
3. Carlos Lemos Antunes, José Cecílio, Hugo Valente, "LMAT_GEOMODEL – The Geometric Configuration Modeller of the Package LMAT_SIMX for Very High Voltage Power Lines", accepted for presentation at 10th Portuguese–Spanish Conference in Electrical Engineering – XCLEEE, to be held in Funchal (Portugal), 5–7 July, 2007
4. Carlos Lemos Antunes, José Cecílio, Hugo Valente, "LMAT_VISUAL.ELF – Software Tool for Visualization and Manipulation of ELF Magnetic/Electric Fields Emanated from Very High Voltage Power Lines", accepted for presentation at 10th Portuguese–Spanish Conference in Electrical Engineering – XCLEEE, to be held in Funchal (Portugal), 5–7 July, 2007

Concerning 2-D Planar Open Boundary Electromagnetic System Computation

Mykhaylo V. Zagirnyak and Yurij A. Branspiz

Abstract 2A possibility of computation of 2-D planar magnetic fields of open boundary electromagnetic systems by means of containing the computation area in a certain square for which the external area is conformally mapped onto the internal area of another square with identical boundary conditions, is demonstrated in the paper.

1 Introduction

Numerical computations of magnetic systems are commonly used in practice of scientific research and engineering calculations. Practical realization of these computations is connected with difficulties resulting from the fact that real electromagnets often have open boundary working areas. Nowadays there is an approach providing the possibility to carry out numerical computations of open boundary electromagnetic system magnetic fields on the basis of containing the computation area in a circle [1].

Such method of computation of magnetic fields with open boundary conditions is grounded on the possibility of conformal mapping of the external area of a certain circle onto the internal area of a circle with the same radius (inversion mapping). In this case the external area magnetic field is mapped onto the internal area magnetic field, the coinciding area boundary is preserved, consequently, the boundary

M.V. Zagirnyak
Kremenchuk Mykhaylo Ostrogradskiy State Polytechnic University, Ukraine, 39614, Kremenchuk, vul. Pershotravneva, 20, KSPU
mzagirn@polytech.poltava.ua

Yu.A. Branspiz
East Ukrainian Dal National University kvartal Molodezhniy, 20-a, 91034, Lugansk, Ukraine
branspiz@mail.ru

M.V. Zagirnyak and Y.A. Branspiz: *Concerning 2-D Planar Open Boundary Electromagnetic System Computation*, Studies in Computational Intelligence (SCI) **119**, 111–116 (2008)
www.springerlink.com

conditions are preserved. It causes preservation of solutions for these areas (correspondent distribution of vector magnetic potential). Thus, the computation includes the second round having a zero value of vector potential in the center and boundary conditions identical to the boundary conditions of the first round wherein the computed magnetic system is situated [1].

2 Problem Statement

In many cases the electromagnet external contour represents a certain rectangle. Then, to carry out the computation of the field created by it, it is more natural to limit the computation area not with a circle, but with a square, the sides of which are parallel to the sides of the electromagnet external contour rectangle.

However, it is possible only when the external area of an arbitrary square can be mapped conformally onto the internal area of a similar square. In this case, with periodic boundary conditions on the corresponding sides of the square enveloping the electromagnet contour, and additional square with a zero value of vector potential in its center, there appears a possibility to carry out the electromagnet computation taking into account the magnetic field extent to infinity. It can be done, for example, applying the same procedure with periodic boundaries which is used in [1], when the computation area is limited with a circle. The proof of this possibility was the problem of the research described in this paper.

3 Conformal Mapping of the External Area of the Square onto its Internal Area

To solve the stated problem it is necessary to find out for what kind of rectangles their external area can be conformally mapped onto their internal area. With this purpose a conformal mapping of the external area of rectangle ABCD (Fig. 1a) onto some rectangle $A^*B^*C^*D^*$ (Fig. 1d) is considered. This conformal mapping can be realized as:

- Conformal mapping of the external area of rectangle ABCD onto upper half-plane (Fig. 1b);
- Corresponding scaling of the obtained upper half-plane (also a conforming mapping, Fig. 1c);
- Conformal mapping of the scaled upper half-plane onto the internal rectangular area (rectangle $A^*B^*C^*D^*$ in Fig. 1d).

In this case conformal mapping of the external area of rectangle ABCD onto the upper half-plane is made by mapping function of the following form:

$$z = \frac{p}{E-(k')^2}\left\{E' - jk^2K\left[Z(\beta) + \frac{\beta}{K}\left(E-(k')^2K\right)\right]\right\}, \tag{1}$$

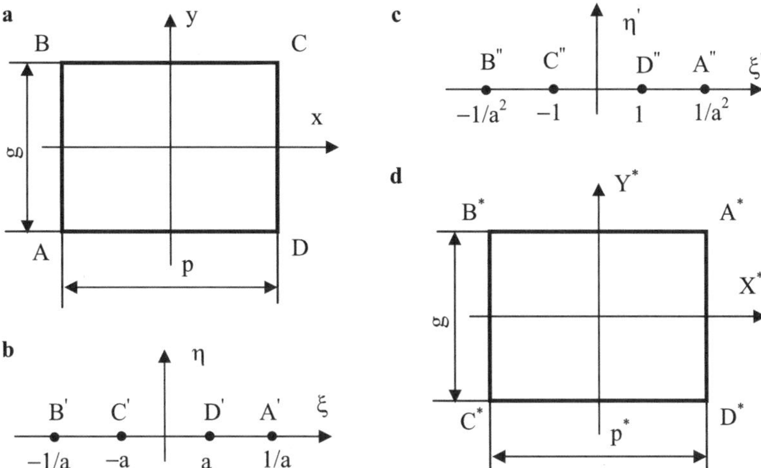

Fig. 1 Used conformal mapping

where z is a complex coordinate of points of the mapped area of rectangle ABCD with the width p and the height g; K, K′ and E, E′ are complete elliptic integrals of correspondingly the first and the second kinds of modules k and $k' = \sqrt{1-k^2}$; Z(β) is Jacobi zeta-function of β argument, which can be defined by the relation

$$k\,\mathrm{sn}(\beta) = -\sin[2\mathrm{arctg}(-t)]. \tag{2}$$

Here t is a complex coordinate of points of the upper half-plane onto which the external area of rectangle ABCD is mapped.

Elliptic integrals modulus k contained in (1) and (2) is defined by relation [2]

$$\frac{p}{g} = \frac{E-(k')^2K}{E'-k^2K'}, \tag{3}$$

which is an equation in modulus k, when the values of width p and height g of the initial rectangle ABCD are known.

This modulus assigns position of points A′, B′, C′ and D′ on the real axis – upper half-plane boundary. It is this half-plane that the external area of rectangle ABCD is mapped onto (Fig. 1b). Namely, relation

$$a = \mathrm{tg}(1/2\arcsin\,k),$$

assigns dimensions a and 1/a, determining position of points C′, D′ and A′, B′, correspondingly (Fig. 1b).

Then, by means of scaling of form

$$T = t/a,$$

the upper half-plane shown in Fig. 1b is mapped onto the upper half-plane shown in Fig. 1c (here T is a complex coordinate of points of this half-plane).

Finally, if it is assumed that $k_1 = a^2$, and Jacobi sn-function of form

$$T = \mathrm{sn}\left(\frac{2K(k_1)}{p^*}w, k_1\right), \tag{4}$$

is used, the upper half-plane shown in Fig. 1b is mapped onto the internal area of rectangle $A^*B^*C^*D^*$ (Fig. 1d) with width p^* and height g^*; the points in this area are defined by complex coordinate w.

In this case dimensions of rectangle $A^*B^*C^*D^*$ are connected by relation [2]

$$\frac{p^*}{g^*} = 2\frac{K(k_1)}{K\left(\sqrt{1-k_1^2}\right)}. \tag{5}$$

This expression, as well as consideration of the fact that relation of dimensions of rectangles ABCD and $A^*B^*C^*D^*$ should coincide (i.e. the equality $p/g = p^*/g^*$ should be true), makes it possible to write down the following equation in modulus k proceeding from the form of right parts of (3) and (5)

$$\frac{E-(k')^2K}{E'-k^2K'} = 2\frac{K(k_1)}{K\left(\sqrt{1-k_1^2}\right)}. \tag{6}$$

Value $k = 1/\sqrt{2}$ is the solution of equation (6). To make sure of it one should substitute $k = 1/\sqrt{2}$ into (6).The mentioned value of modulus k corresponds to relation of the sides of rectangles ABCD and $A^*B^*C^*D^*$

$$\frac{p}{g} = \frac{p^*}{g^*} = 1,$$

which can be proved by direct substitution of modulus k into (3) and (5).

Hence, the external area of a rectangle can be conformally mapped onto the internal area of this rectangle only when this rectangle has equal sides, representing a square.

4 Computation of 2-D Planar Open Boundary Electromagnetic Systems

It is well known that in case of conformal mapping not only geometric correspondence, but also correspondence of boundary conditions for the field problem under consideration are preserved. Then, if boundary conditions for corresponding sides of the squares are preserved when the external area of one square is mapped onto the internal area of another square (Fig. 1a,d), the problem of computation of magnetic field for open boundary area will be solved. In this case the problem of computation of magnetic field for open boundary area presents a joint solution of field problem in a square limiting the computed electromagnetic system, and in another square, with periodic boundary conditions, the center of which has a zero value of vector potential. It should be noted that the latter results from the fact that, due to symmetry, the central point in square A*B*C*D* corresponds to the boundary enveloping the external area of the initial square ABCD on infinity, where the value of vector potential can be assumed equal to 0.

The required correspondence of the boundary conditions on the squares sides is provided by the known procedure of periodic boundary conditions of two computed areas as it is done in [1].

To confirm the described opportunity of computing open boundary electromagnetic systems, using square areas with periodic boundary conditions, computation of E-type electromagnet magnetic field was carried out with the help of FEMM software [1]. Figure 2 is the representation of the correspondent form of magnetic field for two cases: a computed area limited by a square (Fig. 2a) and a computed area limited by a circle (Fig. 2b).

The direct computation showed that all the computed parameters (strength and induction of magnetic field in characteristic points of the system, magnetic flux from various poles surfaces) are determined in both cases (computed area limited by a square or a circle) with the result difference not exceeding 1%. It certifies the acceptability of the offered approach to engineering computation of open boundary electromagnetic systems.

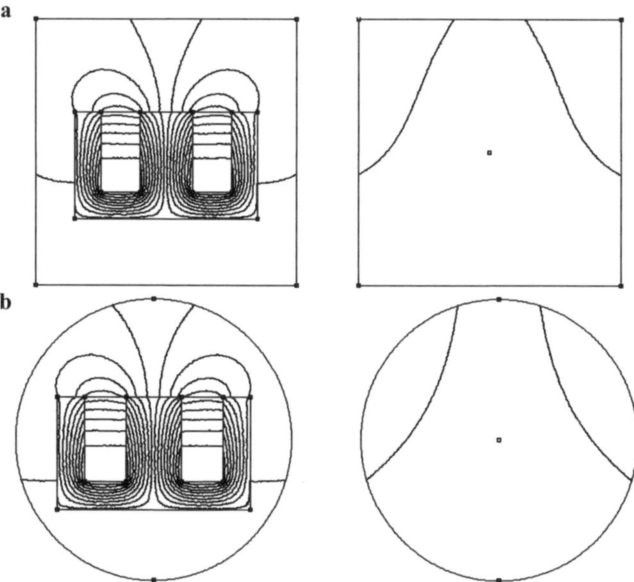

Fig. 2 Numerical computation of E-type electromagnet magnetic field with computed area limited by a square **a** and a circle **b**

5 Conclusion

1. It has been proved that only for a square its external area can be conformally mapped onto the internal area of a square of the same type.
2. Application of periodic boundary conditions makes it possible to perform the computation of 2-D planar magnetic fields with open boundary conditions by means of containing the computation area in a certain square the external area of which is computed in another square with periodic boundary conditions and zero value of vector potential in the center.

References

1. D. Meeker, Finite Element Method Magnetics: User's Manual, http://femm.berlios.de, 2003, p. 79.
2. K.J. Binns, P.J. Lawrenson, C.W.Trowbridge, The Analytical and Numerical Solution of Electric and Magnetic Fields, Wiley, New York, 1992, p. 486.

Quadtree Meshes Applied to the Finite Element Computation of Phase Inductances in An Induction Machine

Jose Roger_Folch, Juan Perez, Manuel Pineda, and Ruben Puche

Abstract In this paper, the quadtree data structure and recent advances on the construction of polygonal finite element Interpolants are applied to the study of an induction machine. Quadtree is a hierarchical data structure that is computationally very attractive for adaptive numerical simulations. Mesh generation, adptive refinement and direct data feeding from CAD drawings in pixel format are straightforward for quadtree meshes. Efficient multigrid methods can be used because of the regularity of the resulting grid. However, finite elements are non-conforming on quadtree meshes due to level-mismatches between adjacent elements, which produce the so called "hanging nodes". Meshfree (Laplace) basis functions have been used to overcome this problem, and a quadtree mesh has been applied to the evaluation of the winding inductances of an induction motor.

1 Introduction

Uniform meshes are not a computationally viable election for solving systems of partial differential equations in which steep gradients or discontinuities must be captured. This is the case that arises in the analys of the air gap field in an induction machine, whose dimensions are very small compared with the machine size, and where there are very steep transitions between the permeability of the iron and the air, of orders of magnitude. Using large elements leads to non-satisfactory results in this region, and the use of small elements in all the domain is impractical. Adaptive refinement strategies automatically adjust mesh resolution only in the area where it is most needed. Today's most advanced refinement techniques are based on natural refinement of elements.

Jose Roger_Folch, Juan Perez, Manuel Pineda, and Ruben Puche
Department of Electrical Engineering, Universidad Politécnica de Valencia, Cno de Vera s/n. 46022 Valencia, Spain
mpineda@die.upv.es, jroger@die.upv.es, juperez@die.upv.es, rupucpa@die.upv.es

J. Roger_Folch et al.: *Quadtree Meshes Applied to the Finite Element Computation of Phase Inductances in An Induction Machine*, Studies in Computational Intelligence (SCI) **119**, 117–124 (2008)
www.springerlink.com

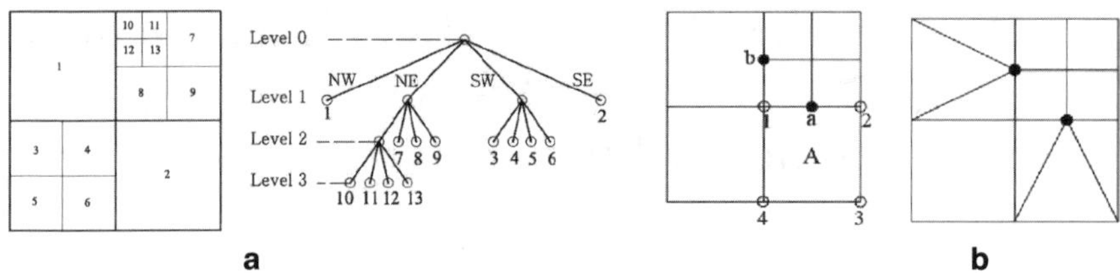

Fig. 1 Quadtree recursive structure **a** and element with a hanging node **b**

The quadtree data structure is a spatial data structure based on the principle of recursive decomposition [1], which provides a simple, fast, and efficient way for both h- and p-refinement. Its efficiency in data storage and fast data retrieval are unmatched. However, hanging nodes, which are generated after each refinement if the adjacent elements are not of the same size, are a significant impediment in using quadtree meshes in finite element methods. In Fig. 1, a quadtree structure is presented with hanging nodes, as well as the traditional way of handling this problem: the use of embedded triangular elements. But this destroys the regularity and simplicity of the mesh, and restricts the maximum difference of level between adjacent elements to 2.

2 Shape Functions for Quadtree Elements with Hanging Nodes

The shape functions used in the presented paper are the Laplace shape functions [2, 3]. They are derived using the concept of natural neighbors: given a set of nodes in the plane, the Voronoi diagram partitions the space into closest-point regions (Voronoi cells). The Delaunay tessellation is the dual of the Voronoi diagram. The natural neighbors of a point p are defined through the Delaunay circumcircles: if p lies within the circumcircle of a Delaunay triangle t, the nodes that define t are neighbors of p.

The Laplace shape function at point p are determined using the Voronoi cell of p: a line is drawn from point p to each of its neighbors. At the mid-point of each of these lines, a perpendicular line is derived. The intersection of all of this perpendicular lines define the new Voronoi of point p. The Laplace function is then defined as

$$\phi_i(\mathbf{x}) = \frac{\omega_i(\mathbf{x})}{\sum_{j=1}^{n} \omega_j(\mathbf{x})}, \qquad \omega_j(\mathbf{x}) = \frac{s_j}{h_j}, \tag{1}$$

where si(x) is the length of the Voronoi edge and $hi(x) = ||x - xi||$ is the Euclidean distance from p to node i.

Figure 2 shows this factors for the case of a regular hexagon.

To obtain the shape functions in a quadtree element with hanging nodes, an affine map between the element and the corresponding regular polygon, with the same

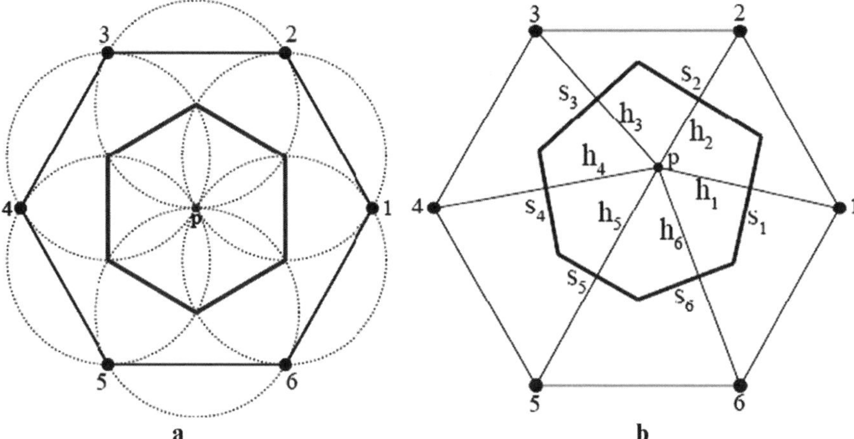

Fig. 2 Laplace shape function: **a** Delaunay circumcircles and Voronoi cell for p in a hexagon; **b** length measures in the definition of the Laplace shape function

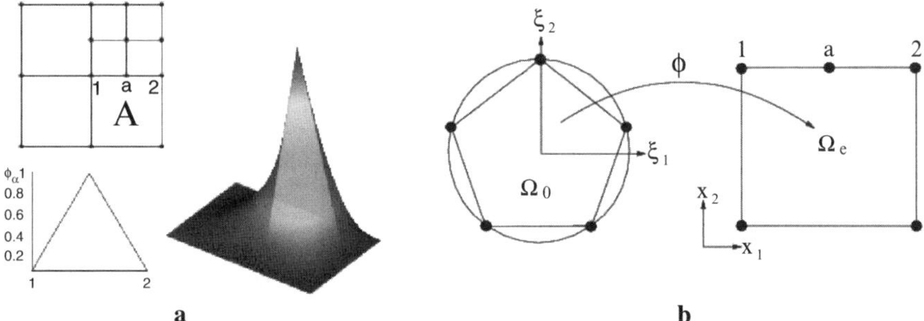

Fig. 3 Conforming (Laplace) shape function in a hanging node, showing linear behavior along edge 1–2 **a** and affine mapping to a regular 5-polygon used to derive it **b**

number of nodes, is established. In this way, the Laplace shape functions must be computed just once for this canonical elements, overcoming the cost of recomputing them at every point of integration. Figure 3 shows the map between a quadtree element with a hanging node and a pentagon. With the use of this method, it is possible to eliminate any restriction regarding the maximum difference of level between adjacent quatree elements, and the additional triangular elements of Fig. 1.

3 System Equations

The following equations system can be written for an induction machine with m stator and n rotor phases with arbitrary layout (that is, even with winding fault conditions like inter-turn short circuits or broken bars):

$$[U_S] = [R_S][I_S] + d[\Psi_S]/dt, \tag{2}$$

$$[0] = [R_r][I_r] + d[\Psi_r]/dt, \tag{3}$$

$$[\Psi_s] = [L_{ss}][I_s] + [L_{sr}][I_r], \tag{4}$$

$$[\Psi_r] = [L_{sr}]^T[I_s] + [L_{rr}][I_r], \tag{5}$$

$$[U_S] = [u_{s1}\ u_{s2} \ldots u_{sm}]^T, \tag{6}$$

$$[I_S] = [i_{s1}\ i_{s2} \ldots i_{sm}]^T, \tag{7}$$

$$[I_r] = [i_{r1}\ i_{r2} \ldots i_{rn}]^T, \tag{8}$$

$$T_e = [I_S]^T \frac{\partial [L_{sr}]}{\partial \theta}[I_r], \tag{9}$$

$$T_e - T_L = J\frac{d\Omega}{dt} = J\frac{d^2\theta}{dt^2}, \tag{10}$$

where $[U]$ is the voltage matrix, $[I]$ is the current matrix, $[R]$ is the resistance matrix, $[\Psi]$ is the flux linkage matrix and $[L]$ is the matrix of inductances. Subscripts s and r stand for stator and rotor. T_e is the electromechanical torque of the machine, T_L is the load torque, J is the rotor inertia, Ω is the mechanical speed and θ is the mechanical angle. To compute (4), (5) and (9), self and mutual phase inductance matrices must be calculated. Due to the presence of the derivatives in (2), (3) and (9), it is necessary to achieve a very good accuracy in this process. End turn and slot leakage inductances are treated as constants in (4) and (5), and saturation of the iron is not considered in this model, as usual in the technical literature [4, 5].

The inductance between two phases, A1 and A2, is calculated in this paper through the following process:

1. Phase A1 is fed with a constant unit current, and its yoke flux is obtained.
2. Flux linkage of phase A2 due to the yoke flux of phase A2 is determined, which corresponds to the mutual inductance between the phases. (If A1 = A2, we get the phase magnetizing self inductance).

The yoke flux generated by phase A1 is obtained by the solving the equation

$$\Delta \mathbf{A} = -\mathbf{J}/\mu \tag{11}$$

in a two-dimensional cross section of the machine, along with the condition $\mathbf{A} = 0$ on the machine boundaries. **A** is the magnetic vector potential (which has only one component in two dimensions, A_z), and **J** is the current density. The flux crossing the surface defined by two normal conductors c1 and c2 at positions (x1,y1), (x2,y2) is the difference of the potential vector at those points times the length of the conductors (see Fig. 4a):

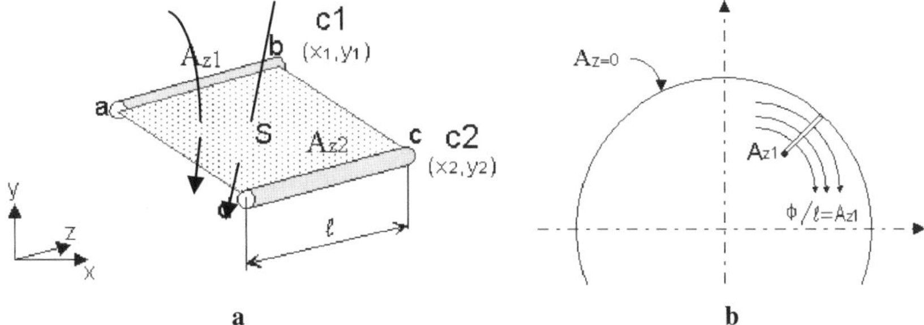

Fig. 4 Flux crossing the surface between two conductors **a** and yoke flux at the position of a conductor **b**

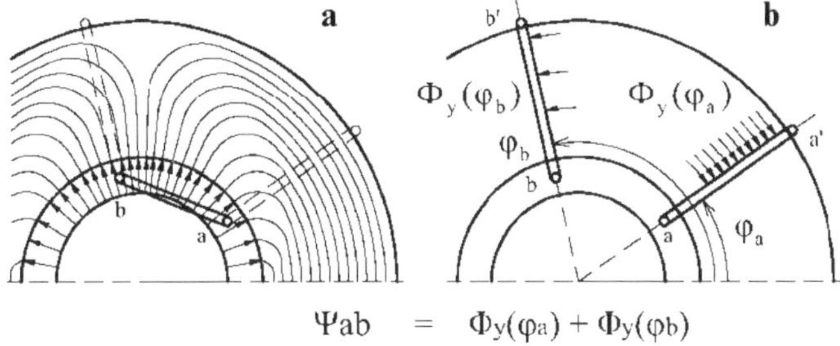

Fig. 5 Flux linkage of an arbitrary coil: (**a**) actual coil (**b**) replaced by two equivalent annular coils

$$\iint_S \vec{B}\,dS = \phi = \iint_S \mathrm{rot}(\vec{A}) \cdot dS = \int_{abcd} \vec{A}\,dl = (A_{z1} - A_{z2}) \cdot 1 \tag{12}$$

If the second conductor is placed on the machine boundary, where $A_z = 0$, then the value of the vector potential at each point is equal to the yoke flux at that point times the conductor's length (see Fig. 4b).

The flux linkage of a phase A_2 is obtained by simply adding up the values of the yoke flux at the yoke sections corresponding to each one of its conductors. Figure 5 shows the basis of this method: the flux linkage Ψ_{ab} of an arbitrary coil a-b can be calculated by replacing the coil by two equivalent annular ones, (a-a′, b-b′) and summing up the yoke flux (that is, the potential vector times the length of the conductors) that crosses them.

So the flux linkage $\Psi_{A2\,A1}$ of phase A_2 due to the yoke flux generated by phase A_1 (which is equal to their mutual inductance $L_{A1\,A2}$ if phase A_1 is fed with a unit current), is computed as:

$$\psi_{A_2A_1} = \ell \cdot \frac{\iint_{\Omega_{A2}} A_z(x,y) \cdot dxdy}{\iint_{\Omega_{A2}} dxdy}. \tag{13}$$

4 Finite Element Computation of Vector Potential Using a Quadtree Mesh

The computation of the vector potential, which value is necessary for solving (13), is carried out using FEM applied to a two-dimensional cross section of the machine perpendicular to its axis. In order to simplify the description of the geometry of the machine, the characteristics of each element (material, current density) and the meshing process, the following approach has been used:

1. Data is fed directly from a raster image of the machine of size $N \times N$ pixels (a $2,024 \times 2,024$ pixels image has been used in the present article), and stored as an $N \times N$ square matrix. The image has two layers, as seen in Fig. 6: material permeability and current density. The color of each pixel in each layer is used as the index of a vector that contain the permeability or the current density of the machine elements.
2. A quadtree mesh is applied directly to the material's layer of the machine, and a recursive algorithm is used to build the quadtree elements of the mesh, with the constraint that the permeability and current densities in all the pixels of each element are the same (see Fig. 7). This mesh is simply stored as an additional layer of the machine image, in which each pixel indicates the level of the quadtree element that it belongs to.
3. The vector potential is approximated in the usual form:

$$\mathbf{A_z}(x,y) = \sum_{i=1}^{N} A_i \cdot \phi_i(x,y) \tag{14}$$

with A_i representing the nodal degrees of freedom. In the FEM context, the nodal degree of freedom A_i corresponds with the vector potential. The nodes used to approximate the vector potential are the pixels corresponding to the quadtree mesh of step 2, and the shape functions Φ_i are those defined by (1). A new layer

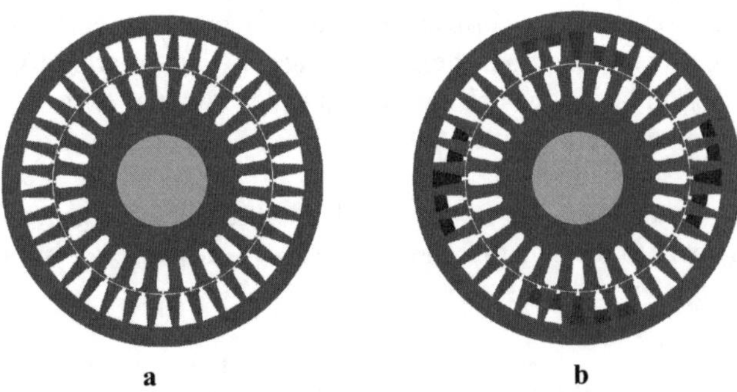

Fig. 6 Layers of the image used to define the machine: **a** material's permeability layer **b** permeability and current density layers superimposed

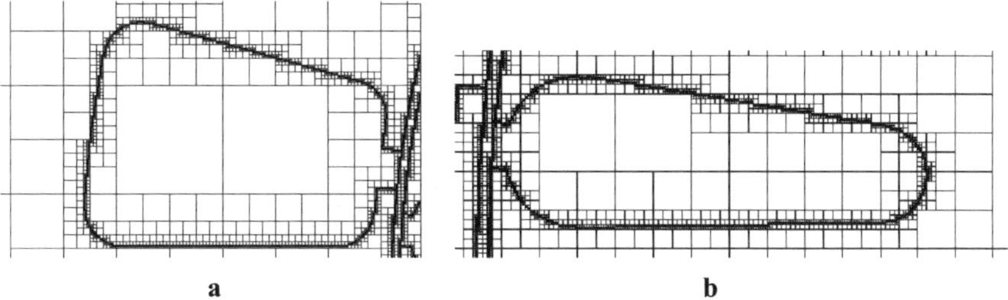

Fig. 7 Quadtree mesh of a stator slot **a** and a rotor slot **b**

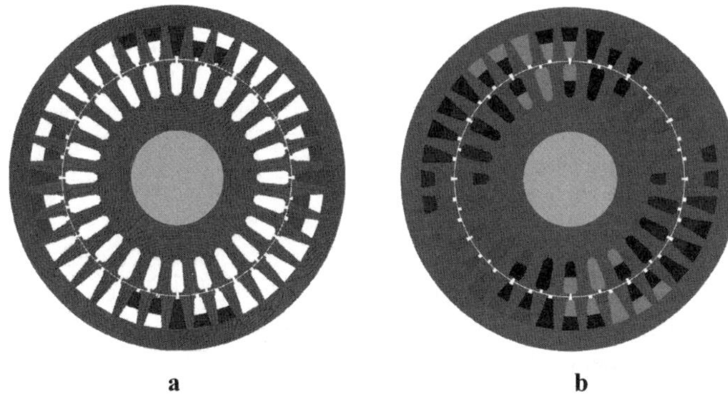

Fig. 8 Layers of the machine image that contains the values of the potential vector and the phase that generates it **a** and layer with the winding distribution of all the phases to compute their mutual inductance **b**

or the machine image is used to store the node values of the vector potential, so that coordinates x,y in (14) are interpreted as pixel coordinates, as seen in Fig. 8a.

4. The linear system obtained by applying the Galerkin method to the weak formulation related to (11) is solved. During this process an adaptative refinement of the mesh is carried out, by simply dividing each element that must be refined in four elements of the next level.
5. The mutual inductance between both phases is computed by applying (16) with the final values of the vector potential. The inductance is computed between the phase that generates the yoke flux and every other phase of the machine. The information about the winding distribution of all the phases is stored in an additional layer of the machine image, as seen in Fig. 7b.

The described method gives the mutual inductance between all the machine phases. It has been applied to an 11 Kw, four poles machine with 36 stator slots and 24 rotor slots. The self inductance of a rotor phase calculated with the theory of ideal magnetic circuits gives a constant value of 0.0714 H for any position of the rotor. The solution obtained by FEM (see Fig. 9) shows the variations due to the different angular positions between rotor and stator.

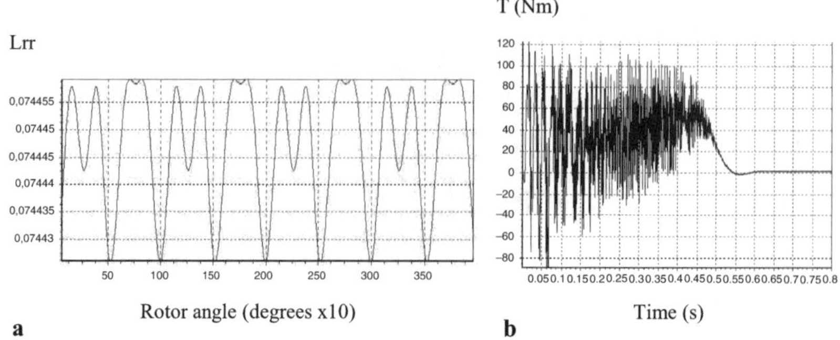

Fig. 9 Self inductance of a rotor phase, Lrr, as a function of the rotor angular position **a** and Torque developed during the start-up of the motor

5 Conclusions

The avaibility of new polygonal finite elements interpolants, based on meshfree Laplace shape functions, which are conforming even for hanging nodes, allow for the use of very simple quadtree meshes without any restriction on the level difference of adjacent elements. This meshes are very easy to construct, extremely efficient in terms of data storage and retrieval, and higly suitable adaptive refinement. They allow for a very simple storage of all de data associated with the machine (geometry, materials, mesh description, current densities and winding distribution) in the form of different layers of a $N \times N$ image of the machine, which drastically reduces and simplifies the required software code. The method has been applied to an induction machine to compute the machine inductances and evaluate its electromagnetic torque during start-up.

References

1. H. Samet, The quadtree and related hierarchical data structure. ACM Comput. Surv., 16(2):187–260, 1984.
2. N. Sukumar and E.A. Malsch, Recent advances in the construction of polygonal finite elements inter-polants. Arch. Comput. Meth. Engng., 13(1):129–163, 2006.
3. A. Tabarraei and N. Sukumar, Adaptive computations on conforming quadtree meshes. Finite Elements Anal. Des., 41:686–702, 2005.
4. X. Luo, Y. Liao, H.A. Toliyat, A. El-Antably, and T. Lipo, Multiple coupled circuit modeling of induction machines. IEEE Trans. Ind. Appl., 31(2):311–318, 1995.
5. S. Nandi, Modeling of induction machines including stator and rotor slot effects. IEEE Trans. Ind. Appl., 40(4): 1058–1065, 2004.

Android Employing 3-D Finite Element Method

Masayuki Mishima, Katsuhiro Hirata, and Hiroshi Ishiguro

Abstract We have been studying the new linear actuator for android using Halbach array of magnets. In this paper, the dynamic performances are computed by the 3-D finite element method and compared with the measurement of the prototype. As the result, both results are in good agreement, and the validity of the analysis is clarified. Moreover, the effectiveness of the proposed actuator is verified through the comparison with the conventional synchronous motor (shaft motor).

1 Introduction

Air servo actuators have been usually applied as driving sources of robots. Figure 1 shows our android. The arm is composed of two air cylinders, and has little backlash, but low response because of the pneumatic control. The authors therefore have proposed the new direct drive electromagnetic linear actuator (see Fig. 2) to solve the above problem. Our proposed actuator is composed of Halbach array [1, 2] of magnets effectively, and confirmed its effectiveness from the viewpoint of the stroke and thrust.

In this paper, the basic structure and the operating principle of the proposed actuator are described, and the dynamic performances of the actuator are computed by the 3-D finite element method. The validity of the computation is confirmed through the measurement of the prototype. Moreover, the effectiveness of the proposed actuator is verified through the comparison with the synchronous motor. As the result, the possibility of applying the actuator to our android is clarified.

Masayuki Mishima, Katsuhiro Hirata and Hiroshi Ishiguro
Department of Adaptive Machine Systems, Osaka University, Yamadaoka, 2-1, Suita-City, Osaka 565-0871, Japan
k-hirata@ams.osaka-u.ac.jp

M. Mishima et al.: *Android Employing 3-D Finite Element Method*, Studies in Computational Intelligence (SCI) **119**, 125–130 (2008)
www.springerlink.com

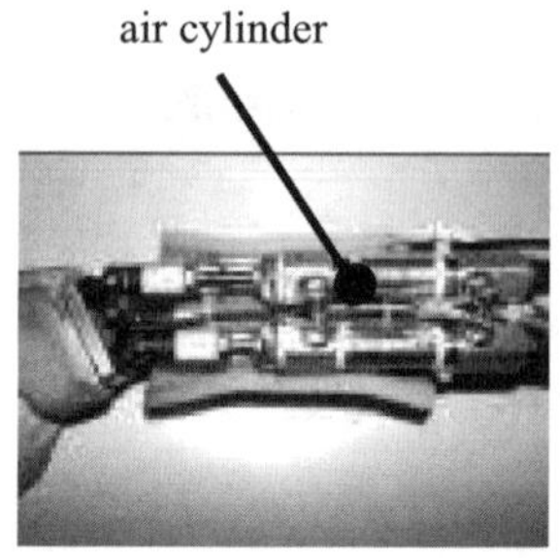

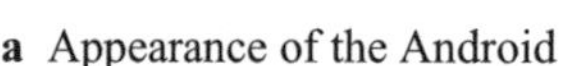

a Appearance of the Android **b** Arm composed of air cylinder

Fig. 1 Photograph of an android

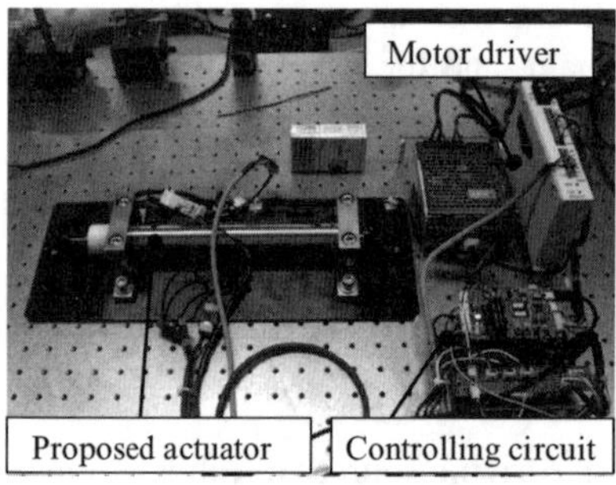

Back yoke
Permanent magnet
Coil
Non-magnetic shaft

a Prototype **b** Magetic structure of the actuator

Fig. 2 Proposed actuator system

2 Analyzed Model and Operating Principle

Figure 2 shows the prototype of the proposed linear actuator and the magnetic structure. The mover is mainly composed of a magnet block and non-magnetic shaft. The magnet block consists of Halbach array magnets, which can generate the high magnetic flux, especially radial component of flux along the outside of the magnet block that varies periodically. Magnetization of the magnet (NdFeB) is 1.4 T.

The stator is composed of three-phase coils and a back yoke. The back yoke may cause the non-linearity of the magnetic field, however, it generates the high magnetic flux on the coils.

When these coils are excited, they are forced to move by the Lorenz force. The mover is driven by the reaction force while the stator is fixed (see Fig. 3). The mover can be freely controlled by switching the three-phase currents.

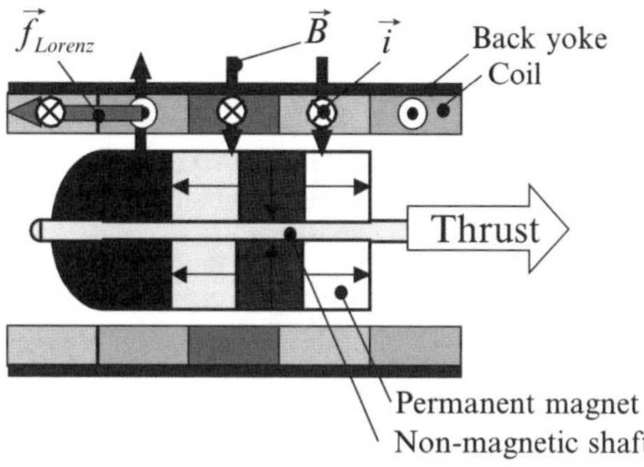

Fig. 3 Operating principle

3 Analysis Method

The equations of the magnetic field and the electric circuit are coupled using the 3-D FEM, which are given by the magnetic vector potential $\boldsymbol{A}$ and the exciting current I_0 as follows [3]:

$$\mathrm{rot}(\nu\ \mathrm{rot}\,\boldsymbol{A}) = J_0 + \nu_0\ \mathrm{rot}\,\boldsymbol{M} \tag{1}$$

$$E = V_0 - RI_0 - \frac{d\psi}{dt} = 0 \tag{2}$$

$$J_0 = \frac{n_c}{S_c} I_0 \boldsymbol{n}_s \tag{3}$$

Where ν is the reluctivity, $\boldsymbol{J}_0$ is the exciting current density, ν_0 is the reluctivity of the vacuum, $\boldsymbol{M}$ is the magnetization of permanent magnet, V_0 is the applied voltage, R is the effective resistance, Ψ is the interlinkage flux of exciting coil, n_c and S_c are the number of turns and the cross-sectional area of the coil respectively, and n_s is the unit vector along with the direction of exciting current.

The motion of the mover is described as follows.

$$M\frac{d^2z}{dt^2} + D\frac{dz}{dt} = F \mp F_s \tag{4}$$

Where M is the mass of the mover, z is the displacement of the mover, D is the viscous damping coefficient, F is the electromagnetic force, and F_s is the dynamic friction force.

4 Analyzed and Measured Results

Dynamic performances are computed when the mover is moved from 0 to 8 mm by three-phase excitation of sinusoidal input voltage. Tables 1 and 2 show the dicretization data and analyzed conditions. The measurement of the prototype is completed

Table 1 Discretization data and CPU time

Number of element	275,520
Number of node	69,785
Number of edge	391,336
Number of unknown variables	391,283
Number of time steps	413
Total CPU time [h]	24.2

Table 2 Analyzed conditions

Input volt [V]	5
Number of turns [Turns]	525
Resistance [Ω]	5
Mass of mover [g]	238
Friction force [N]	0.93
Viscous damping coefficient [N(s m)¯1]	1.0

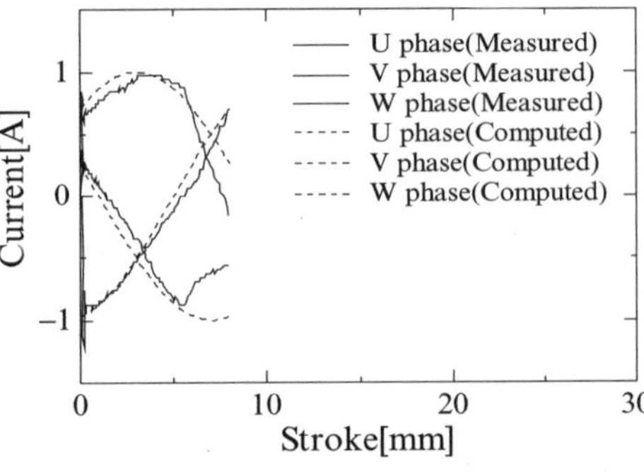

Fig. 4 Current waveforms

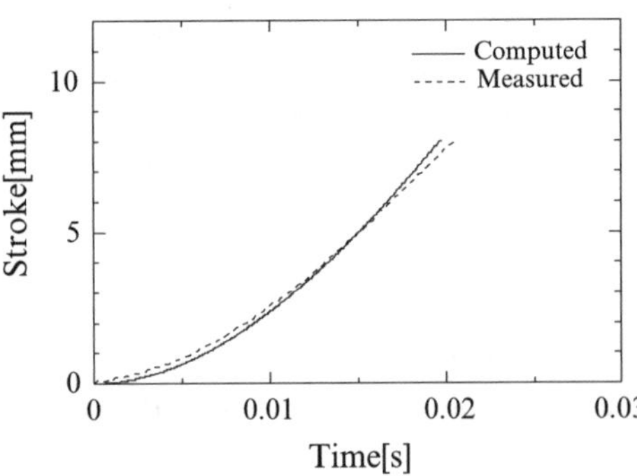

Fig. 5 Response characteristics

under the same conditions as the analysis. Figure 4 shows the comparison between computed and measured waveforms of three-phase currents. As shown, both results are in good agreement though the V-phase current of measurement is different from the computation because of the experimental error. Figure 5 shows the comparison between computed and measured time variations of the stroke. As shown, both results are in good agreement. The maximum velocity is $0.71\,\mathrm{m\,s^{-1}}$. The magnetic

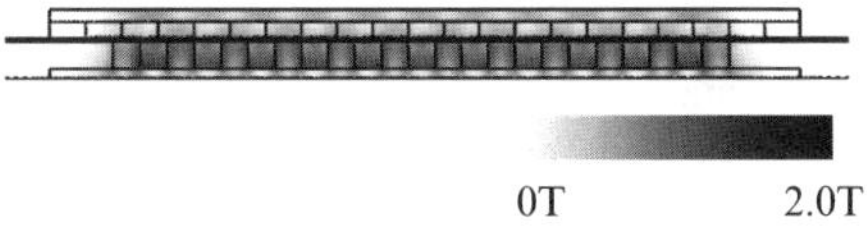

Fig. 6 Magnetic flux density distribution

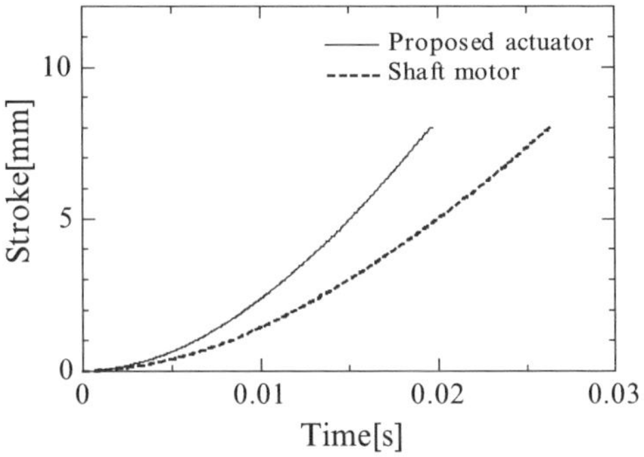

Fig. 7 Comparison with shaft motor

flux density distribution is shown in Fig. 6 (stroke: 0 mm). It shows that the maximum magnetic flux density of each coil is approximately 0.7 T. It is found that the magnetic density of the back yoke is 1.7 T (not saturated).

5 Comparison with Conventional Synchronous Motor

In order to confirm the effectiveness of the Halbach array magnets, this actuator is compared with the linear synchronous motor (so called shaft motor). The mover of the shaft motor has magnets of opposite magnetization direction side by side, and generates the high magnetic flux above the boundary of these magnets. The operating principle is the same as the proposed actuator. Under the same size and analyzed conditions, their dynamic performances are calculated. The computed results of both models are shown in Fig. 7. From this figure, the response time of the proposed actuator and the shaft motor is 0.0192 and 0.0273 s, respectively (approximately 30% decrease).

6 Conclusions

In this paper, dynamic performances of the new linear actuator whose mover has Halbach array of magnets were computed. The validity of the dynamic computation was verified through the comparison with the measured results. The effectiveness of the actuator was confirmed from the viewpoint of response by the comparison with the conventional synchronous motor.

This research was supported by "Special Coordination Funds for Promoting Science and Technology: Yuragi Project" of the Ministry of Education, Culture, Sports, Science and Technology, Japan.

References

1. K. Halbach, Application of permanent magnets in accelerators and electron storage rings, Journal of Applied Physics, vol. 57, pp. 3605–3608, 1985.
2. Z.O. Zhu, D. Howe, Halbach permanent magnet machines, and applications: a review, IEEJ Proceedings of Electrical Power Applications, vol. 148, no. 4, pp. 299–308, 2001.
3. Y. Kawase and S. Ito, New practical analysis of electrical and electronic apparatus by 3-D finite element method, Morikita Publishing Co., Tokyo, Japan, 1997.

An Optimized Support Vector Machine Based Approach for Non-Destructive Bumps Characterization in Metallic Plates

Matteo Cacciola, Giuseppe Megali, and Francesco Carlo Morabito

Abstract Within the framework of non-destructive testing techniques, it is very important to quickly and cheaply recognize flaws into the inspected materials. Moreover, another requirement is to carry out the inspection in an automatic way, with a total departure from the inspector's experience, starting from the experimental measurements. In this case, a further problem is represented by the fact that many open problems within the electromagnetic diagnostic are inverse ill-posed problems. This paper just studies a method for the analysis of metallic plates, with the aim of bumps detection and characterization starting from electromagnetic measurements. The ill-posedness of the inverse problem has been overcame by using an optimized heuristic method, i.e., the so called support vector regression machines.

1 Introduction

In recent years, non-destructive testing (NDT) techniques have more and more increase their importance for the material inspection. The relevant methodologies and techniques justify a deep interest in ensuring safety in industrial equipments, in the field of the production of energy and in biomedical environment, therefore involving several scientific disciplinary fields. Integrity of materials has a great importance in many industrial applications, where a trade-off is necessary between the necessity of material re-use in order to minimize the functioning expenses and the safety requirement for human beings. Qualitative information, such as location and extension of defects, can be very important for technicians. These are still now open questions, and can be resumed in a single problem: how to obtain a whole characterization of a defect starting from non-destructive measurements. Moreover, they

M. Cacciola, G. Megali, and F.C. Morabito
Università Mediterranea degli Studi di Reggio Calabria, Via Graziella Feo di Vito, 89100 Reggio Calabria, Italy
matteo.cacciola@unirc.it, giuseppe.megali@unirc.it, morabito@unirc.it

M. Cacciola et al.: *An Optimized Support Vector Machine Based Approach for Non-Destructive Bumps Characterization in Metallic Plates*, Studies in Computational Intelligence (SCI) **119**, 131–138 (2008)
www.springerlink.com

often are ill-posed problems, since flaws having different characteristics can generate similar electromagnetic measurements. The resulting theory is devoted to cope with the problem of finding general, approximate, stable and unique solutions and is referred to as regularization theory of ill-posed problems [1]. In recent years, support vector machines (SVMs) [2] have been extensively used for the solution of complex problems which show some cognitive aspects. In particular, support vector regression machines (SVRMs) [3] have been exploited for regression problems as a regularization method [4], since several engineering problems can be posed as search problems whose directions of search can be determined either in deterministic or stochastic ways. The main advantage is the feasibility to arrange the inspected problem as an identification process in which an interpolating heuristic system estimates some features of the defect in terms, e.g., of location and size. In this paper, we preliminarily evaluate a system for detecting bumps in a metallic plate by electric potentials as well as a suitable SVRM. The latter has been used in order to regularize the ill-posedness of the inspected problem, and has been optimized in order to use the minimal number of features.

2 The Case of Study: An Overview

In our work, we deal with an electromagnetic problem in which a thin electrically grounded metallic plate has a semi-spherical bump on its surface. It has been geometrically described in [5] and has been recently re-considered in order to refine the computational method of defect characterization [6]. Our main aim is to approximate the location as well as the radius of the bump by using a point-wise charge (i.e., a source with very small dimensions) and a set of pseudo-sensors able to measure the electric potential, just starting from the measured voltages. Because of plate's dimensions are far greater than the other distances (e.g., distance between the plate and the inducing charge), and since the material composing the plate is realistically assumed as isotropic, it is possible to consider the same plate as an infinite perfect electric conductor (IPEC) laying on the xy-plane. The point-wise charge q is located along the z-axis, at a quote z_q above the IPEC. If the IPEC has no bumps, then a suitable set of pseudo-sensors will detect a voltage measurement symmetrical to the z-axis, with a peak in correspondence to the location of the q charge (see Fig. 1a). On the contrary, if the plate has a sort of defect such as a semi-spherical bump, then a charge -q', having an opposite sign than q, will be induced within the bump according to the Van De Graaff theory [7]:

$$q' = q\left(\frac{r_b}{b^*}\right); b^* = \sqrt{x_b^2 + y_b^2 + z_q^2}, \tag{1}$$

where $(x_b,\ y_b)$ is the couple of the coordinates of the bump's center, and r_b is its radius. Thus, a sensible variation of the electric potential can be measured by the pseudo-sensors in correspondence to the bump (Fig. 1b).

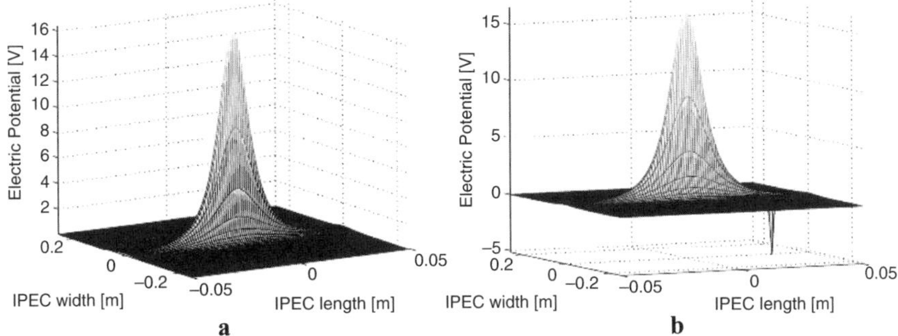

Fig. 1 The voltages measured by a set of pseudo-sensors in correspondence of an IPEC with **a** no bumps, **b** a bump located at (0.0339, −0.0129) with a radius of 0.26 mm

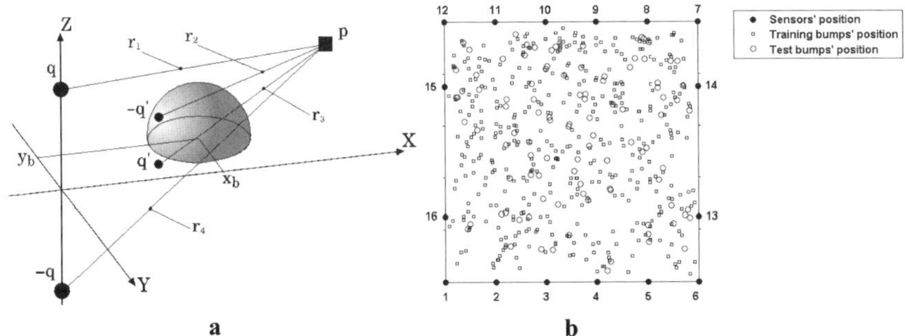

Fig. 2 **a** A graphical description of the geometry of the in-study problem; **b** location of pseudo-sensors and centers of bumps used to collect the patterns for training and testing the SVRMs

The geometry of the problem is depicted in Fig. 2a. The electric potential V measured by a pseudo-sensor located on $\mathbf{p} \in \Re^3$ spatial point, at a quote equal to z_q, is given by the following formulation, whereas the location of pseudo-sensors is depicted in Fig. 2b [8]:

$$V(\mathbf{p}) = \frac{1}{4\pi\varepsilon_0}\left(\frac{q}{r_1} - \frac{q'}{r_2} + \frac{q'}{r_3} - \frac{q}{r_4}\right), \tag{2}$$

where r_1 and r_2 are the distance between the ith sensor and the position of the inducing charge q and the induced charge $-q'$ respectively, whilst r_3 and r_4 are the distance between the ith sensor and the images of $-q'$ and q, respectively. The ill-posedness of the analyzed problem has been proven in [6] and is due to its non-linearity.

3 SVRMs: Theory and Its Usage in the In-Study Case

The SVRMs have historically been used even to regularize ill-posed inverse problems by means of a sort of "learning by sample" technique. In order to explain the mathematical framework in which SVRMs are defined, let us consider the problem to approximate the set of data $D = \{(\mathbf{s}_1, y_1), \ldots, (\mathbf{s}_l, y_l)\}$, $\mathbf{s}_i \in \mathfrak{R}^n$, $y_i \in \mathfrak{R}$, by a linear function $f(\mathbf{s}) = < \mathbf{w}, \mathbf{s} > +b$, where $<,>$ denotes the inner product in S, i.e., the space of the input patterns. This problem can be solved by selecting the optimal regression function as the minimum of the functional [9]

$$\Phi\left(\mathbf{w}, \xi^-, \xi^+\right) = \frac{1}{2}\|\mathbf{w}\|^2 + C\left(\nu\varepsilon + \frac{1}{l}\sum_{i=1}^{l}\left(\xi_i^- + \xi_i^+\right)\right), \tag{3}$$

where C is a constant determining the trade-off between minimizing training errors and minimizing the model complexity term $\|\mathbf{w}\|$; ξ^-, ξ^+ are slack variables representing upper and lower constraints on the output of the system, respectively; $0 \leq \nu \leq 1$ is a user defined constant trading-off the tube size ε against model complexity and slack variables. If the in-study problem is non-linear, a non-linear transformation, accomplished by a kernel function $K(\mathbf{s}_i, \mathbf{s}_j)$ [9], maps D into a high-dimensional space where the linear regression can be carry out. In order to implement a SVRM-based estimator, a loss function must be used; in our approach, it has been exploited the Vapnik ε-insensitive loss function [9] $|\mathrm{y} - \mathrm{f}(\mathbf{x})|_\varepsilon = \max\{0, |\mathrm{y} - \mathrm{f}(\mathbf{x})| - \varepsilon\}$. A number of 500 patterns has been collected by exploiting a set of 16 sensors, located on the xy-plane as Fig. 2b draws at a quote equal to z_q, and randomly varying the center (x_b, y_b) as well as the radius r_b of the bump. The features are the measured V at the sensors' locations. Figure 2b shows the bumps' centers used to both train and test suitable SVRMs. The collected patterns have been split into a training set (300 patterns) and a test set (200 patterns). Three multiple input single output SVRMs have been implemented in order to estimate x_b (SVRM_X), y_b (SVRM_Y) and r_b (SVRM_R), respectively, and their performances have been evaluated by using the Willmott's index of agreement (WIA): $0 \leq \mathrm{WIA} \leq 1$ (see (4)). The closer WIA is to 1, the more affordable the SVRM's estimation.

$$WIA = 1 - \frac{\sum_{i=1}^{n}(\hat{y}_i - y_i)^2}{\sum_{i=1}^{n}\left(|y_i - \bar{y}| + |\hat{y}_i - \bar{y}|\right)^2}. \tag{4}$$

Here, y_i is the ith pattern, $\bar{y}$ is the average value of the whole set of test patterns, and $\hat{y}_i$ is the estimation of the ith pattern carried out by the trained SVRM. Table 1 sums the characteristics of each SVRM as well as the obtained performances. Since this phase of the algorithm's implementation has been already developed in [6], let us focus the attention on the algorithm's optimization.

Table 1 Summary of the settings for the best performing SVRMs

	Kernel	C (cost parameter)	ε (loss parameter)	WIA
SVRM_X	Radial basis function ($\sigma = 1$)	100	10^{-4}	0.8354
SVRM_Y	Polynomial (2nd order)	0.1	10^{-3}	0.8365
SVRM_R	Polynomial (2nd order)	10^{-3}	10^{-4}	0.6645

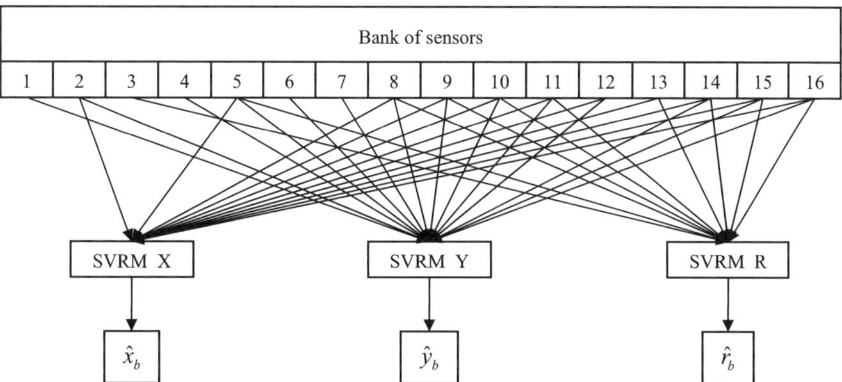

Fig. 3 The block schema of the optimized SVRM based system implemented to automatically estimate the characteristics of bumps in an IPEC in a real-time way

4 Optimization of Implemented Technique

In order to improve the performances of the implemented SVRMs, the so called leave-one-out cross validation procedure [10] has been exploited. It consists on removing a sensor measure at each time, ad thus re-training and re-testing the single SVRM. The procedure is iterated until the minimal number of sensors showing the maximal WIA is found. In this way, for each considered SVRM, the method tends to take away the measurements obtained by some sensors, so becoming a sort of leave-multi-out cross validation. The results of the optimization is depicted in Fig. 3. As Table 2 describes, the so optimized SVRMs show increased performances with a smaller complexity, thus without incurring into the so called "curse of dimensionality problem". The different performances have been evaluated in terms of WIA as well as considering other statistical parameters, giving information about the systems' reliability, i.e., the root mean squared error (RMSE), the root relative squared error (RRSE), the mean absolute error (MAE) and relative absolute error (RAE) [11]:

$$RMSE = \sqrt{\frac{\sum_{i=1}^{n} (\hat{y}_i - y_i)^2}{n}}; \; RRSE = \sqrt{\frac{\sum_{i=1}^{n} (\hat{y}_i - y_i)^2}{\sum_{i=1}^{n} (y_i - \bar{y})^2}}; \; MAE = \frac{\sum_{i=1}^{n} |\hat{y}_i - y_i|}{n}; \; RAE = \frac{\sum_{i=1}^{n} |\hat{y}_i - y_i|}{\sum_{i=1}^{n} |y_i - \bar{y}|}. \tag{5}$$

Table 2 Performances of the original and optimized SVRMs

		SVRM_X	SVRM_Y	SVRM_R
WIA	Original	0.8354	0.8365	0.6645
	Optimized	0.8378	0.8464	0.7268
RMSE	Original	0.0197	0.0089	0.0006
	Optimized	0.0188	0.0088	0.0005
RRSE	Original	0.7181	0.6890	1.0429
	Optimized	0.6973	0.6616	0.8606
MAE	Original	0.0229	0.0119	0.0020
	Optimized	0.0229	0.0119	0.0020
RAE	Original	0.6345	0.5772	0.7937
	Optimized	0.6290	0.5739	0.7884

Let us denote how the RMSE has the same measure as the evaluated quantity, and is based on the absolute error. For a perfect fit, $\hat{y}_i = y_i$ and RMSE $= 0$. So, the RMSE index ranges from 0 to infinity, with 0 corresponding to the ideal. A related unitless measure is the RRSE: for a perfect fit, the numerator is equal to 0 and RRSE $= 0$. So, the RRSE index ranges from 0 to infinity, with 0 corresponding to the ideal. Concerning the MAE and RAE, in the former case for a perfect fit $\hat{y}_i = y_i$ and MAE $= 0$. So, the MAE index ranges from 0 to infinity, with 0 corresponding to the ideal; whilst, in the latter case, for a perfect fit, the numerator is equal to 0 and RAE $= 0$. So, the RAE index ranges from 0 to infinity, with 0 corresponding to the ideal. Let us remark how the RMSE is more sensitive to outliers in the data than the MAE.

Practically speaking, the optimized SVRMs do not consider the following set of sensors: $\{1, 3, 4, 6, 7\}$ for SVRM_X; $\{3, 13\}$ for SVRM_Y and $\{1, 2, 4, 6, 7, 12\}$ for SVRM_R, as described by Fig. 4.

5 Discussion About Results and Conclusions

Since their introduction into the scientific environment, SVRMs have been largely used as a computational intelligence technique able to solve complex regression problems. Nowadays, the extensive usage of this kinds of techniques allows to regularize inverse problems starting from a "learning by sample" approach. In this paper, the SVRMs' performances have been evaluated within the context of a NDT inverse problem, i.e., the characterization of a semispherical bump laying on a metallic plate in terms of both location and size. The leave-multi-out cross validation based optimization allows us to obtain increased precision in the characterization of bumps, improving the performances obtained by other heuristic techniques, such as artificial neural networks (presented in [6]), with the furthermore advantage of a low time consuming for training as well as for testing procedure. The proposed approach, therefore, seems very interesting for a real-time working in real-world applications.

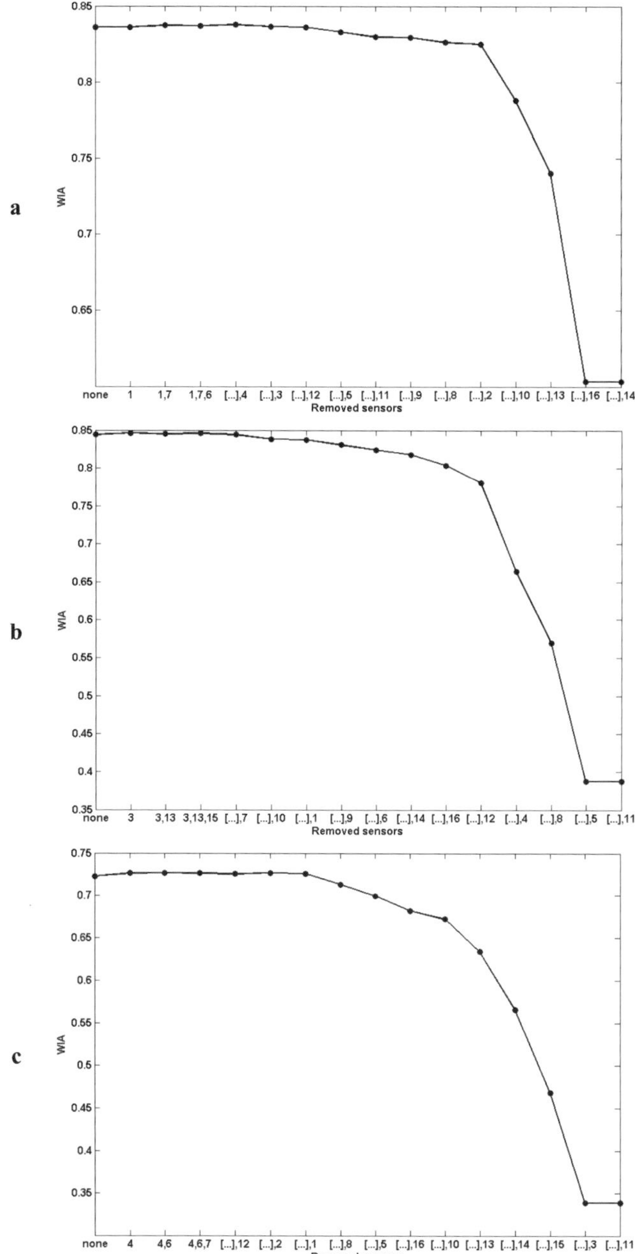

Fig. 4 Results of the leave-one-out cross validation for the estimation of: **a** x_b; **b** y_b; **c** r_b

References

1. M. Bertero, T.A. Poggio, V. Torre, Ill-posed problems in early vision, Proceedings of the IEEE, no. 76, pp. 869–889, 1988.
2. V.N. Vapnik, The Nature of Statistical Learning Theory. New York, USA: Springer Verlag, 1995.
3. A.J. Smola, Regression estimation with support vector learning machines, Master Thesis, Technische Universitat, Munchen, Germany, 1996.
4. M. Cacciola, F. La Foresta, F.C. Morabito, M. Versaci, Advanced Use of Soft Computing and Eddy Current Test to Evaluate Mechanical Integrity of Metallic Plates. NDT& E International, vol. 40 no. 5, pp. 357–362, 2007.
5. F.C. Morabito, M. Campolo, Location of Plural Defects in Conductive Plates via Neural Networks. IEEE Transactions on Magnetics, vol. 31, no. 3, pp. 1765–1768, 1995.
6. M. Cacciola, M. Campolo, F. La Foresta, F.C. Morabito, M. Versaci, A kernel based learning by sample technique for defect identification through the inversion of a typical electric problem, in Lecture Notes in Artificial Intelligence, Special Issue of Joint Conference WIRN2007-KES2007, vol. 4694, Part III, pp. 243–250, 2007.
7. R.J. Van De Graaff, A 1,500,000 Volt electrostatic generator, Physical Review, no. 38, pp. 1919–1920, 1931.
8. E. Durand, Electrostatique, Tome II. Paris: Masson, 1966.
9. B. Schölkopf, A. Smola, Learning with Kernels. New York, USA: MIT Press, 2002.
10. J. Weston, Leave-one-out support vector machines, Proceedings of the Sixteenth International Joint Conference on Artificial Intelligence, Stockholm, Sweden, July 31–August 6, pp. 727–733, 1999.
11. H. Ian, F. Eibe, I.H. Witten, Data Mining: Practical Machine Learning Tools and Techniques. New York: Morgan Kaufmann, 2005.

The Back Reconstruction of Signals by the NMR Techniques

Eva Kroutilova, Miloslav Steinbauer, Pavel Fiala, Jarmila Dedkova, and Karel Bartusek

Abstract This article deals with the reverse reconstruction results obtained from the numerical simulation of MR signals by various techniques, which will be usable for the experimental results verification. We solved the effect of changes of magnetic fields in MR tomography. The paper will describe the magnetic resonance imaging method applicable mainly in MRI and MRS in vivo studies.

1 Numerical Analysis

The numerical modeling was realized using the finite element method together with the Ansys system. As the boundary condition, there was set the scalar magnetic potential φ_m by solving Laplace's equation [1]

$$\Delta\varphi_{\mathrm{m}} = \mathrm{div}\mu\,(-\mathrm{grad}\varphi_{\mathrm{m}}) = 0 \tag{1}$$

together with the Dirichlet boundary condition

$$\varphi_{\mathrm{m}} = const. \text{ on the areas } \Gamma_1 \text{ a } \Gamma_2 \tag{2}$$

and the Neumann boundary condition

$$\mathbf{u}_{\mathrm{n}} \cdot \mathrm{grad}\ \varphi_{\mathrm{m}} = 0 \text{ on the areas } \Gamma_3 \text{ a } \Gamma_4. \tag{3}$$

Eva Kroutilova, Miloslav Steinbauer, Pavel Fiala, and Jarmila Dedkova
Brno University of Technology, Faculty of Electrical Engineering and Communication, Department of Theoretical and Experimental Electrical Engineering, Kolejni 2906/4, 612 00 Brno, Czech Republic, http://www.utee.feec.vutbr.cz/EN/index.htm
kroutila@feec.vutbr.cz, steinbau@feec.vutbr.cz, fialap@feec.vutbr.cz, dedkova@feec.vutbr.cz

Karel Bartusek
Institute of Scientific Instruments, Academy of Sciences of the Czech Republic, Královopolská 147, 612 64 Brno, Czech Republic, http://www.isibrno.cz
bar@isibrno.cz

E. Kroutilova et al.: *The Back Reconstruction of Signals by the NMR Techniques*, Studies in Computational Intelligence (SCI) **119**, 139–145 (2008)
www.springerlink.com

The continuity of tangential elements of the magnetic field intensity on the interface of the sample region is formulated by the expression

$$\mathbf{u}_{\mathrm{n}} \times \operatorname{grad} \varphi_{\mathrm{m}} = 0. \tag{4}$$

The description of the quasi-stationary model MKP is based on the reduced Maxwell's equations

$$rot\mathbf{H} = \mathbf{J}. \tag{5}$$

$$div\mathbf{B} = 0, \tag{6}$$

where **H** is the magnetic field intensity vector, **B** is the magnetic field induction vector, **J** is the current density vector. For the case of the static magnetic irrotational field, (5) is reduced to the expression (7).

$$rot\mathbf{H} = 0. \tag{7}$$

Material relations are represented by the equation

$$\mathbf{B} = \mu_0 \mu_{\mathrm{r}} \mathbf{H}, \tag{8}$$

where μ_0 is the permeability of vacuum, $\mu_r(\mathbf{B})$ is the relative permeability of ferromagnetic material. The closed area Ω, which will be applied for solving (6) and (7), is divided into the region of the sample Ω_1 and the region of the medium Ω_2. For these, there holds $\Omega = \Omega_1 \cup \Omega_2$. For the magnetic field intensity **H** in area Ω there holds the relation (7). The magnetic field distribution from the winding is expressed with the help of the Biot–Savart law, which is formulated as:

$$\mathbf{T} = \frac{1}{4\pi} \int_{\Omega} \frac{\mathbf{J} \times \mathbf{R}}{|\mathbf{R}|^3} d\Omega, \tag{9}$$

where **R** is the distance between a point in which the magnetic field intensity **T** is looked for and a point where the current density **J** is assumed. The magnetic field intensity **H** in the area can be expressed as

$$\mathbf{H} = \mathbf{T} - \operatorname{grad}\phi_m, \tag{10}$$

where **T** is the preceding or estimated magnetic field intensity, ϕ_m is the magnetic scalar potential. The boundary conditions are written as

$$\mathbf{u}_{\mathrm{n}} \cdot \mu \, (\mathrm{T} - grad\phi_m) = 0 \text{ on the areas } \Gamma_3 \text{ and } \Gamma_4, \tag{11}$$

where $\mathbf{u}_n$ is the normal vector, $\Gamma_{\mathrm{Fe}-0}$ is the interface between the areas Ω_{Fe} and $\Omega_0 \cup \Omega_{\mathrm{W}}$. The area Ω_0 is the region of air in the model, the area Ω_{W} is the region with the winding. The continuity of tangential elements of the magnetic field intensity on the interface of the area with ferromagnetic material is expressed

$$\mathbf{u}_{\mathrm{n}} \times (\mathrm{T} - grad\phi_m) = 0. \tag{12}$$

By applying the relation (10) in the relation (11) we get the expression

$$\operatorname{div}\mu_0\mu_r\mathrm{T}-\operatorname{div}\mu_0\mu_r\operatorname{grad}\phi_\mathrm{m}=0. \tag{13}$$

The equation can be discretized (13) by means of approximating the scalar magnetic potential

$$\varphi_\mathrm{m}=\sum_{j=1}^{NN}\varphi_j W_j(x,y,z)\ \text{pro}\ \forall(x,y,z)\subset\Omega, \tag{14}$$

where φ_j is the value of the scalar magnetic potential in the j-th node, W_j the approximation function, NN the number of nodes of the discretization mesh. By applying the approximation (14) in the relation (13) and minimizing the residues according to the Galerkin method, we get the semidiscrete solution

$$\sum_{j=1}^{NN}\varphi_j\int_\Omega \mu\ \operatorname{grad}W_i\cdot\operatorname{grad}W_j\mathrm{d}\Omega=0,\qquad i=1,\ldots NN. \tag{15}$$

The system of equations (15) can be written briefly as

$$[k_{ij}]\cdot[\varphi_i]^T=0,\qquad i,j\in\{1,\ldots NN\}. \tag{16}$$

The system (16) can be divided into

$$\mathbf{K}\begin{bmatrix}\mathbf{U}_\mathrm{I}\\ \mathbf{U}_\mathrm{D}\end{bmatrix}=\begin{bmatrix}0\\0\end{bmatrix}, \tag{17}$$

where $\mathbf{U}_\mathrm{I}=[\varphi_1,\ldots,\varphi_\mathrm{NI}]^\mathrm{T}$ is the vector of unknown internal nodes of the area Ω including the points on the areas Γ_3 and Γ_4. $\mathbf{U}_\mathrm{D}=[\varphi_1,\ldots,\varphi_\mathrm{ND}]^\mathrm{T}$ is the vector of known potentials on the areas Γ_1 and Γ_2 (the Dirichlet boundary conditions). NI in the index marks the number of internal nodes of the discretization mesh, ND is the number of the mesh boundary nodes. Then, the system can be written further in four submatrixes

$$\begin{bmatrix}\mathbf{k}_{11}\ \mathbf{k}_{12}\\ \mathbf{k}_{21}\ \mathbf{k}_{22}\end{bmatrix}\begin{bmatrix}\mathbf{U}_\mathrm{I}\\ \mathbf{U}_\mathrm{D}\end{bmatrix}=\begin{bmatrix}0\\0\end{bmatrix} \tag{18}$$

and this yields the system with the introduced boundary conditions, which is solved in the MKP as

$$\mathbf{k}_{11}\mathbf{U}_\mathrm{I}+\mathbf{k}_{12}\mathbf{U}_\mathrm{D}=0. \tag{19}$$

The coefficients $\mathrm{k_{ij}}$ of the submatrix k are non-zero only when the element of mesh contains both the i and the j nodes. The contribution of the element e to the coefficient k_{ij} is

$$k_{ij}^e=\int_{\Omega^e}\mu^e\ \operatorname{grad}W_i^e\cdot\operatorname{grad}W_j^e\ \mathrm{d}\Omega,\qquad e=1,\ldots NE, \tag{20}$$

where Ω^e is the area of the discretization mesh element, μ^e is the permeability of the selected element medium, NE is the number of the discretization mesh elements. The matrix k elements are then the sums of contributions of the individual elements.

$$k_{ij} = \sum_{e=1}^{NE} k_{ij}^{e}. \tag{21}$$

The system of equations (16) can be solved with the help of standard algorithms. The scalar magnetic potential value is then used for evaluating the magnetic field intensity according to (10).

2 The Boundary Conditions

The boundary conditions $\pm\varphi/2$ were set to the model edges, to the external left and right boundaries of the air medium. The excitation value $\pm\varphi/2$ was set using again the relation (21).This is derived for the assumption that, in the entire area, there are no exciting currents, therefore there holds for the rot $\boldsymbol{H} = \mathbf{0}$ and the field is irrotational. Consequently, for the scalar magnetic potential φ_m holds

$$H = -grad\ \varphi_m. \tag{22}$$

The potential of the exciting static field with intensity $\mathbf{H}_0$ is by applying (23)

$$\varphi_m = \int \vec{H}_0 \cdot \vec{u}_z dz = H_0 \cdot z, \tag{23}$$

where

$$H_0 = \frac{B}{\mu_0 \cdot \mu_r}. \tag{24}$$

Then

$$\pm\frac{\varphi}{2} = \frac{B \cdot z}{2\mu_0} = \frac{4,7000T \cdot 90mm}{2\mu_0}, \tag{25}$$

where z is the total length of the model edge.

3 Geometrical Model

Figure 1 describes the sample geometry for the numerical modeling. On both sides, the sample is surrounded by the referential medium. During the real experiment, the reference is represented by water, which is ideal for obtaining the MR signal.

As shown in Fig. 1, in the model there are defined four volumes with different susceptibilities. The materials are defined by their permeabilites: material No. 1 – the medium outside the cube (air), $\chi = 0$, material No. 2 – the cube walls

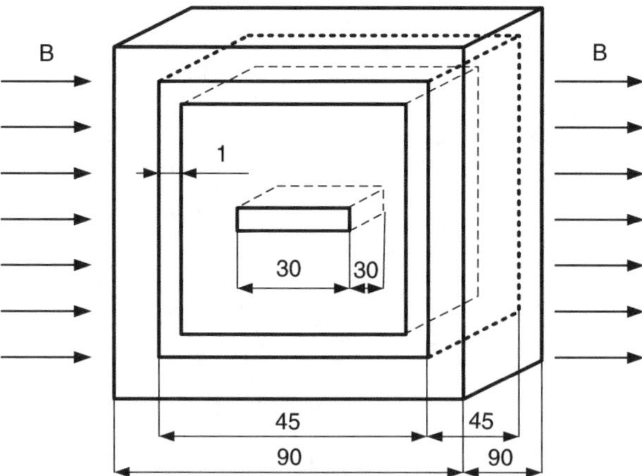

Fig. 1 The sample geometry for numerical modeling

(sodium glass), $\chi = -11,67.10^{-6}$, material No. 3 is the sample material (sodium glass), $\chi = -11,67.10^{-6}$, quartz glass, $\chi = -8,79.10^{-6}$, the simax glass (commercial name), $\chi = -8,82.10^{-6}$, material No. 4 is the medium inside the cube (water with nickel sulfate solution NiSO4, $\chi = -12,44.10^{-6}$). The permeability rate was set with the help of the relation $\chi = 1 + \mu$.

4 Numerical Model

For the sample geometry according to Fig. 1, the geometrical model was built in the system. In the model there was applied the discretization mesh with 168,948 nodes and 159,600 elements, type Solid96 (Ansys) [2, 3]. The boundary conditions (25) were selected for the induction value of the static elementary field to be $B_0 = 4,7000\,T$ in the direction of the z coordinate (the cube axis) – corresponds with the real experiment carried out using the MR tomograph at the Institute of Scientific Instruments, ASCR Brno (Figs. 2 and 3).

5 Experimental Verification

The experimental measuring was realized using the MR tomograph at the Institute of Scientific Instruments, ASCR Brno. The tomograph elementary field $B_0 = 47,000\,T$ is generated by the superconductive solenoidal horizontal magnet produced by the Magnex Scientific company. The corresponding resonance frequency for the 1H cores is 200 MHz.

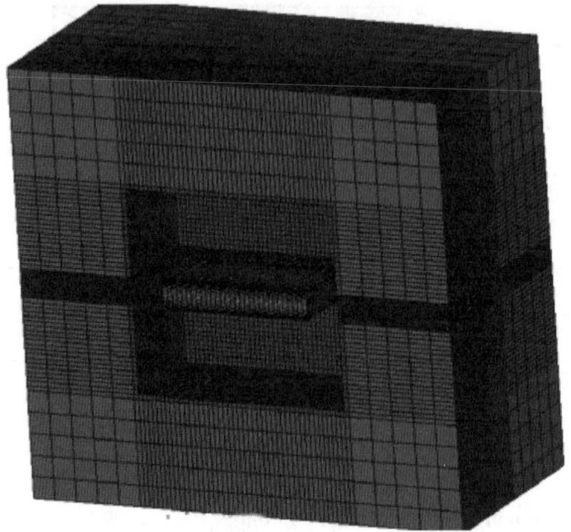

Fig. 2 The model with the elements in the system Ansys

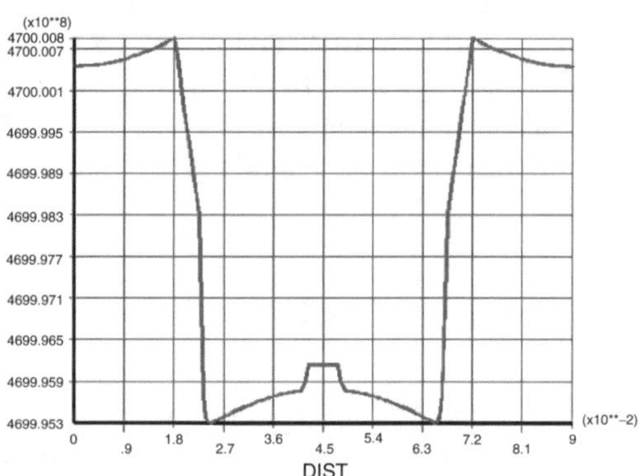

Fig. 3 The distribution of the magnetic induction module in the section of the sample for sodium glass with $\chi = -11{,}67.10^{-6}$

6 The Comparison of Results: Numerical Modeling and Measuring

The numerical modeling and analysis of the task have verified the experimental results and, owing to the modificability of the numerical model, we have managed to advance further in the experimental qualitative NMR image processing realized at the ISI ASCR. The differences between measured and simulated values aren't higher than 20% (Fig. 4).

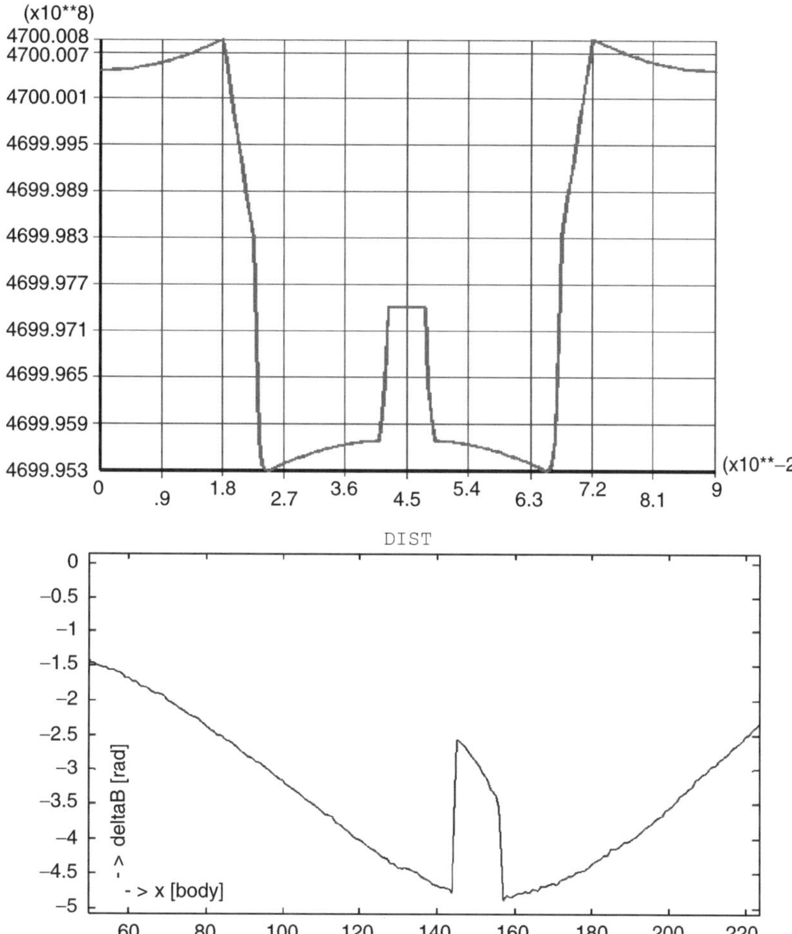

Fig. 4 The comparison of the results from numerical modeling and measuring for the quartz glass ($\Delta B = 17\,\mu T$)

Acknowledgements The research described in the paper was financially supported by research plans GAAV B208130603, MSM 0021630516 and GA102/07/0389.

References

1. Fiala, P., Kroutilová, E., Bachorec, T., Modelování elektromagnetických polí, počítačová cvičení. vyd. Brno: VUT v Brně, FEKT, Údolní 53, 602 00, Brno, 2005, pp. 1–69.
2. Steinbauer, M., Měření magnetické susceptibility technikami tomografie magnetické rezonance. vyd. Brno: VUT v Brně, FEKT, Údolní 53, 602 00, Brno, 2006.
3. Ansys User's Manual. Huston (USA): SVANSON ANALYSYS SYSTEM, Inc., 1994–2006. List and number all bibliographical references at the end of the paper.

3D Edge Element Calculations of Electrical Motor with Double Cylindrical Rotor

Rafal Wojciechowski, Cezary Jędryczka, Andrzej Demenko, and Ernest A. Mendrela

Abstract The brushless permanent magnet motor with cylindrical double rotor is investigated. The calculations have been performed using the 3D edge element method (EEM). The authors developed computer software for 3D electromagnetic field computations. Selected results of electromagnetic torque characteristics, for the different motor dimensions are compared and discussed.

1 Introduction

In the paper, the 8-pole permanent magnet motor with double cylindrical rotor (PMDRM) is considered. The structure of motor is presented in Fig. 1. The analyzed PMDRM consists of toothless toroidal stator, double cylindrical rotor with rare-earth magnets and 48 coils made of stranded copper wire. Coils are wound in a Gramme's winding arrangement. The 3D edge element method (EEM) has been applied for calculation of magnetic field in the motor. For motor design and optimisation, a special computer program has been elaborated. The authors used their experience got in developing such a software before [1,4,5]. In the presented approach the motor has been subdivided into curved rectangular parallelepipeds (Fig. 2). The element edges are parallel to the axis of a cylindrical coordinate system r, z, ψ. The trace of single element in the r, ψ plane is a curved rectangular of the same angular length β, see Fig. 2.

R. Wojciechowski, C. Jędryczka, and A. Demenko
Poznań University of Technology, Piotrowo 3, 60-965 Poznań, Poland
rafal.wojciechowski@doctorate.put.poznan.pl,
cezary.jedryczka@doctorate.put.poznan.pl, andrzej.demenko@put.poznan.pl

E.A. Mendrela
Louisiana State University, Electr. Eng. Building 102, Baton Rouge, LA 70803, USA
ermen@ece.lsu.edu

R. Wojciechowski et al.: *3D Edge Element Calculations of Electrical Motor with Double Cylindrical Rotor*, Studies in Computational Intelligence (SCI) **119**, 147–153 (2008)
www.springerlink.com

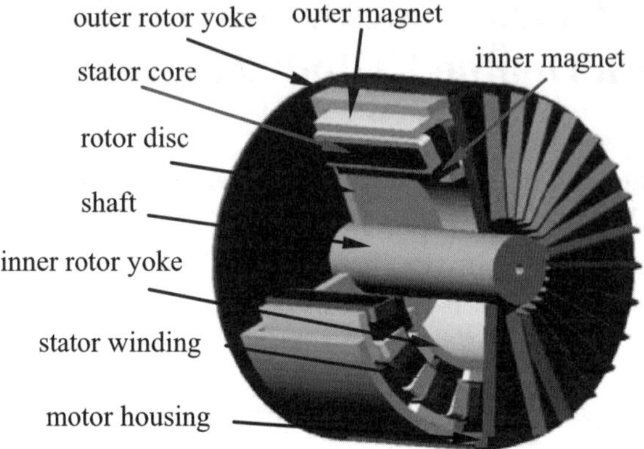

Fig. 1 Construction of PMDR motor

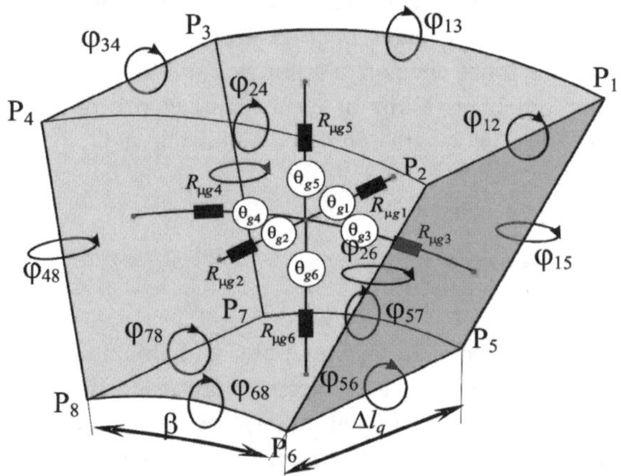

Fig. 2 Network model of curved rectangular parallelepiped – reluctance network

2 Reluctance Motor Model

In the software preparation there is an advantage to describe the EEM equations using the notation of equivalent circuits. The edge element equations represent the loop equations of reluctance network (RN). The branches of RN connect the centres of the elements. The vector of edge values of $\boldsymbol{A}$ represents the loop fluxes φ in the loops around edges. The loop equations that represent EEM equations for magnetic field can be written as follows:

$$\boldsymbol{k}_e\,\boldsymbol{R}_{\mu gd}\boldsymbol{k}_e^T\varphi = \boldsymbol{k}_e\theta_{gd}. \tag{1}$$

Here $\boldsymbol{R}_{\mu gd}$ is the matrix of branch reluctances for the disjoint set of elements, matrix $\boldsymbol{k}_e$ is the transposed loop matrix for reluctance network, θ_{gd} is the vector of branch magnetomotive forces (*mmfs)* for the disjoint set of elements. The components of vector θ_{gd} in the region with permanent magnets (PMs) are defined using function that describe magnetizing vector $\boldsymbol{T}_m$ [2]. In considered motor coils are made of stranded conductors. In the winding region, the *mmfs* are described by the facet values of conduction currents. The *ith* component of loop $mmfs$, $\boldsymbol{k}_e\theta_{gd}$, is equal to the product of current and numbers of conductors z_i passing through the *ith* loop of RN. In the considered winding region, the skin effect is negligible. The problem is simplified by neglecting the eddy current in the shaft and in PMs.

The magnetic core of the machine is laminated [1]. In order to model this core in 3D it is assumed that reluctivity ν_z in the direction parallel to shaft axis z differs from the reluctivity $\nu_{r,\psi}$ in the direction orthogonal to axis z, i.e. the core reluctivity is considered to be orthogonally anisotropic. The values of ν_z and $\nu_{r,\psi}$ have been obtained from the formulas that describe the equivalent reluctivity of the system composed of two reluctances: the reluctance for the flux which penetrates the ferromagnetic sheets and the reluctance for the flux that penetrates the isolation. This approach gives

$$\nu_{r,\psi} = \nu_0 \nu_{Fe} \left(k_{nd}\nu_0 + (1-k_{nd})\ \nu_{Fe}\right)^{-1} \approx \nu_{Fe}/k_{nd}, \tag{2}$$

$$\nu_z = k_{nd}\nu_{Fe} + (1-k_{nd})\ \nu_0 \approx (1-k_{nd})\ \nu_0, \tag{3}$$

where k_{nd} is the stacking factor, $\nu_{Fe} = \nu(\boldsymbol{B})$ is the reluctivity of isotropic ferromagnetic sheet.

The electromagnetic torque for RN is obtained from finite difference approximation of the magnetic energy derivative vs. the rotor displacements [1,3,6]. The finite difference approximation gives

$$T(\alpha) = -\frac{W(\alpha+\beta) - W(\alpha-\beta)}{2\beta}, \tag{4}$$

where $W(\alpha \pm \beta)$ is the magnetic energy for position $\alpha \pm \beta$, α is an considered rotor position. In the 3D calculations of $T(\alpha)$, the circular band of length l is placed inside the air gap; where l is the EE model length in the direction of z axis parallel to the shaft. The band is subdivided into m layers of thickness Δl_q, see Fig. 2. In the considered motors, the two circular bands are formed: (a) band under the outer PMs and (b) band over the inner PMs. For each band with EEM formula (4) is expressed as follows:

$$T(\alpha) = \frac{1}{4\beta}\left\{\sum_{q=1}^{m} R_{bq} \sum_{i=1}^{n}\left(\phi^2_{\psi q,i-1} - \phi^2_{\psi q,i+1}\right) + \sum_{q=1}^{m-1} R_{zq} \sum_{i=1}^{n}\left(\phi^2_{zq,i-1} - \phi^2_{zq,i+1}\right)\right\}. \tag{5}$$

Here $\phi_{\psi i\pm 1,q}$, is the value of flux in branch P_{bq} for position $\alpha \pm \beta$, $\phi_{zi\pm 1,q}$ is the value of flux in branch P_{zq} for position $\alpha \pm \beta$ and R_b, R_z are the branch reluctances in the band – see Fig. 3.

The branch fluxes have been expressed by loop fluxes (see Fig. 4), then (5) can be transformed into the following form

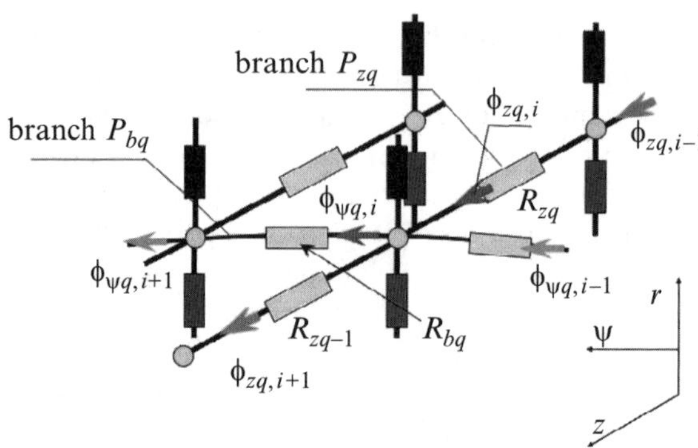

Fig. 3 Fragment of RN in upper air gap

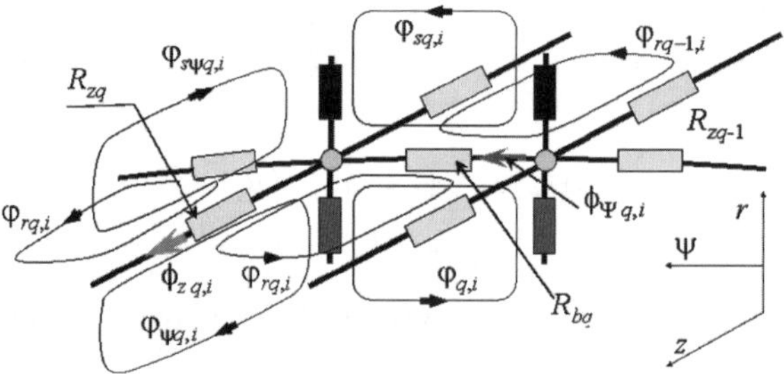

Fig. 4 Part of the band with branches P_{bq} and P_{zq}

$$T(\alpha) = \frac{1}{2\beta}\left\{\sum_{q=1}^{m} R_{bq}\sum_{i=1}^{n}\phi_{\psi q,i}\left(\varphi_{q,i-1}-\varphi_{q,i+1}\right)+\sum_{q=1}^{m-1} R_{zq}\sum_{i=1}^{n}\phi_{zq,i}\left(\varphi_{q,i-1}-\varphi_{q,i+1}\right)\right\}. \tag{6}$$

As a result we obtain equation that represents the stress tensor formula for torque calculation using RN and edge element method.

An important quantity in the PM machines torque analysis is the torque pulsation factor ε. The authors used the torque pulsation factor which is defined in [3]

$$\varepsilon = \frac{T_{max} - T_{min}}{T_{av}} \cdot 100\ \%, \tag{7}$$

where T_{max} T_{min}, T_{av} is the maximum, the minimum and the average electromagnetic torque $T(\alpha)$ functions, for given value of internal power load angle δ.

3 Results

The method presented above has been used in the analysis of two types of PM-DRM: (a) motor with magnets mounted on the cylindrical rotors (without "ferromagnetic keys" between magnets), (b) motor with ferromagnetic keys between magnets. Motor with ferromagnetic keys can be considered as a motor with magnets in the slots. The considered region has been subdivided into $\approx 400,000$ elements per pole pair. To find the optimal motor parameters the calculations have been performed for the different values of angular length γ_m of permanent magnets. The other dimensions of analyzed motor are assumed to be identical. The influence of the angular length γ_m on average electromagnetic torque has been examined. The average value of electromagnetic torque as a function of γ_m is shown in Fig. 5, the calculation have been performed for internal power angle $\delta = 81°$. For each value of γ_m torque waveforms and torque pulsation factor ε have been also analysed. Here, the results of torque calculations for motor supply from 3 phase balanced system are presented. Figures 6 and 7 show the calculated torque waveforms for magnets of $\gamma_m = 18°$, $\gamma_m = 20°$, and $B_r = 1.2\,\mathrm{T}$. For presented torque waveforms the percentage of higher harmonics related to the average value of torque are shown in Figs. 8 and 9. For selected angles γ_m values of torque pulsation factor ε are presented in Table 1.

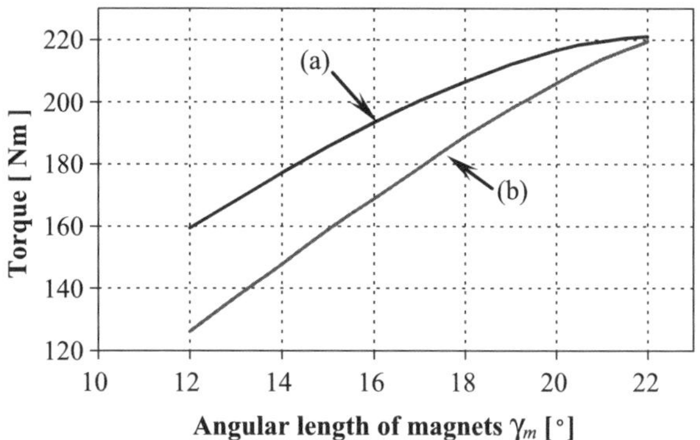

Fig. 5 Average torque as a function of γ_m

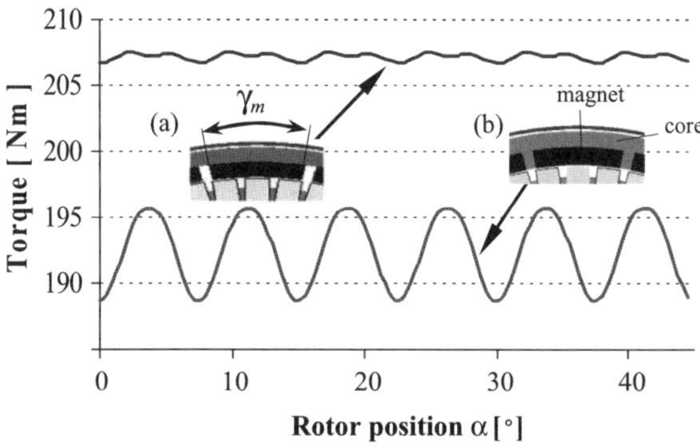

Fig. 6 Torque pulsations for $\gamma_m = 20°$ (3 phase balanced supply, $\delta = 81°$)

Fig. 7 Share of higher harmonics T_ν in functions $T(\alpha)$ shown in Fig. 8

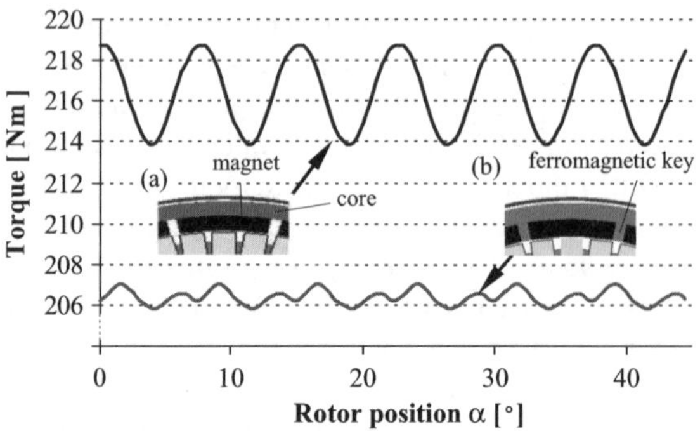

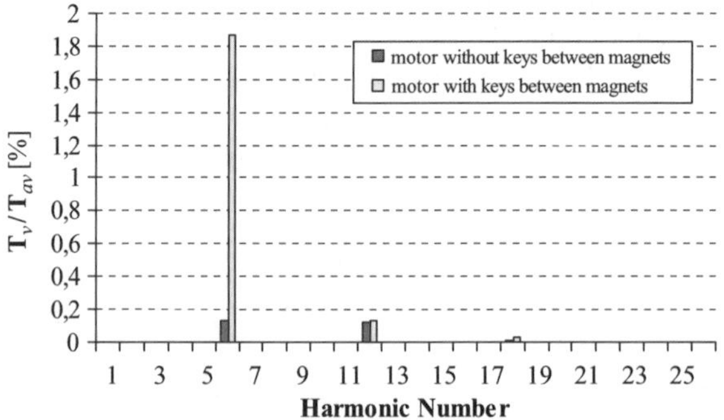

Fig. 8 Torque pulsations for $\gamma_m = 18°$ (3 phase balanced supply, $\delta = 81°$)

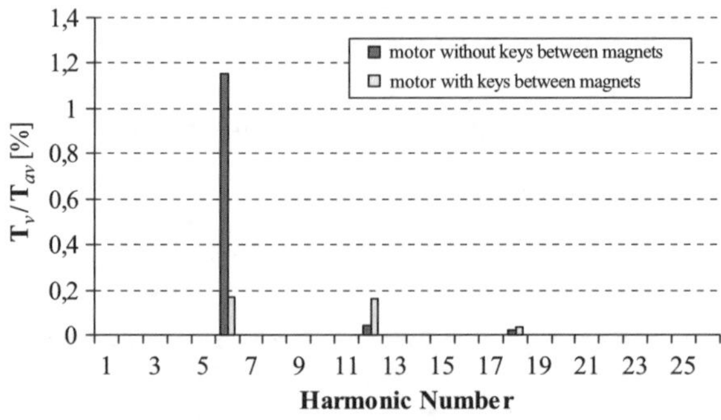

Fig. 9 Share of higher harmonics T_ν in functions $T(\alpha)$ shown in Fig. 6

Table 1 The values of torque pulsation factor ε for consider motors

	$\gamma_m [°]$	16	17	18	19	20	21	22
ε %	Motor type (a)	2.79	1.58	0.41	1.09	2.25	3.16	3.68
	Motor type (b)	6.32	5.23	3.66	1.83	0.55	2.03	3.41

4 Conclusion

The paper presents the edge element software that has been specially prepared for 3D calculations of permanent magnet motor with double cylindrical rotor. The right-hand side of the considered EE equations is expressed by the edge values of $\boldsymbol{T}$. This approach guarantees a good convergence of ICCG procedure of solving EE equations for ungaged formulation. The proposed edge element method of torque calculation conforms to the band structure of applied edge element network in the air gap region. Therefore the formula of torque calculation gives very accurate results. For the internal power angle of torque equal to zero (angle determined analytically) the calculated torque is equal to zero with 12 decimal-place accuracy.

The advantage of the considered PMDRM is that the cogging torque is negligible. The torque pulsations are only caused by non-sinusoidal distribution of magnetomotive force in the air gaps. Even for the motor with ferromagnetic keys between the magnets the torque pulsations are very small, less than 0.6% of average value of torque. It is interesting to notice that in the motor without ferromagnetic keys the torque pulsation factor ε is minimal for the angular length of magnet equal to 18°. However, for motor with ferromagnetic keys this factor reaches minimum for the angular length equal to 20°. For this type of motor the average value of electromagnetic torque is smaller than for its counterpart without ferromagnetic keys.

The presented method and elaborated software can be successfully used as CAD system for detailed analysis and optimisation of the motor of special constructions.

References

1. A. Demenko, 3D edge element analysis of permanent magnet motor dynamics, IEEE Trans. Magn., vol. 34, no. 5, pp. 3620–3623, September 1998
2. A. Demenko, D. Stachowiak, Representation of permanent magnets in 3D finite element description of electrical machines, XV Symposium Micromachines & Servosystems, Soplicowo, September 2006, Poland
3. D. Stachowiak, Edge element analysis of brushless motors with inhomogeneously magnetized permanent magnets, COMPEL – Int. J. Comput. Math. Electrical Electronic Eng., vol. 23, no. 4, 2004, pp. 1119–1128
4. E.A. Mendrela, M. Jagieła, Analysis of torque developed in axial flux, single-phase brushless DC motor with salient-pole stator, IEEE Trans. Energy Conv., vol. 19, no. 2, pp. 271–277, June 2004
5. E. Mendrela, M. Łukaniszyn, K. Macek-Kamińska, Disc-type brushless DC motors, Polish Academy of Science, 2002 (in Polish)
6. J. Coulomb, G. Meunier, Finite element implementation of virtual work principle for magnetic or electric force and torque computation, IEEE Trans. Magn., vol. 20, no. 5, pp. 1894–1896, September 1984

Sensitivity Analysis of Electromagnetic Quantities by Means of FETD and Semi-Discrete Method

Konstanty M. Gawrylczyk and Mateusz Kugler

Abstract The paper deals with numerical aspects of the sensitivity analysis in time domain by means of adjoint model method. The excitation shape in adjoint model allows to apply semi-discrete method for solution in time-domain, which should be more effective for some problems, than the finite element method with time-stepping scheme used until now.

1 Introduction

The semi-discrete finite-element method (FEM) is applied to solve the diffusion equation for the magnetic vector potential of the electromagnetic field. This numerical technique is a variant of the conventional FEM. Using Tellegen's method [5] for sensitivity evaluation, the second, the so called adjoint model, has to be solved. For specific excitation of this model, the solution on time coordinate may be handled analytically. Proposed method seems to be simpler and more economical than the conventional, fully-discrete, time-stepping version. In this work we consider only axial-symmetric models in linear and izotropic medium.

2 Forward Problem Formulation

Analyzing field diffusion into conducting region, the following non-uniform equation may be established:

$$[\mathrm{K}]\{\mathrm{U}(t)\} + [\mathrm{M}]\left\{\frac{\partial}{\partial t}\mathrm{U}(t)\right\} = \{\mathrm{i}(t)\}, \tag{1}$$

Konstanty M. Gawrylczyk and Mateusz Kugler
Szczecin University of Technology, Department of Electrical and Computer Engineering, Piastów 17, PL-70-310 Szczecin, Poland
kmg@ps.pl, mkugler@ps.pl

K.M. Gawrylczyk and M. Kugler: *Sensitivity Analysis of Electromagnetic Quantities by Means of FETD and Semi-Discrete Method*, Studies in Computational Intelligence (SCI) **119**, 155–162 (2008)
www.springerlink.com

where [**K**] and [**M**] are the stiffness and mass matrices of finite elements containing the material parameters and geometric properties of the simulated model, $\{\mathbf{U}(t)\}$ is the vector of the desired node values (modified magnetic vector potentials of nodes and $\{\mathbf{i}(t)\}$ is the excitation vector. $\{\mathbf{U}(0)\}$ is the initial condition vector. In most cases of field penetration into conducting region, this vector should be set to zero. Knowledge of modified magnetic vector potential distribution allows to calculate voltage induced in measurement coil as the time function.

3 Sensitivity Analysis in the Time Domain

The Tellegen's sensitivity equation may be derived from the Lorenz reciprocity theorem. Two systems have to be analyzed: original and adjoint [1]. The adjoint one has the same topology and material parameters, and differs from the original only with the excitation and boundary conditions. Both are analyzed on the same area Ω, but for different times t and τ. The time τ is reversed to t, it means $\tau = \xi - t$, where ξ denotes the time, while the sensitivity is evaluated.

For the tasks of electric field sensitivity versus electric conductivity γ, the sensitivity equation simplifies to:

$$\int_0^{\xi} \iint_{\Omega} J^{+}(\tau) \cdot \delta E(t) \mathrm{d}\Omega \mathrm{d}t = \int_0^{\xi} \iint_{\Omega} E^{+}(\tau) \cdot \delta\gamma \cdot E(\tau) \mathrm{d}\Omega \mathrm{d}t, \tag{2}$$

where E is the only non-zero component of electric intensity vector, perpendicular to the plane of analysis, U modified magnetic vector potential, and J excitation current density. The variables denoted with $(^{+})$ relate to the adjoint system, the other one to the original. The sensitivity equation shows the changes in δE caused by conductivity variation $\delta\gamma$. The adjoint model allows one to calculate the changes of field value for the assumed area on the whole. This area depends on the excitation of the adjoint model.

4 Adjoint Model Excitation

In the case of transient analysis, the shape of excitation has to be chosen [2]. From the sensitivity analysis point of view, the right choice of adjoint model excitation leads to simplification of the left-hand side (2). We propose an application either unit step impulse (Fig. 1).

The assumed excitations are not realizable physically, they are acting only in virtual, adjoint system.

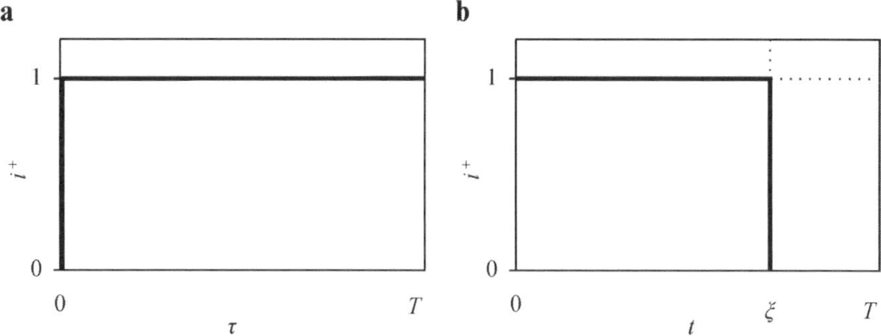

Fig. 1 Unit step impulse as adjoint model excitation: **a** in the backward time τ domain, **b** in the current time t domain

5 Finite-Element-Time-Domain Method

By introducing linear time-elements, we obtain well known generalized time stepping scheme theta [3]. The time function is approximated with linear shape functions, continuous for every time element. In the case of first order time-elements, the following two level scheme can be shown:

$$\left[\Theta\,[\mathbf{K}]+\frac{1}{\Delta t}\,[\mathbf{M}]\right]\{U_j\}=\left\{\left(\frac{1}{\Delta t}\,[\mathbf{M}]-(1-\Theta)\,[\mathbf{K}]\right)\{U_{j-1}\}\right.$$
$$\left.+\,(1-\Theta)\left\{\mathbf{i}_j\right\}+\Theta\left\{\mathbf{i}_{j-1}\right\}\right\}, \tag{3}$$

where $\{\mathbf{U}_j\}$ and $\{\mathbf{i}_j\}$ is discretized for time steps $j\cdot\Delta t$ with $j=1..n$. For faster sensitivity evaluation, the time step Δt should remain unchanged during the whole analysis. Depending on the assumed value of parameter Θ, from the range $<0,1>$, different stepping schemes may occur. The unconditional stability is guaranteed for Θ from the range of $<0.5,\ 1>$.

The time dicretization schemes force a very special form of discretized unit-step. The excitation remains linear inside the time element and continuous in their nodes. The numerical representation of unit-step impulse is decribed in [2].

6 Semi-Discrete Method

The assumed excitation shape of the adjoint model (Fig. 1) allows one to apply the semi-discrete method. Then, the vector of nodal potentials $\{\mathbf{U}(t)\}$ may be handled analytically [4]. Dividing the matrices in (1) on conducting region with potential distribution $\{\mathbf{U}_1\}$ and non-conducting region described by potential $\{\mathbf{U}_2\}$ containing sources $\{\mathbf{i}_2\}$ we obtain the following system of differential equations:

$$\begin{bmatrix} \mathbf{K}_{11} & \mathbf{K}_{12} \\ \mathbf{K}_{21} & \mathbf{K}_{22} \end{bmatrix} \begin{Bmatrix} \mathrm{U}_1(t) \\ \mathrm{U}_2(t) \end{Bmatrix} + \begin{bmatrix} \mathbf{M}_{11} & 0 \\ 0 & 0 \end{bmatrix} \begin{Bmatrix} \frac{\partial}{\partial t}\mathrm{U}_1(t) \\ 0 \end{Bmatrix} = \begin{Bmatrix} 0 \\ \mathrm{i}_2 \end{Bmatrix}, \tag{4}$$

where for izotropic media the matrices $[\mathbf{K}_{11}]$, $[\mathbf{K}_{22}]$ i $[\mathbf{M}_{11}]$ are banded and symmetric, however $[\mathbf{K}_{21}] = [\mathbf{K}_{12}]^{\mathrm{T}}$ are general matrices.

$$\begin{cases} [\mathbf{K}_{11}]\{\mathrm{U}_1(t)\} + [\mathbf{K}_{12}]\{\mathrm{U}_2(t)\} + [\mathbf{M}_{11}]\left\{\frac{\partial}{\partial t}\mathrm{U}_1(t)\right\} = \{0\} \\ [\mathbf{K}_{21}]\{\mathrm{U}_1(t)\} + [\mathbf{K}_{22}]\{\mathrm{U}_2(t)\} = \{\mathrm{i}_2\} \end{cases}. \tag{5}$$

Deriving $\{\mathbf{U}_2\}$ from the second equation:

$$\{\mathrm{U}_2(t)\} = [\mathbf{K}_{22}]^{-1}(\{\mathrm{i}_2\} - [\mathbf{K}_{21}]\{\mathrm{U}_1(t)\}), \tag{6}$$

we can eliminate it in the first one, obtaining equation for conducting region:

$$[\mathrm{K}_{11}]\{\mathrm{U}_1(t)\} + [\mathrm{K}_{12}][\mathbf{K}_{22}]^{-1}(\{\mathrm{i}_2\} - [\mathbf{K}_{21}]\{\mathrm{U}_1(t)\}) + [\mathbf{M}_{11}]\left\{\frac{\partial}{\partial t}\mathrm{U}_1(t)\right\} = \{0\}, \tag{7}$$

or in simplified form:

$$[\mathbf{K}_{\mathrm{c}}]\{\mathrm{U}_1(t)\} + [\mathbf{M}_{11}]\left\{\frac{\partial}{\partial t}\mathrm{U}_1(t)\right\} = \{\mathrm{i}_{\mathrm{c}}\}, \tag{8}$$

where

$$[\mathbf{K}_{\mathrm{c}}] = [\mathbf{K}_{11}] - [\mathbf{K}_{12}][\mathbf{K}_{22}]^{-1}[\mathbf{K}_{21}], \tag{9}$$

and

$$[\mathrm{i}_{\mathrm{c}}] = -[\mathbf{K}_{12}][\mathbf{K}_{22}]^{-1}[\mathrm{i}_2]. \tag{10}$$

Equation (8) may be handled analytically as the system of differential equations of first order. With characteristic equation

$$[\mathbf{K}_{\mathrm{c}}] + [\mathbf{M}_{11}] \cdot [\mathrm{s}] = [0], \tag{11}$$

we obtain the particular solution:

$$[\mathbf{K}_{\mathrm{c}}]\left\{\mathrm{U}_{1f}(t)\right\} = \exp\left(-t[\mathbf{M}_{11}]^{-1}[\mathbf{K}_{11}]\right)\{\mathrm{C}\} \tag{12}$$

with constant $\{\mathbf{C}\}$. This constant may be determined, while the solution $\{\mathbf{U}_{1s}\}$ for steady state is known:

$$[\mathbf{K}_{\mathrm{c}}]\{\mathrm{U}_{1s}(t)\} = \{\mathrm{i}_{\mathrm{c}}\}. \tag{13}$$

Equation (13) acts in the case of unit step excitation. For harmonic excitation the steady-state solution may be derived from the frequency-domain analysis. Taking into account initial conditions:

$$\left\{\mathrm{U}_{1f}(0)\right\} + \{\mathrm{U}_{1s}(0)\} = \{\mathrm{U}_1(0)\}, \tag{14}$$

the constant $\{C\}$ equals:

$$\{C\} = \{U_1(0)\} - [\mathbf{K}_c]^{-1}\{\mathbf{i}_c\}. \tag{15}$$

Assuming initial condition $U_1(0) = 0$ and homogenous boundary condition for adjoint model, the analytical form for the solution in time-domain is obtained:

$$[\mathbf{K}_c]\{U_1(t)\} = \{i_c\} - \exp\left(-t[\mathbf{M}_{11}]^{-1}[\mathbf{K}_{11}]\right)\{\mathbf{i}_c\}. \tag{16}$$

Equation (16) contains the exponential function of inversed mass matrix $[\mathbf{M}_{11}]$ and stiffness matrix $[\mathbf{K}_{11}]$. The matrix $[\mathbf{K}_c]$ contains inverse of $[\mathbf{M}_{22}]$. So, despite of band nature of [**M**] and [**K**] the calculations have to be performed using the general form of matrices. The formula (16) allows to calculate potentials for any time of analysis, inverting the matrices only once.

7 Test Example

The simple model of non-destructive eddy-current testing probe (Fig. 2) consists of two (exciting and measurement) coils over an infinite conducting plate.

The problem is assumed to be axial symmetric and can be analyzed using 2-D formulation (Fig. 3). The Dirichlet boundary condition is given for all boundaries of

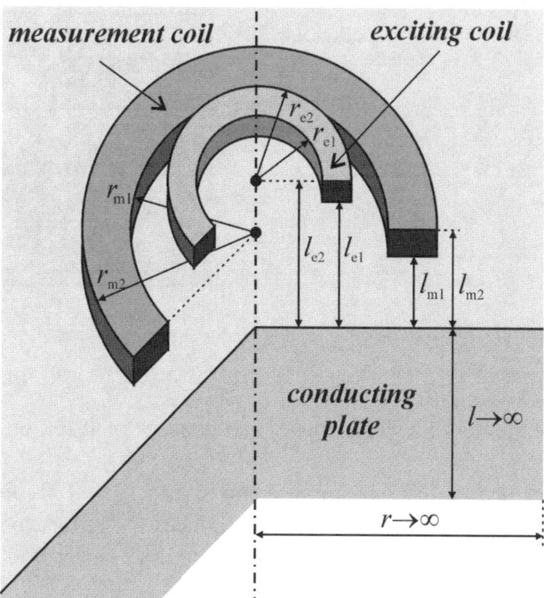

Fig. 2 Coils over infinite conducting plate

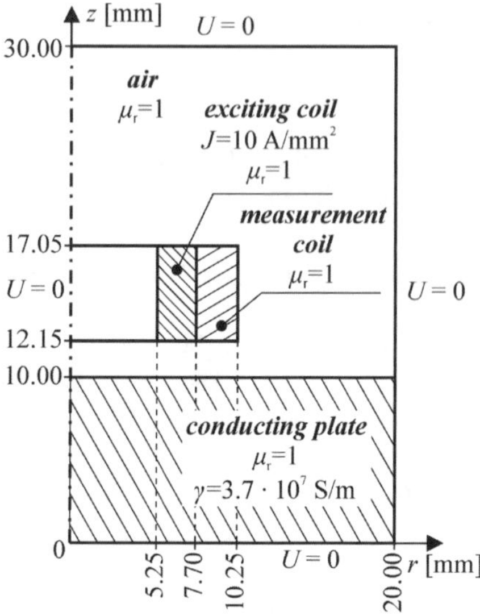

Fig. 3 Analyzed model

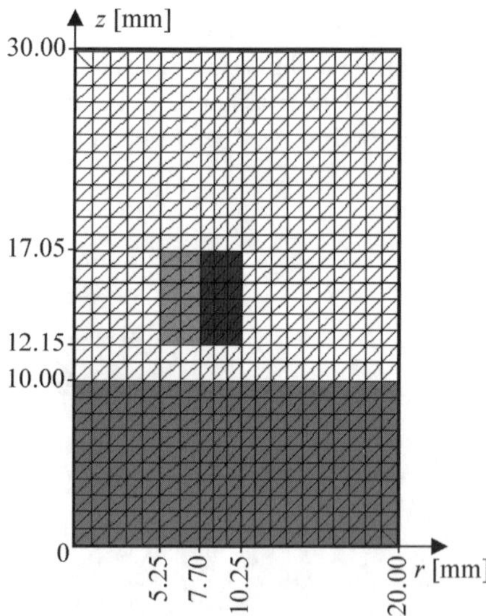

Fig. 4 Finite element mesh

model. The finite element mesh for this model consists of 1,200 elements and 651 nodes (Fig. 4).

The coil is driven with unit step impulse current of $i_{\max} = 0.3\text{A}$, exciting electromagnetic field. Due to Faraday's law voltage induces in the measurement coil and depends on material properties of the plate. Using the Fourier transformation this voltage can be analytically calculated [6].

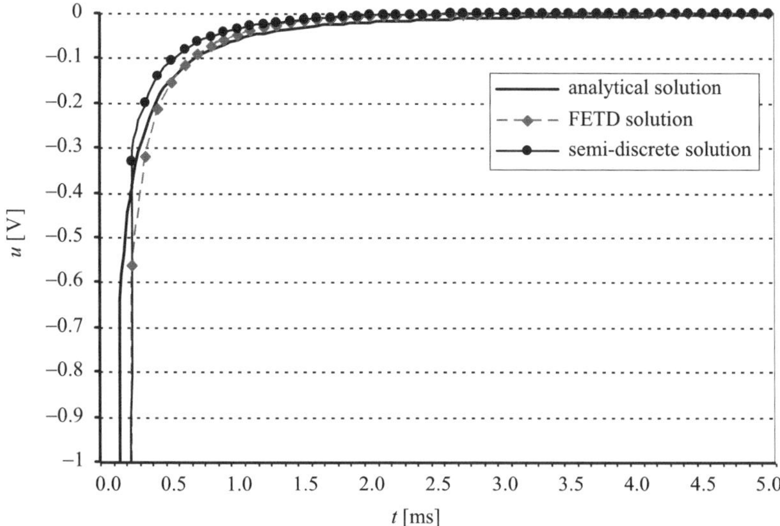

Fig. 5 Comparison of induced voltage for different solutions

The voltage induced in measurement coil was compared for analytical, finite-elements-time-domain (FETD) and semi-discrete methods (Fig. 5). The approximate solutions converge very well to exact, analytical solution.

8 Conclusions

The specific shape of the adjoint model excitation is made possible using the aim of sensitivity analysis in electromagnetic fields the semi-discrete method instead of the typical finite element algorithm. The semi-discrete method allows one to calculate analytically the distribution of field quantity for any given time. This fact is very important from the sensitivity analysis point of view. Evaluating sensitivity with time-stepping schemes for the original model and for the adjoint model in backward time results in some special conditions, and practically causes the need to analyze using constant time step. However, while solving inverse problems by means of the gradient method, the useful information is delivered not only by the first time steps, but also by the advanced time points. So, in the case of typical time-stepping algorithm we have to meet a compromise between the time step length and the number of steps. The aforementioned problem vanishes while using the semi-discrete method.

References

1. K.M. Gawrylczyk, M. Kugler, *Time domain sensitivity analysis of electromagnetic quantities utilizing FEM for the identification of material conductivity distributions*, COMPEL, vol. 25, no. 3, 2006, pp. 589–598.
2. K.M. Gawrylczyk, M. Kugler, Adjoint models in the time domain sensitivity analysis utilizing FEM for the identification of material conductivity distribution, *14th International Symposium on Theoretical Electrical Engineering, Szczecin, Poland*, June 20–23, 2007.
3. O.C. Zienkiewicz, K. Morgan, *Finite Elements and Approximation*, New York: Wiley, 1983.
4. A.R. Mitchell, R. Wait, *The Finite Element Method in Partial Differential Equations*, New York: Wiley, 1977.
5. B.D.H. Tellegen, *A General Network Theorem with Applications*, Philips Research Reports, vol. 7, 1952, pp. 259–269.
6. V.O. Haan, P.A. Jong, *Analytical expressions for transient induction voltage in a receiving coil due to a coaxial transmitting coil over a conducting plate*, IEEE Transactions on Magnetics, vol. 40, no. 2, March 2004, pp. 371–378.

Inverse Problems in Magnetic Induction Tomography of Low Conductivity Materials

Ryszard Palka, Stanislaw Gratkowski, Piotor Baniukiewicz, Mieczyslaw Komorowski, and Kreysztof Stawicki

Abstract The paper deals with the computational problems typical for magnetic induction tomography (MIT) of low conductivity materials. The forward problem is solved in order to define and verify the solution of the inverse problem. The paper focuses on the formulations and solutions of both. Different formulations and simplifications are discussed, depending on the physical properties of the exciter and examined object.

1 Introduction

Magnetic induction tomography (MIT) is a relatively new, non-invasive method which uses eddy currents phenomenon for reconstructing spatial distribution of the electrical conductivity in the examined object. It can be used for diagnostics of objects with wide spectra of physical properties, ranging from molten metals and other conductive fluids, through magnetic environments with low electrical conductivity, to non-magnetic weak-conducting objects like saline solutions representing certain body tissues [1–3]. The measurement system consists of the set of exciters and receivers, sensing the changes in the excited field, due to the different distribution of conductivity in the body. In certain implementations of such systems there may be only one exciter and (or) receiver, which is (are) rotated consecutively around the object. The system discussed in the paper is presented schematically in Fig. 1. Figure 2 shows the picture of the measurement system built by the authors. The exciter is driven by a sinusoidal current of frequency 100 kHz. The exciter consists of a coil with a ferrite core and conducting shield. The conducting shield has been optimized in order to shield sensitive electronic equipment as well as focus the main part of the magnetic flux in the vicinity of the receiver, which ensures sufficient values

R. Palka, S. Gratkowski, P. Baniukiewicz, M. Komorowski, and K. Stawicki
Szczecin University of Technology, Sikorskiego 37, 70-313 Szczecin, Poland
rpalka@ps.pl

R. Palka et al.: *Inverse Problems in Magnetic Induction Tomography of Low Conductivity Materials*, Studies in Computational Intelligence (SCI) **119**, 163–170 (2008)
www.springerlink.com

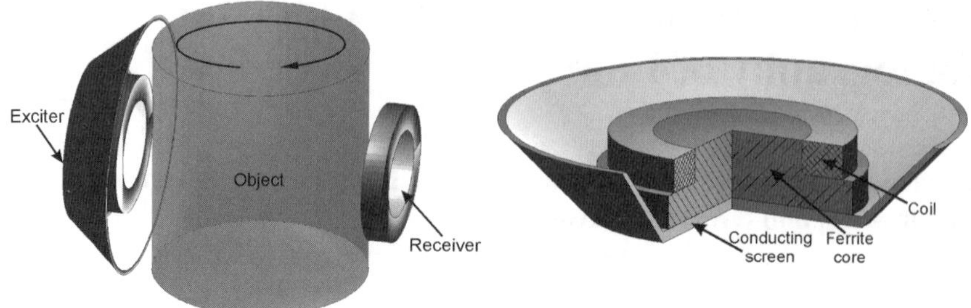

Fig. 1 Schematic diagram of the measurement system and the excitation unit in details (*on the right*)

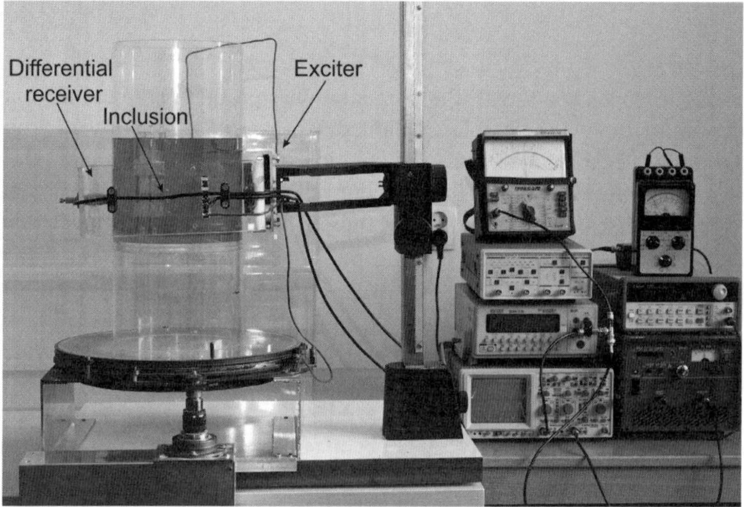

Fig. 2 The measurement system

of the measured signals. Eddy currents induced in the object are the source of the secondary magnetic field, opposing the source field and therefore modifying the measurements taken from the receivers. There is no need for physical contact with the surface of the object being tested. The object can be scanned either manually or with the aid of a mechanical device, or can be rotated as shown in Fig. 1. In general the magnetic field produced by the eddy currents excited in a uniformly conductive object and the field produced by the excitation coil represent unwanted signals. These signals can be eliminated by using differentially connected signal coils.

2 Forward Problem

Different formulations can be used to solve the forward problem, i.e., to find the eddy current distribution for a given conductivity distribution and then the magnetic field outside the object in the plane of the signal coil. We have used the Φ–$\mathbf{A}$

formulation. In a linear, nonmagnetic, isotropic conductive medium one can find the following equation for these two potentials (provided that the condition $\nabla \cdot \mathbf{A} = 0$ is adopted) [4]:

$$\nabla \cdot [(\sigma + \varepsilon\, j\omega)\nabla\Phi] = -j\omega\mathbf{A} \cdot \nabla(\sigma + \varepsilon\, j\omega), \tag{1}$$

where: σ – conductivity, ε – permittivity, ω – angular frequency. If $\omega\varepsilon \ll \sigma$, (1) can be written as follows:

$$\nabla \cdot (\sigma\nabla\Phi) = -j\omega\mathbf{A} \cdot \nabla\sigma. \tag{2}$$

Under the assumption that the object does not influence the excitation field, the vector magnetic potential in (2) may be replaced by the primary magnetic potential $\mathbf{A}_\mathrm{p}$, computed without the presence of the object. In this case the final equation and the relevant boundary condition have the forms:

$$\nabla \cdot (\sigma\nabla\Phi) = -j\omega\mathbf{A_p} \cdot \nabla\sigma, \qquad \frac{\partial\Phi}{\partial n} = -j\omega\mathbf{A_p} \cdot \mathbf{n}. \tag{3}$$

The primary magnetic vector potential can easily be calculated for a given configuration of the exciter by using the standard axi-symmetrical FEM. After solving the above equations for the scalar potential, the induced current density in the conductive object can be calculated. Formulation of the forward problem continues by applying Biot–Savart's law. The object is divided into polyhedrons of any shape with a uniform current density $\mathbf{J}^\mathrm{e}(\mathbf{r}')$. It can be shown that the magnetic field at the point $\mathbf{r}$ is given by (S_k refers to the facets of the polyhedron e) [5]:

$$\mathbf{B}(\mathbf{r}) = \frac{\mu_0}{4\pi}\sum_e\sum_k \mathbf{J^e} \times \mathbf{n_k} \int_{S_k} \frac{dS}{\left|\mathbf{r} - \mathbf{r}'\right|}. \tag{4}$$

The finite element mesh and exemplary results of calculations of the magnetic flux density distribution (secondary field only) in the plane of the signal coil are shown in Fig. 3.

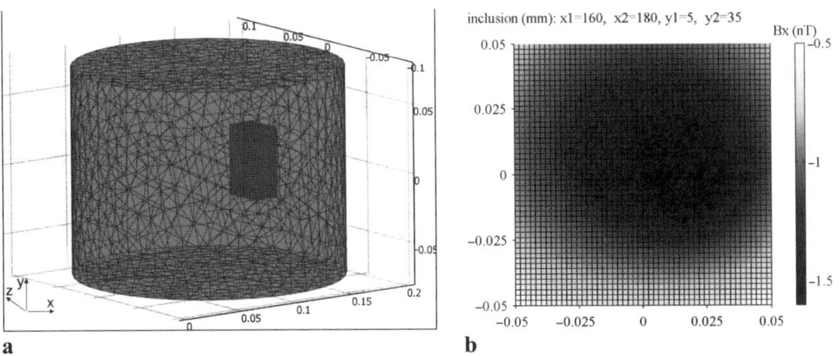

Fig. 3 The finite element mesh and the x-component of the magnetic flux density in the plane of the signal coil; conductivity of the inclusion is equal to $100\,\mathrm{S\,m^{-1}}$ while the conductivity of the background equals $0.8\,\mathrm{S\,m^{-1}}$

3 Application of An Artificial Neural Network for Solving the Inverse Problem

For solving the inverse field problem of reconstruction of the conductivity distribution, an artificial neural network (ANN) has been applied. The ANN is a system composed of artificial neurons organized in layers which uses a mathematical or computational model for information processing based on a connectionist approach to computation. The inverse problems, which can be met in the MIT are usually ill-posed and have large number of unknowns to be determined. The ANN, thanks to their flexibility and possibility to model very complicated systems, can be successfully used for solving such problems. The ANN approximates the whole measurement system, which can be treated as a "black box" that transforms the spatial distribution of measured quantity (e.g., conductivity) to the output signal values (e.g., voltage or current) or vice versa. Fitting the ANN to the measurement system, a transfer function is determined during a learning stage. The learning of the ANN is carried out using a "knowledge base" that contains signals obtained from measurements performed on the calibration specimens. These results should cover the whole input space of possible shapes and properties of inclusions. The main advantages of proposed formulation of the problem are that the sensor parameters, excitation parameters and the geometry of the system are not important in point of view ANN-based algorithm. In the approach to the inverse problem proposed by the authors, the artificial neural network was used for modeling the whole measurement system (Fig. 4). The system is identified by input–output data pairs, where the output stands for the spatial distribution of the conductivity, whereas the input stands for the measurement results obtained as a result of scanning the specimen in $y-z$ plane (Fig 5). The signals obtained during measurements are used as the ANN input.

The main assumption is that the sensor is only sensitive to the inclusion located close to it. Thus, it is not necessary to analyze at the same time the signals

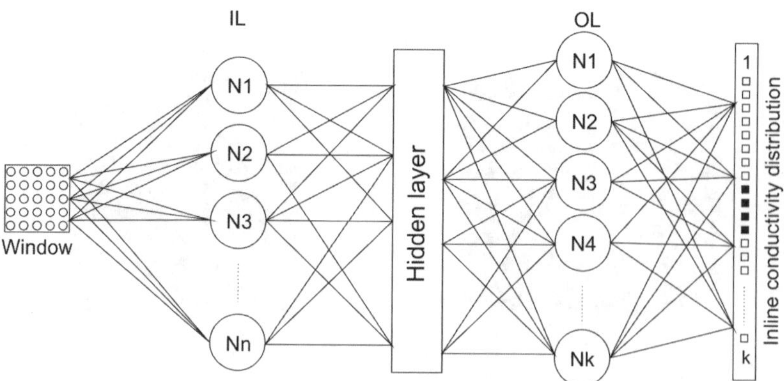

Fig. 4 The block scheme of the measurement signal inversion using ANN. IL – input layer, OL – output layer, N – neurons, k – number of output space discretization points

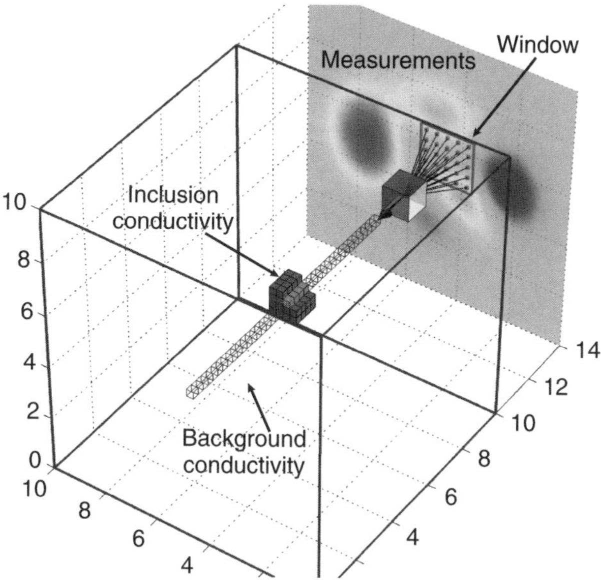

Fig. 5 The ANN system used for the conductivity identification

measured over the whole specimen. This leads to the idea of rectangular window moved through the measurement plane. The window size corresponds to the distance from which the inclusion affects the sensor. Figure 5 shows the basis of proposed method. The specimen is discretized and approximated by cubic elements with constant conductivity. The window moves in the $y-z$ plane with a defined step depending on the measurement accuracy. The distribution of the conductivity along the x-axis is computed by the ANN for each position of the window. The number of elements, which approximate the specimen along the x-axis, equals to the number of ANN outputs. Thus, particular network output responds to the conductivity of one cubic element. The wire cube stands for the estimated conductivity of the specimen, whereas the solid cube stands for the estimated conductivity of the inclusion.

4 Reconstruction of the Conductivity Distribution Via Biot–Savart's Law

The reconstruction of the conductivity distribution within the object under investigation by the use of the external field values in the close neighborhood of the specimen, known from scanning measurements, constitutes a typical inverse problem and belongs to the class of the improperly posed tasks [6, 7]. This problem can be replaced by the equivalent problem of the determination of the current distribution. Figure 6 shows the region under consideration where the external flux densities can

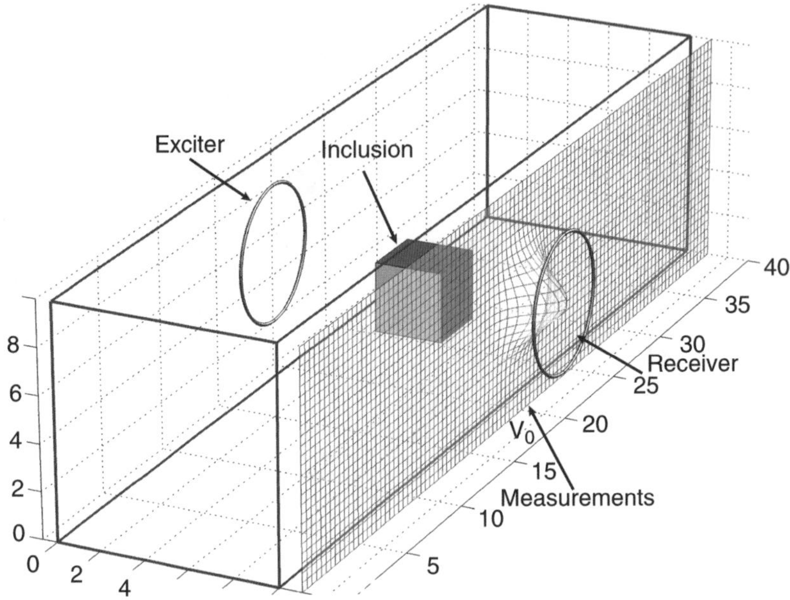

Fig. 6 Object under consideration and the measurement plane V_0

be defined/measured (the possible region of measurements V_0 has been indicated). In order to define the mathematical model for the inverse task, the whole conducting region has been discretized into N identical sub-cubes with three individual current sheet values J_k surrounding each cube. The magnetic field generated by each cube in any point of the region under consideration can be symbolically expressed in the following form (the Biot–Savart's law):

$$\mathbf{B}(x,y,z) = \mathbf{f}(x,y,z)J_{k1} + \mathbf{g}(x,y,z)J_{k2} + \mathbf{h}(x,y,z)J_{k3}. \tag{5}$$

Detailed formulas describing all components of the magnetic field generated by current sheets are given in [7]. The resulting magnetic field in any point of the external region can be determined as the sum of the magnetic field of all sub-domains. The mapping: current density distribution $\rightarrow$ magnetic field distribution leads to an overdetermined, linear and ill-conditioned set of $3P$ equations (P – number of evaluation points) with $3N$ unknown surface current densities $(P >> N)$:

$$\mathbf{WJ} = \mathbf{B}, \tag{6}$$

where $\mathbf{W}$ is a real $m \times n$ matrix $(m = 3P,\ n = 3N)$.

The minimal least squares solution of (6) is the vector of the minimum Euclidean length which minimizes the length of the residual vector $\mathbf{R} = \mathbf{B} - \mathbf{WJ}$ [6, 7]. The matrix $\mathbf{W}$ in (6) can be factorised by the singular value decomposition as:

$$\mathbf{W} = \mathbf{QDP^T}, \tag{7}$$

where **Q** is an $m \times m$ orthogonal matrix, **P** is an $n \times n$ orthogonal matrix and **D** is $n \times n$ diagonal matrix with singular values of **W** arranged in decreasing order. Finally the minimal least squares solution of this problem is given by

$$\mathbf{J} = \mathbf{P}\mathbf{S}^{-1}\mathbf{Q}^{\mathrm{T}}\mathbf{B}^{\mathrm{T}}, \tag{8}$$

where $\mathbf{S} = diag(sv_1, sv_2, \ldots, sv_k)$, with sv_k being the last singular value greater than zero.

The practical implementation of the above algorithm can be very difficult because of the possible instabilities. Therefore the matrix **W** should be scaled in order to prevent the least squares problem being unnecessarily sensitive to both measurement and numerical calculation inaccuracies. The numerical algorithm has been intensively tested for different external field distribution (both – measured and calculated) and for extremely large equation systems. As the convergence criterion the total error between the primary (measured) B_M and final (calculated) field distribution B_C has been used:

$$\varepsilon = \int_{\mathrm{V}_0} (B_M - B_C)\,dV. \tag{9}$$

The value of above quality criterion together with the calculation time, rank of the equation matrix and their singular values enable the proper choice of the meshing of the conducting region and the size, position and discretization of the measurements regions. An example of the reconstruction of the current density region has been shown in Fig. 7.

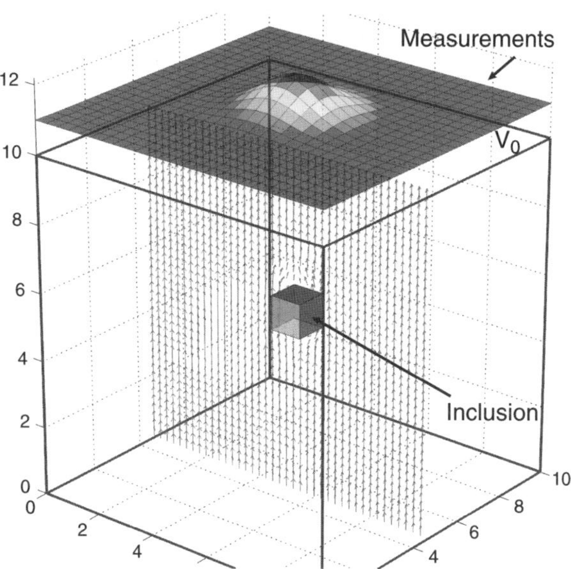

Fig. 7 Identified current density regions with the flux density distribution

5 Conclusions

In this paper the forward and inverse problems of magnetic induction tomography are discussed. In order to find the scalar potential distribution for an object of arbitrary conductivity distribution a 3-D finite element formulation was employed. The secondary magnetic field, due to the induced currents in the object was calculated in the plane of the signal coil using the Biot–Savart law. Next, the inverse problems were formulated. Two algorithms have been proposed and tested: first using the artificial neural network and second using the inversion of the Biot–Savart law.

Acknowledgement This work was supported by the Ministry of Education and Science, Poland, under grant 3 T10A 033 30 (2006–2009).

References

1. M. Soleimani, W. Lionheart, A. Peyton, X. Ma and S. Higson, A three-dimensional inverse finite-element method applied to experimental eddy-current imaging data, IEEE Transactions on Magnetics, vol. 42, pp. 1560–1567, 2006
2. J. Rosell, R. Casañas and H. Scharfetter, Sensitivity maps and system requirements for magnetic induction tomography using a planar gradiometer, Physiological Measurement, vol. 22, pp. 121–130, 2001
3. K. Hollaus, C. Magle, R. Merwa and H. Scharfetter, Numerical simulation of the eddy current problem in magnetic induction tomography for biomedical applications by edge elements, IEEE Transactions on Magnetics, vol. 40, pp. 623–626, 2004
4. N.G. Gencer and M.N. Tek, Electrical conductivity imaging via contactless measurements, IEEE Transactions on Medical Imaging, vol. 18, pp. 617–627, 1999
5. S. Pissanetzky and Y. Xiang, Analytical expressions for the magnetic field of practical coils, Compel, vol. 9, pp. 117–121, 1990
6. R. Pałka, Synthesis of magnetic fields by optimization of the shape of areas and source distributions, Archiv für Elektrotechnik, vol. 75, pp. 1–7, 1995
7. R. Pałka, H. May and W.-R. Canders, Nondestructive quality testing of high temperature superconducting bulk material used in electrical machines and magnetic bearings, Optimization and Inverse Problems in Electromagnetism. Kluwer, Dordrecht, pp. 303–312, 2003

Adaptive Plasma Identification from Internal and External Measurements

Alessandro Formisano, Raffaele Martone, and Vincenzo Cutrupi

Abstract A key aspect of the controlled thermonuclear fusion is the plasma identification process, that is the determination of the relevant plasma parameters from measurements taken during the fusion event. One of the classical approaches to plasma identification is based on a representation of the plasma current in terms of simpler, equivalent, currents, whose values are determined by best fitting different sets of measurements. The possibility of combining different diagnostic systems is here analyzed to assess possible improvements of the identification process: in particoular, the integration of diagnostics providing information about magnetic field both in external and internal plasma region improves the quality of the reconstruction process and gives information about the inner plasma current profile. Under suitable hypotheses, the resulting inverse problem can be formulated as the pseudo-inversion of a matrix linking the unknown equivalent currents to the various measurements.

1 Introduction

Present experiment to realize thermonuclear fusion reactors are mainly oriented towards devices called "Tokamaks" [1], the most important being the ITER project [2]. In Tokamaks, a current is induced in an ionized gas, and confined using high intensity magnetic fields. Consequently, one of the most relevant aspects of the process is the estimation of plasma parameters such as the plasma *magnetic contour profile*, or *internal inductance* and *poloidal beta* (eventually related to the internal plasma density current distribution), in order to control the plasma evolution. This "diagnostic" process is typically performed starting from a number of available measurements taken by probes of various kind. In this paper, three different kinds of

A. Formisano, R. Martone, and V. Cutrupi
Dip. di Ingegneria dell'Informazione, Seconda Università di Napoli, I-81031 Aversa, Italy
Vincenzo.Cutrupi@unina2.it

A. Formisano et al.: *Adaptive Plasma Identification from Internal and External Measurements*, Studies in Computational Intelligence (SCI) **119**, 171–177 (2008)
www.springerlink.com

measurement will be used to study the identification problem, with the aim to identify shape and position of the plasma. The first set of measurements is the external magnetic measurements (EMm) [3], that provide the values of the magnetic field and flux in suitable measurement points located out of the vessel. Two additional sets of internal measurements will be considered here, namely Integral Polarimetric measurements (IPm) [4] and motional stark effect measurements (MSEm) [5].

Note that the parameters estimate must be available in real time, in order to opportunely provide to the control loop the required information. At this purpose, many different approaches have been proposed in literature, the most relevant being:

- *Multifilament approach*. The plasma is discretized as a set of *equivalent* currents (EC), chosen in such a way to best fit the available measurements [6];
- *Multipolar expansion approach*. The relevant plasma magnetic moments are reconstructed from magnetic measurements, and then used to estimate relevant parameters [7];
- *Neural networks*. The plasma parameters are extracted using artificial neural networks (or other soft-computing techniques) from the data sets of available experiments [8].

The multifilament approach will be used here. Such an approach is based on the assumption that plasma current can be treated as a set of equivalent currents. In its classical version, one of the most diffused approach to plasma identification, the equivalent currents are simple circular filamentary coils. The inverse problem, in the assumption of absence of magnetic materials, consists then in the inversion of a linear system, whose coefficients matrix is the Green matrix linking the unknown equivalent currents to the various measurements, and the known term is composed by the measurement values.

2 Mathematical Formulation

The main disadvantage of the multifilament approach is due to the poor quality of current representation in the inner plasma region, due to the presence of singularities when field and source points overlap. To overcome this problem, it's possible to adopt a different representation base for the plasma current; as a matter of fact, in this paper the plasma will be treated as a set of massive coils of rectangular cross section. Analytical expressions for the vector potential and magnetic field of such coils are present in literature [9].

Under the hypothesis of no ferromagnetic materials, the trial (i.e. *computed*) Emms $\mathbf{m}$ depend linearly on the plasma currents vector $\boldsymbol{i}_p$ (unknown) and on the external currents $\mathbf{i}_{\text{ext}}$:

$$\mathbf{m}(\mathbf{i}_p, \mathbf{i}_{ext}) = \mathbf{M}_p \mathbf{i}_p + \mathbf{M}_{ext} \mathbf{i}_{ext}, \tag{1}$$

where $\mathbf{M}_\mathrm{p}$ and $\mathbf{M}_\mathrm{ext}$ are the Green matrices describing the contributions to EMm due to $\boldsymbol{i}_p$ and $\boldsymbol{i}_{ext}$ respectively, computed using analyitical expressions, as anticipated above.

Anyway, EMm only are not able to provide information about the inner current density distribution, and the use of internal diagnostic measurements such as the IPm and the MSEm is necessary.

Polarimetric measurements are based on the Faraday rotation of the polarization plane of an electromagnetic wave induced by an external magnetic field:

$$p(r) = k\lambda^2 \int_l n_e(r,z) B_{tl}(r,z)\, dl, \tag{2}$$

where l is a segment of the probing beam (a *chord*) laying within the plasma cross section, B_{tl} is the component of induction field B along the line l, λ is the probing wavelength and k is a dimensional constant; finally n_e is the plasma electron density (supposed known in this analysis). This kind of diagnostic can be used then to estimate line integrals of plasma magnetic field in the inner plasma region itself.

The motional Stark effect (MSE) is a diagnostic technique for measurement of the internal poloidal field B_p profile in neutral beam heated Tokamak plasmas. The technique relies the measurement of the polarization angle (also called pitch angle) γ_p:

$$\gamma_p = \tan^{-1}\left(\frac{B_p}{B_t}\right), \tag{3}$$

where B_t is the toroidal component of the field. The knowledge of γ_p provides a significant constraint on the q profile. The numerical value of $q(r)$ gives the number of times a magnetic field line will traverse the angular toroidal length of the plasma before returning to its initial position. From the mathematical point of view $q(r)$ is so defined:

$$q(r) = \frac{rB_t}{RB_p}, \tag{4}$$

where R is the major radius of the torus. In particular way, this value will usually be low with respect to the magnetic axis of the plasma $q(r=0)$, because this field line exhibits little poloidal variation.

Under suitable hypotheses, also Ipms and MSEms can be linearly related to currents in the following way:

$$\mathbf{p}(\mathbf{i}_p, \mathbf{i}_{ext}; l) = \mathbf{P}_p \mathbf{i}_p + \mathbf{P}_{ext} \mathbf{i}_{ext}, \tag{5}$$

$$\mathbf{s}(\mathbf{i}_p, \mathbf{i}_{ext}; \mathbf{x}_{MSE}) = \mathbf{S}_p \mathbf{i}_p + \mathbf{S}_{ext} \mathbf{i}_{ext}, \tag{6}$$

where $\mathbf{p}$ and $\mathbf{s}$ are the trial values of IPms, and MSEms, respectively, l is the length of the polarimetric chord(s), $\mathbf{x}_\mathrm{MSE}$ are the coordinates of the points where the MSEms are provided, and finally $\mathbf{P}_\mathrm{p}$, $\mathbf{P}_\mathrm{ext}$, $\mathbf{S}_\mathrm{p}$ and $\mathbf{S}_\mathrm{ext}$ are the (Green) matrices describing the contributions to IPm and MSEm due to $\boldsymbol{i}_p$ and $\boldsymbol{i}_{ext}$, respectively.

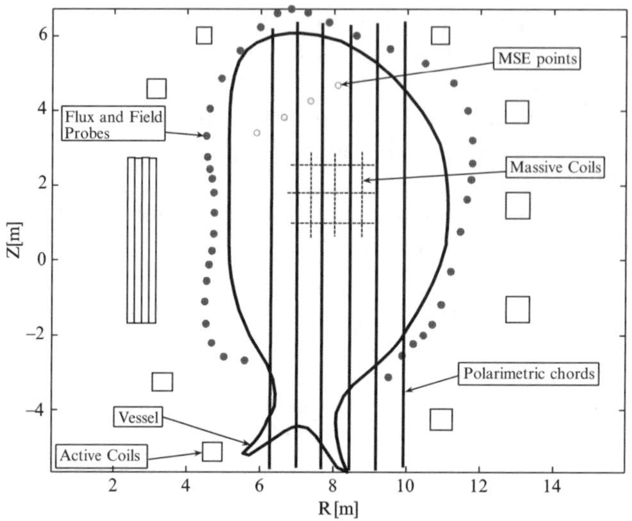

Fig. 1 Poloidal field measurements in a Tokamak (ITER-like) poloidal scheme

The values of l and $\mathbf{x}_{MSE}$ as well as the chords and Stark points positions impact on the **P** and **S** matrices, and different choices may be optimal for different plasma configurations. Obviously, the assembly of the Green Matrices ($\mathbf{M}_{\mathrm{p}}$, $\mathbf{M}_{\mathrm{ext}}$, $\mathbf{P}_{\mathrm{p}}$, $\mathbf{P}_{\mathrm{ext}}$, $\mathbf{S}_{\mathrm{p}}$ and $\mathbf{S}_{\mathrm{ext}}$) is more complex than the equivalent Green matrices obtained in the case of multifilament approach, as field and source points may possibly overlap. As an example, a sketch of the present design for ITER [2] poloidal cut plane is reported in Fig. 1, showing the location of foreseen EMm, of the chords for IPm, and of the points $\mathbf{x}_{\mathrm{MSE}}$.

Polarimetric, Stark and Magnetic measurements can be combined in the plasma reconstruction procedure by adding to the linear equation (1) the further sets of linear equations (5) and (6). The resulting augmented linear system of equations becomes:

$$\mathbf{A}\mathbf{i}_p = \mathbf{b}, \tag{7}$$

where **b** represents the vector of measurements filtered by the contribution of known external currents. A further row is eventually added to **A** and **b**, to impose the total plasma current, as measured by the Rogowsky coil [1]. Finally, the (truncated) pseudo-inverse of the global matrix **A** is computed to determine the unknown plasma currents.

3 Numerical Results

To assess the effectiveness and robustness of the proposed approach, a sample ITER equilibrium has been considered, namely one falling into the class of X-point configurations (i.e. the magnetic flux density vanishes in one point inside the plasma

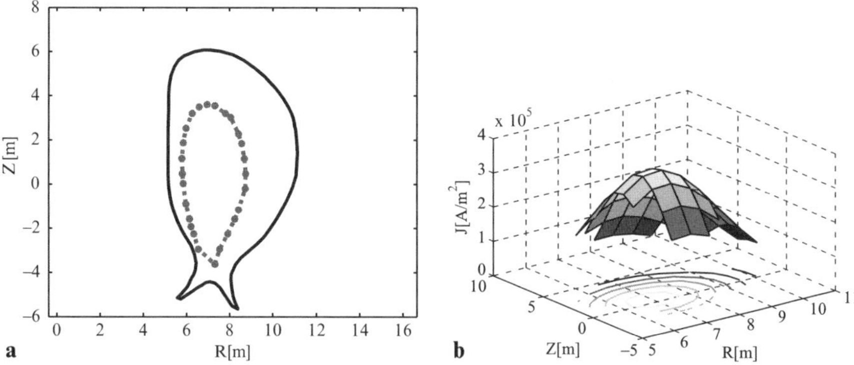

Fig. 2 **a** Reference plasma magnetic contour (*dashed line*). **b** Reference internal current density map

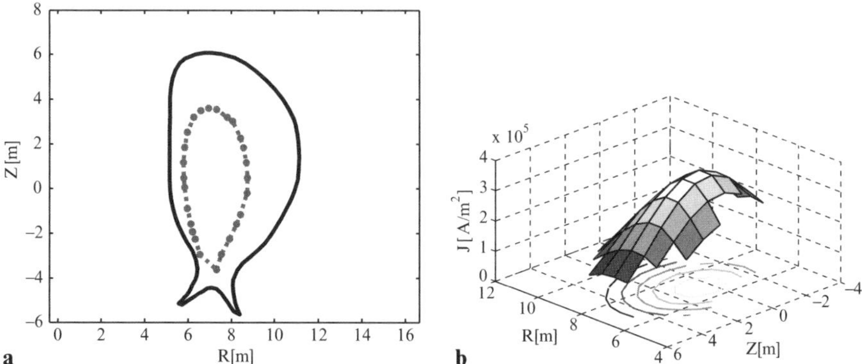

Fig. 3 **a** Plasma magnetic contour (*dashed line*) tackled using only EMms. **b** Internal current density map tackled using only EMms

chamber). The set of measurement used consists of 40 magnetic probes (21 flux-loops, 19 pick-up coils), six vertical polarimetric chords spaced between 5.8 and 10.4 m, and finally four stark points. Reference plasma boundary and current distribution are reported in Fig. 2a,b, respectively.

The reconstruction of plasma contour (Figs. 3a and 4a) and internal current density map (Figs. 3b and 4b) has been tackled using first only EMms (Fig. 3), and then using the full set of EMms, IPms and MSEms (Fig. 4). From Fig. 4b it's possible to see how the capability of polarimetric and stark measurements helps in the identification of the inner plasma current profile, while they don't improve the reconstruction of plasma contour (Figs. 3a and 4a).

A further improvement in the procedure is to adaptively cancel from the set of equivalent currents those falling outside the estimated plasma contour. In Fig. 5a the currents (thick points) falling inside the reconstructed plasma magnetic contour are shown, while in Fig. 5b the reconstructed current density map with 61% less unknowns is shown.

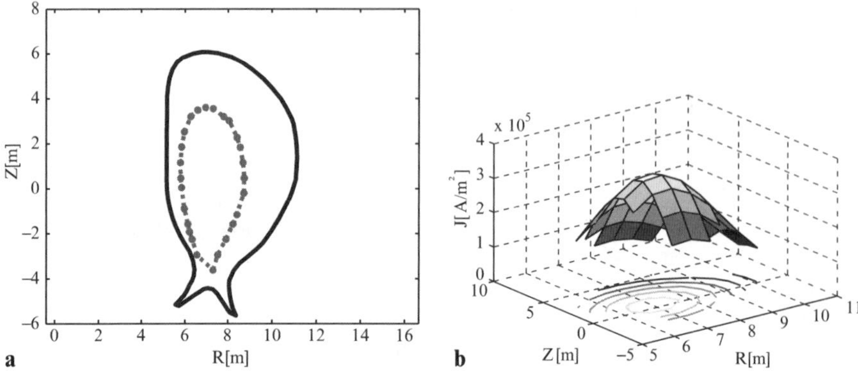

Fig. 4 **a** Plasma magnetic contour (*dashed line*) tackled using the full set of EMms, IPms and MSEms. **b** Internal current density map tackled using the full set of EMms, IPms and MSEms

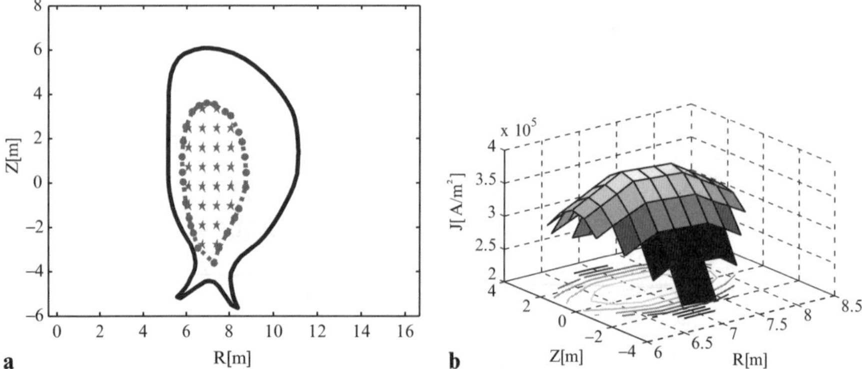

Fig. 5 **a** Plasma magnetic contour (*dashed line*) and currents (*thick stars*) falling inside the reconstructed plasma magnetic contour tackled using the full set of EMms, IPms and MSEms. **b** Internal current density map tackled using the only currents falling inside the reconstructed plasma magnetic contour

4 Conclusions

The introduction of Polarimetric and Stark measurements in the plasma identification procedures allows to improve the reconstruction accuracy (in terms of inner current density distribution and of inner local magnetic flux). In particular way, from the numerical simulations performed, it can be deduced the convenience of different typologies of measurements. Note that the increased effort necessary to assemble Green matrices starting from massive equivalent currents (to prevent singularities due to filamentary currents when field and source points overlap) has an impact only in the preliminary matrices assembly step, while does not impact the efficiency in the course of real-time identification.

Acknowledgements This work has been partly sponsored by Association ENEA-Euratom-CREATE.

References

1. J. Wesson, Tokamaks, Third Edition, Oxford Science Publications, 2004
2. http://www.iter.org/
3. B. J. Braams, The interpretation of Tokamak magnetic diagnostics, *Plasma Physics and Controlled Fusion*, vol. 33, no. 7 (1991), pp. 715–748
4. S. L. Pruntly, M. C. Sexton, E. Zilli, High resolution far infrared polarimeter for plasma diagnostics, *Springler Link*, vol. 13, No. 5 (1992), pp. 591–608
5. B. W. Rice, K. H. Burrell, L. L. Lao, Effect of plasma radial electric field on motional stark effect measurements and equilibrium reconstruction, *Nuclear Fusion*, vol. 37, No. 4 (1997)
6. D. W. Swain, G. H. Neilson, An efficient technique for magnetic analysis of non-circular, high-beta Tokamak equilibria, *Nuclear Fusion*, vol. 22, no. 8 (1982), pp. 1015–1030
7. F. Alladio, F. Crisanti, M. Marinucci, Analysis of axisymmetric MHD equilibria by toroidal multipolar expansion, *Nuclear Fusion*, vol. 22 (1986)
8. E. Coccorese, C. Morabito, R. Martone, Identification of non-circular plasma equilibria using a neural network approach, *Plasma Physics and Controlled Fusion*, vol. 33, no. 7 (1991), pp. 715–748
9. L. K. Urankar, Vector potential and magnetic field of current carrying finite arc segment in analytical form, part III: exact computation for rectangular cross section, *IEEE Transaction on Magnetics*, vol. 18, no. 6 (1982)

Analytical and Numerical Solutions of Temperature Transients in Electrochemical Machining

Daiga Žaime, Janis Rimshans, and Sharif Guseynov

Abstract A new semi-implicit difference scheme based on our propagator method is elaborated for solution of initial-boundary value problem of electrochemical machining. The scheme is unconditionally monotonic and has truncation errors of the first order in time and of the second order in space. Analytical and numerical solutions for temperature distribution in 2D electrolytic cell are presented.

1 Formulation of the Problem

The effects of dissociation under applied electrical field cause a variety of technological and theoretical issues such as micromechanical processes in electro osmotic pumps [1], electrochemical dissolution of silicon [2] and electrochemical machining [3]. Our aim is to obtain analytical and numerical solutions of temperature distributions for relatively simple machining model-anodic dissolution of metal [3, 4].

Unsteady numerical simulations for temperature distribution in the electrolytic cell have been performed in [3]. High role of simulation software was shown in prediction of heating of electrolyte in thermal boundary layer, estimation of thermal limits of the system, prediction of heat accumulation during the separated current pulses. Two different time scales for temperature transients were found, related to the convective transport and thermal dynamics of the electrodes. The analytical solution has been obtained on the electrodes on the assumption of fully convective transport, when processes can be described as 1D problem. We aim to obtain analytical solution for the full parabolic problem in 2D, and to expand effective

D. Žaime, J. Rimshans, and Sh. Guseynov
Institute of Mathematical Sciences and Information Technologies, University of Liepaja, 14 Liela, Liepaja LV3401, Latvia
daigazaime@inbox.lv

D. Žaime et al.: *Analytical and Numerical Solutions of Temperature Transients in Electrochemical Machining*, Studies in Computational Intelligence (SCI) **119**, 179–186 (2008)
www.springerlink.com

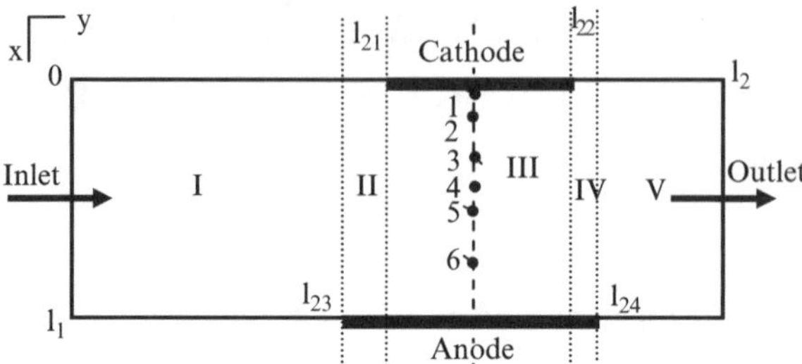

Fig. 1 The electrolytic cell geometry

numerical propagator method [5] with extended numerical stability calculations for temperature and potential distribution.

Actual initial-boundary value problem for temperature T and potential U in 2D electrolytic cell (Fig. 1) is given as follows [3]:

$$\frac{\partial T(x_1,x_2,t)}{\partial t} = \frac{k}{\rho C_p}\Delta_2 T(x_1,x_2,t) - (\overrightarrow{v}, \nabla T(x_1,x_2,t)) + rT(x_1,x_2,t) + P_{bulk}(x_1,x_2,t), \tag{1}$$

$$0 < |r| \ll 1, \quad x_n \in (0, l_n),\ n = 1,2;\ t > 0,$$

$$\varepsilon \frac{\partial U(x_1,x_2,t)}{\partial t} = \sigma \Delta_2 U(x_1,x_2,t), \quad 0 < \varepsilon \ll 1, \quad x_n \in (0,l_n),\ n = 1,2; t > 0, \tag{2}$$

where $\overrightarrow{v} = (v_1,\ v_2)$ is velocity vector, $P_{bulk} = \frac{J^2}{\sigma \rho C_p}$ is Joule heating and $J = -\sigma \nabla U$ is local current density, with the respective initial, Dirichlet and Neumann boundary conditions [3, 4] at boundary lines (Fig. 1) for T and U. Isothermal electrodes are assumed, which is valid for the considered case of short current pulse phenomena, when thermal effects are negligible. Then for the potential U at the electrodes, specific boundary conditions are used, taking into account density of reaction current, expressed through the polarization resistance and the onset of polarization [4].

Rewriting (1)–(2) in such a new notation: $u = T,\ a^2 = k/(\rho C_p),\ \alpha_i = -v_i,\ i = 1,2,\ \beta = r,\ f = P_{bulk};\ w = U,\ a_1^2 = \sigma$, and introducing a substitution for u:

$$u(x_1,x_2,t) = e^{-\frac{\alpha_1 \cdot x_1 + \alpha_2 \cdot x_2}{2 \cdot a^2} - \frac{\alpha_1^2 + \alpha_2^2 - 4 \cdot a^2 \cdot \beta}{4 \cdot a^2} \cdot t} \cdot \vartheta(x_1,x_2,t), \tag{3}$$

we will obtain following formulation of initial-boundary value problem:

$$D \equiv \{(x_1,x_2,t) : 0 \le x_n \le l_n, (n = 1;2), t \ge 0\}, \tag{4}$$

$$\vartheta_t = a^2 \Delta_2 \vartheta + F(x_1,x_2,t),\ 0 < x_n < l_n,\ (n = 1;2),\ t > 0, \tag{5}$$

$$\text{with function } F(x_1,x_2,t) = e^{\frac{\alpha_1 \cdot x_1 + \alpha_2 \cdot x_2}{2 \cdot a^2} + \frac{\alpha_1^2 + \alpha_2^2 - 4 \cdot a^2 \cdot \beta}{4 \cdot a^2} \cdot t} \cdot f(x_1,x_2,t), \tag{6}$$

$$\varepsilon w_t = a_1^2 \Delta_2 w, \qquad 0 < \varepsilon \ll 1, \tag{7}$$

and, respectively initial and boundary conditions.

2 Analytical and Numerical Methods for Solution

Both analytical and numerical methods of solution of the initial-boundary value problem (4)–(7) have been considered. Analytical solution is found for 2D problem (4)–(7), by subdividing cell region to subproblems I, II, III, IV and V, as it is shown in Fig. 1. For each subproblem the Green function method is applied in order to obtain solution in considered region, by assuming that Green function is known at the border of adjacent regions. Full description of the method requires a lot of space, and here we will describe in details only the solution ϑ of the first subproblem. The solution for the function w has been found in a similar way.

2.1 Analytical Method

Definition of Subproblem I.

It is necessary to find a function $V^{(1)} = V^{(1)}(x_1,\ x_2,\ t)$ on the following formulation:

$$V_t^{(1)} = a^2 \Delta_2 V^{(1)} + F(x_1,x_2,t),\ 0 < x_1 < l_1,\ 0 < x_2 < l_{23},\ t > 0, \tag{8}$$

$$V^{(1)}(x_1,x_2,t)|_{t=0} = \vartheta_0(x_1,x_2),\ 0 \le x_1 \le l_1,\ 0 \le x_2 \le l_{23}, \tag{9}$$

$$\frac{\partial V^{(1)}(x_1,x_2,t)}{\partial x_1}|_{x_1=0} - \frac{\alpha_1}{2a^2} V^{(1)}(x_1,x_2,t)|_{x_1=0} = 0,\ 0 \le x_2 \le l_{23},\ t \ge 0, \tag{10}$$

$$\frac{\partial V^{(1)}(x_1,x_2,t)}{\partial x_1}|_{x_1=l_1} - \frac{\alpha_1}{2a^2} V^{(1)}(x_1,x_2,t)|_{x_1=l_1} = 0,\ 0 \le x_2 \le l_{23},\ t \ge 0, \tag{11}$$

$$V^{(1)}(x_1,x_2,t)|_{x_2=0} = \vartheta_3(x_1,t), V^{(1)}(x_1,x_2,t)|_{x_2=l_{23}} = \vartheta_4(x_1,t),\ 0 \le x_1 \le l_1,\ t \ge 0. \tag{12}$$

In addition, it is assumed that the respective consistency conditions are also satisfied.

Subproblem I is a mixed initial-boundary value problem, and it can be solved by creating the corresponding Green function [6]:

$$V^{(1)}(x_1,x_2,t)=\int_0^{l_1}d\xi_1\int_0^{l_{23}}G^{(1)}(x_1,x_2;\xi_1,\xi_2;t)\vartheta_0(\xi_1,\xi_2)d\xi_2$$

$$+a^2\int_0^t d\tau\int_0^{l_1}\frac{\partial G^{(1)}(x_1,x_2;\xi_1,\xi_2;t-\tau)}{\partial\xi_2}|_{\xi_2=0}\vartheta_3(\xi_1,\tau)d\xi_1$$

$$-a^2\int_0^t d\tau\int_0^{l_1}\frac{\partial G^{(1)}(x_1,x_2;\xi_1,\xi_2;t-\tau)}{\partial\xi_2}|_{\xi_2=l_{23}}\vartheta_4(\xi_1,\tau)d\xi_1$$

$$+\int_0^t d\tau\int_0^{l_1}d\xi_1\int_0^{l_{23}}G^{(1)}(x_1,x_2;\xi_1,\xi_2;t-\tau)F(\xi_1,\xi_2,\tau)d\xi_2, \quad (13)$$

where the Green function is given as follows:

$$G^{(1)}(x_1,x_2;\xi_1,\xi_2;t)\equiv\frac{1}{l_1l_{23}}\left\{\sum_{n=1}^{\infty}A_n\left(\cos(\Theta_n^{(1)}x_1)+\frac{\alpha_1}{2a^2\Theta_n^{(1)}}\sin(\Theta_n^{(1)}x_1)\right)\right.$$

$$\left.\times\left(\cos(\Theta_n^{(1)}\xi_1)+\frac{\alpha_1}{2a^2\Theta_n^{(1)}}\sin(\Theta_n^{(1)}\xi_1)\right)e^{-a^2(\Theta_n^{(1)})^2t}\right\}$$

$$\times\left\{-\sum_{m=1}^{\infty}\sin\left(\frac{\pi m x_2}{l_{23}}\right)\sin\left(\frac{\pi m\xi_2}{l_{23}}\right)e^{-\frac{a^2\pi^2m^2t}{l_{23}^2}}\right\},$$

$$A_n=\frac{8a^4\left(\Theta_n^{(1)}\right)^2}{\alpha_1\left[\alpha_1l_1+4a^2\right]+4a^4\left(\Theta_n^{(1)}\right)^2l_1},$$

and $\Theta_n^{(1)}$ are positive roots of the following transcendent equation:

$$tg(\Theta^{(1)}l_1)=\frac{4a^2\alpha_1\Theta^{(1)}}{4a^4(\Theta^{(1)})^2-\alpha_1^2}. \quad (14)$$

Up to now function $\vartheta_4(x_1,t)$ in (12) on the right line $d_1\equiv\{(x_1,t):0\le x_1\le l_1,$ $x_2=l_{23},\,t\ge 0\}$ of the first subproblem region and all $\vartheta_n(x_1,t),\ n=5,6,7$ on the border lines of subproblems for $V^{(2)}$, $V^{(3)}$, $V^{(4)}$ and $V^{(5)}$ are unknown and we should give a system of equations to determine these functions $\vartheta_n(x_1,t),\ n=4,5,6,7$. Firstly, such type of equation for subproblem I follows from the continuity condition:

$$\frac{\partial V^{(1)}(x_1,x_2,t)}{\partial x_2}|_{x_2=l_{23}-0}=\frac{\partial V^{(2)}(x_1,x_2,t)}{\partial x_2}|_{x_2=l_{23}+0} \quad (15)$$

From (15) we have following integral equation in respect to ϑ_4 and ϑ_5:

$$\int_0^t d\tau \int_0^{l_1} K^{(1)}(x_1,\xi_1;t,\tau)\vartheta_4(\xi_1,\tau)d\xi_1$$

$$-\int_0^t d\tau \int_0^{l_1} \frac{\partial^2 G^{(2)}(x_1,x_2;\xi_1,\xi_2;t-\tau)}{\partial\xi_2\partial x_2}|_{\xi_2=l_{21};x_2=l_{23}+0}\vartheta_5(\xi_1,\tau)d\xi_1$$

$$= F_1(x_1,t),\ K^{(1)}(x_1,\xi_1;t,\tau) \equiv \frac{\partial^2 G^{(1)}(x_1,x_2;\xi_1,\xi_2;t-\tau)}{\partial\xi_2\partial x_2}|_{\xi_2=l_{23};x_2=l_{23}-0}$$

$$+\frac{\partial^2 G^{(2)}(x_1,x_2;\xi_1,\xi_2;t-\tau)}{\partial\xi_2\partial x_2}|_{\xi_2=l_{23};x_2=l_{23}+0},$$

$$F_1(x_1,t) \equiv \frac{1}{a^2}\int_0^t d\tau \int_0^{l_1} d\xi_1 \left\{ \int_0^{l_{23}} \frac{\partial G^{(1)}(x_1,x_2;\xi_1,\xi_2;t-\tau)}{\partial x_2}|_{x_2=l_{23}-0}F(\xi_1,\xi_2,\tau)d\xi_2 \right.$$

$$\left. -\int_{l_{23}}^{l_{21}} \frac{\partial G^{(2)}(x_1,x_2;\xi_1,\xi_2;t-\tau)}{\partial x_2}|_{x_2=l_{23}+0}F(\xi_1,\xi_2,\tau)d\xi_2 \right\}$$

$$+\int_0^t d\tau \left\{ \int_0^{l_1} \frac{\partial^2 G^{(1)}(x_1,x_2;\xi_1,\xi_2;t-\tau)}{\partial\xi_2\partial x_2}|_{\xi_2=l_{23};x_2=l_{23}-0}\ \vartheta_3(\xi_1,\tau)d\xi_1 \right.$$

$$\left. +\int_{l_{23}}^{l_{21}} \frac{\partial^2 G^{(2)}(x_1,x_2;\xi_1,\xi_2;t-\tau)}{\partial\xi_1\partial x_2}|_{\xi_1=l_1;x_2=l_{23}+0}\vartheta_2(\xi_2,\tau)d\xi_2 \right\}$$

$$+\frac{1}{a^2}\int_0^{l_1} d\xi_1 \left\{ \int_0^{l_{23}} \frac{\partial G^{(1)}(x_1,x_2;\xi_1,\xi_2;t)}{\partial x_2}|_{x_2=l_{23}-0}\vartheta_0(\xi_1,\xi_2)d\xi_2 \right.$$

$$\left. -\int_{l_{23}}^{l_{21}} \frac{\partial G^{(2)}(x_1,x_2;\xi_1,\xi_2;t)}{\partial x_2}|_{x_2=l_{23}+0}\vartheta_0(\xi_1,\xi_2)d\xi_2 \right\}.$$

The remaining three continuity conditions on the boundary lines of subproblems give us complete system of the integral equations for the functions $\vartheta_n(x_1,\ t),\ n = 4,5,6,7$. They can be found by applying Tikhonov's regularization method for the solution of ill-posed problem [7–11] for this system of equations. Then the solution of the full problem is function:

$$\vartheta(x_1,x_2,t)=\begin{cases} V^{(1)}(x_1,x_2,t); & (x_1,x_2,t)\in[0,l_1]\times[0,l_{23}]\times[0,+\infty), \\ V^{(2)}(x_1,x_2,t); & (x_1,x_2,t)\in[0,l_1]\times[l_{23},l_{21}]\times[0,+\infty), \\ V^{(3)}(x_1,x_2,t); & (x_1,x_2,t)\in[0,l_1]\times[l_{21},l_{22}]\times[0,+\infty), \\ V^{(3)}(x_1,x_2,t); & (x_1,x_2,t)\in[0,l_1]\times[l_{22},l_{24}]\times[0,+\infty), \\ V^{(4)}(x_1,x_2,t); & (x_1,x_2,t)\in[0,l_1]\times[l_{24},l_2]\times[0,+\infty). \end{cases} \quad (16)$$

The obtained analytical solution (16) has been used here to estimate the limit of time step predictions for the propagator numerical scheme.

2.2 Numerical Method

A new semi-implicit difference scheme based on propagator method [5] is elaborated for solution of 2D problem (1)–(2) equations. The method exploits an approach of a non-standard representation of the time derivative, by applying the derivative to the solution given as the product of two functions, where one of them is a propagator function. Propagator function is chosen in a non-local way, and with respect to the solution, a new finite volume difference scheme is elaborated. For the problem (1)–(2) on the grid $\omega=\omega_\tau\times\omega_h$ semi-implicit propagator difference scheme can be written as follows:

$$a^2(\Lambda T^{l+1})_{ij}-\frac{(T)^{l+1}_{ij}}{\tau}=-\exp\left(\frac{\Omega^l_{ij}}{T^l_{ij}}\right)\frac{(T)^l_{ij}}{\tau},\quad 1\le i\le N_x-1,$$
$$1\le j\le N_y-1,\quad l=1,2,3,\ldots, \quad (17)$$

$$\Omega^l_{ij}=-v_1\frac{T^l_{i+1/2j}-T^l_{i-1/2j}}{h^*_i}-v_2\frac{T^l_{ij+1/2}-T^l_{ij-1/2}}{h^*_i}+(f)^l_{ij},$$
$$h^*_i=0.5\,(h_{i+1}+h_i),\ h^*_j=0.5\left(h_{j+1}+h_j\right), \quad (18)$$

$$(\Lambda U^{l+1})_{ij}=0,\quad 1\le i\le N_x-1,\qquad 1\le j\le N_y-1,\quad l=1,2,3,\ldots, \quad (19)$$

where $(\Lambda\Phi^{l+1})_{ij}=\frac{1}{h^*_i}\left(\frac{\Phi^{l+1}_{i+1j}-\Phi^{l+1}_{ij}}{h_{i+1}}-\frac{\Phi^{l+1}_{ij}-\Phi^{l+1}_{i-1j}}{h_i}\right)+\frac{1}{h^*_j}\left(\frac{\Phi^{l+1}_{ij+1}-\Phi^{l+1}_{ij}}{h_{j+1}}-\frac{\Phi^{l+1}_{ij}-\Phi^{l+1}_{ij-1}}{h_j}\right)$ is the standard five point central difference operator for non-regular grid.

Schemes (17)–(19) are unconditionally monotonic, i.e., especially useful for temperature distribution simulations, and has truncation errors of the first order in time and of the second order in space.

Numerical calculations are conducted for electrochemical cell geometry shown in the Fig. 1, with dimensions and parameters given from [4]: $l_1=0.28mm$,

$l_2 = 75.6mm$, $l_{21} = 45.097mm$, $l_{22} = 47.6mm$, $l_{23} = 32.6mm$, $l_{24} = 62.6mm$. On the anode and cathode polarization behavior is taken in the linearized form [4]: $U_\xi = V_\xi - Q_\xi - J_\xi R_\xi$, $\xi = a, c$, $V_c = 0$. Rectangular pulsed currents with $J_\xi = 10^6\,Am^{-2}$ and on the peak voltage $V_a = 7V$ and a period of 0.1s are applied. Conditions on the other boundaries for potential U and temperature T are the same as in [4]. Temperature distribution in time of several x direction points 1–6 (see Fig. 1) are shown in Fig. 2. A case of small overage flow velocity t.i. $v_2 = 1.5 \cdot 10^{-4}\,m/s$ and $v_1 = 0\,m/s$ is assumed. On the given electrochemical cell parameters [4] the Reynolds number is very high $\text{Re} = l_2 \rho C_p / k \cong 5.3 \cdot 10^5$.

Estimated stability restrictions [5] of the difference scheme (17–19) for the allowed time step is $\tau_0 \leq \frac{2 l_2^2 \rho C_p}{(\text{Re } v_2)^2 k}$, and for the given Reynolds number and overage flow velocity $\tau_0 \leq 12.8\,\text{s}$. For the case of velocities substantially large than considered, allowed time step τ_0 is very small and numerical calculations, applying propagator difference scheme (17–19) became ineffective. Such essentially convective dominated transport phenomena will be considered in additional paper, in which we will propose limiting expressions for analytical solutions (17), when $a^2 \to 0$, and convective case of the propagator difference scheme (17–19).

The calculations for nano-electrochemical cell for the case of large velocity, when $v_2 = 1.5m/s$, are shown in Fig. 3. Now, all previously defined dimensions have been multiplied by 10^{-3}. As the result the Reynolds number also becomes 10^{-3} times smaller, and does not cause essential restrictions on allowed time step. Pulsed current period was chosen equal to $10^{-4}\,\text{s}$. Other cell parameters are fitted the same as in the previous case. As it shown in Fig. 3, temperature distribution in nano-electrochemical cell shows less inertial behavior and has more sharp pulse boundaries in comparison to temperature distributions in 'macro-electrochemical' cell, Fig. 2.

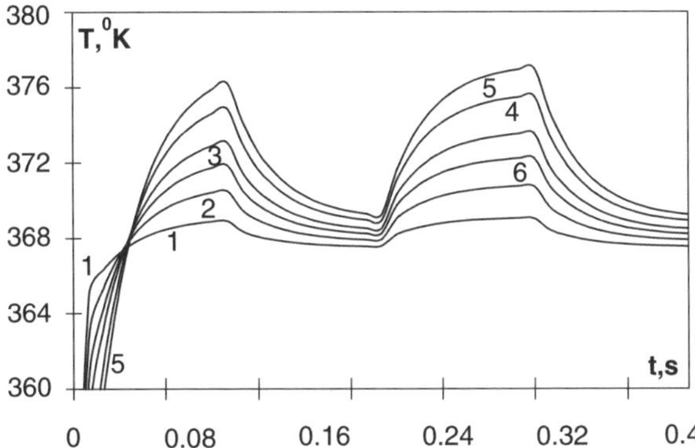

Fig. 2 Temperature distribution in time, at several x points 1–6, for $x = 0.01$; 0.02; 0.04; 0.06; 0.09; 0.17 mm; $l_1 = 0.28\,\text{mm}$, $l_2 = 75.6\,\text{mm}$

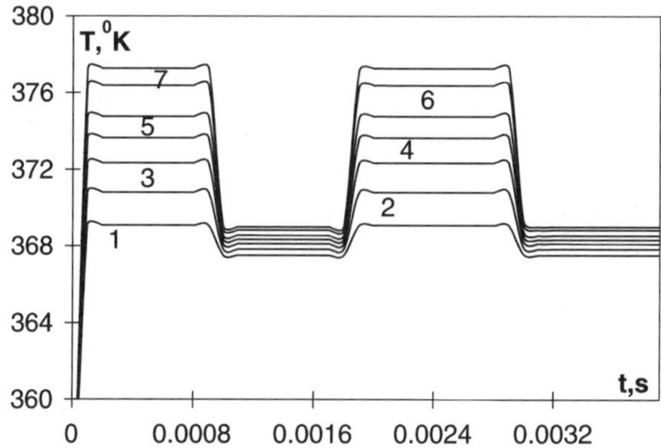

Fig. 3 Temperature distribution in time, at several x points 1–7, $x = 9{,}9\ 10^{-6}$; $2\ 10^{-5}$; $3\ 10^{-5}$; $4\ 10^{-5}$; $5\ 10^{-5}$; $7\ 10^{-5}$; $1\ 10^{-4}$; $l_1 = 0.00028\,\text{mm}$, $l_2 = 0.0756\,\text{mm}$

References

1. A. Ramos, H. Morgan, N.G. Green, A. Castellanos, AC electric-field induced flow in microelectrodes, Journal of Colloid and Interface Science, 217(2):420–422, 1999
2. L. Volker, Electrochemistry of Silicon, Wiley, New York, 2002
3. N. Smets, S. Van Damme, D. De Wilde, G. Weyns, J. Deconinck, Calculation of temperature transients in pulse electrochemical machining (PECM), Journal of Applied Electrochemistry, 37:315–324, 2007
4. S. Van Damme, G. Nelissen, B. Van Den Bossche, J. Deconinck, Numerical model for predicting the efficiency behaviour during pulsed electrochemical machining of steel in $NaNO_3$, Journal of Applied Electrochemistry, 36:1–10, 2005
5. J. Rimšāns, D. Žaime, Propagator numerical method for solving 2D ADR equation, In: Proceedings of the 11th International Conference Mathematical Modelling and Analysis – MMA 2006, Jūrmala, Latvia, pp. 56, 2006
6. H.S. Carslaw, J.C. Jaeger, Conduction of Heat in Solids, the Second Edition, Clarendon Press, Oxford, 1959
7. A.N. Tikhonov, V.Ya. Arsenin, Methods for Solutions of Ill-Posed Problems (in Russian), Moscow, Nauka, 1986
8. Sh.E. Guseynov, M. Okruzhnoya, Quasi-optimal regularization parameter choice for the first kind operator equations (in Russian), Journal of Transport and Telecommunication, 6(3):471–486, 2005
9. A.F. Verlan, V.S. Sizikov, Integral Equations: Methods, Algorithms, Programs (in Russian), Kiev, Naukova Dumka, 1986
10. Sh.E. Guseynov, V.I. Dmitriev, Investigation of the Resolvability and Detailedness of Solutions of Magnetotelluric Sounding Inverse Problems (in Russian), Moscow State University Bulletin, Series 15, Computational Mathematics and Cybernetics, 1:17–25, 1995
11. V.A. Morozov, Regularized Methods for Solutions of Ill-Posed Problems (in Russian), Moscow, Nauka, 1987

Optimal Frame Coherence for Animation in Global Illumination Environment

Piotr Napieralski

Abstract This paper presents an algorithm for animation in global illumination environment for dynamic scenes illuminated by arbitrary, static light sources. Global illumination techniques are difficult to apply for the generation of photorealistic animations because of their high computational costs. These costs are so huge that they take many weeks of computation even on expensive rendering farms, using traditional frame-by-frame rendering systems. Reducing high computational and time costs is of significant practical importance.

1 Introduction

Three-dimensional visualization of electromagnetic fields requires fast processing of individual scenes. Electromagnetic field distribution changes very quickly in function of time. Another problem with visualization of electromagnetic fields is the necessity to distinguish the shape of the simulated machine from the field calculations results. Very often the results are presented in form of scale of colors. The observer needs to have the possibility to analyze the phenomenon inside the examined volume. To correctly extract the information important for the researcher, one of the possibilities is to use the texture of the modelized objects. In such an approach the question of visualization of fields is analogous to photorealistic visualization. The method proposed in the paper permits to decrease the time of the individual scenes rendering. The presented method is useful in the visualization connected with the animation of rotating machines and with the animation of the quick changed electromagnetic field distribution. The physically based simulation of all light scattering in a synthetic model is called global illumination. The goal of global illumination

Piotr Napieralski
Institute of Computer Science, Technical University of Lodz, University of Littoral, Côte d'Opale, France
napieral@ics.p.lodz.pl

P. Napieralski: *Optimal Frame Coherence for Animation in Global Illumination Environment*, Studies in Computational Intelligence (SCI) **119**, 187–195 (2008)
www.springerlink.com

is to simulate all reflections of light in a model and enable an accurate prediction of the intensity of the light at any point in the model. The input to a global illumination simulation is a description of the geometry and the materials, as well as the light sources. It is the work of the global illumination algorithm to compute how light leaving the light sources interacts with the scene. Over the past 25 years or so, several global illumination algorithms have been developed for solving the rendering equation [1, 2] and finding the best way to scene illumination. Mathematically, the equation is expressed by the formula (1). L_o, L_e, L_r are respectively the outgoing radiance, the emitted radiance and the reflected radiance.

$$L_o(x,\omega) = L_e(x,\omega) + L_r(x,\omega). \tag{1}$$

The final rendering equation is:

$$L_o(x,\omega) = L_e(x,\omega) + \int_{\Omega} f_r(x,\omega',\omega) L_i(x,\omega')(\omega' \cdot n) d\omega' \tag{2}$$

Global Illumination algorithms like radiosity, Monte Carlo path tracing, and Metropolis light transport are numerical methods for solving this rendering equation. All these techniques are difficult to apply to the generation of photorealistic animations because of their high computational costs. In recent years, principles and various kinds of implementation and theoretical analyzes of computer animation have been proposed and published to reduce photorealistic computer animation costs [3, 4]. I propose Spatio-Temporal Coherence to find all samples in pixel which can be re-used in next frames. Photo-realistic rendering algorithms are based on stochastic tools, which provide appropriate algorithms for the simulation of light transport. In the rendering of production quality animation, global illumination computations in this framework are divided into some stages. The first stage is the preparation of scene and animation. This process provides definition of objects geometry, light sources parameters as well as definition of camera (position in all animation frames and parameters required for rendering). To define a camera path, I create several key-frames points of camera parameters, each with a different camera position. These key-frames are put at periodic intervals in a key-frame list and in between is generated a spline, or curved line, that passes through all of these key-frame camera positions. The following stage is a pre-processing phase. This stage is responsible for optimal frame segregation for future rendering [5], all instance of camera and scenes are directly connected with all animation frames. To obtain maximal rendering speed and high value of samples reutilization the process of simple rendering is turned only with direct light computation. This resolves a problem of order of sequences, better for optimal samples reutilization. The last stage is the Rendering phase, where animation will be rendered and saved in output disk file. This approach uses all methods based on raytracing algorithms extended to make the global illumination computation more efficient for dynamic environments.

2 The Presentation of Methods

Animations for non-interactive media, such as video and film, are rendered much more slowly. Non-real time rendering allows for the limited processing power in order to obtain higher image quality. Rendered frames are stored on a hard disk, then, when required, transferred to other media such as motion picture film. These frames are then displayed sequentially at high frame rates, typically 24, 25, or 30 frames per second, to achieve the illusion of movement. Image of the photo-realistic quality is often the desired outcome. To achieve these purposes, several different, often specialized, rendering methods have been developed [6–8]. My approach finds all samples in pixel which can be re-used in next frames. The first computed sample that can be called native sample will be re-used in pixel in next frames. The re-used sample can be called recycled sample (Fig. 1). I want to find the best method to re-use native sample as far as it is possible. However I intend that rendering frames will be storage on disk only in the final process. To not overload memory, operation of samples recycling will be stopped when the sample is not visible in the current pixel.

When computing a native sample for a given point in the image plane of one frame the following operation is realized. The ray originating at the eye position and passing through is traced until its intersection point with the nearest object is found and available in following frames. Then the shading computation at this point is performed and the sample value contribution to following frames is added. To derive a recycled sample for a neighboring frames based on this native sample, point is re-projected to determine its position in the image plane of next frames. When the camera is moving, or point represents a dynamic object, the position in next frames might be different of the corresponding position in the previous frame (Fig. 2).

There are three situations when the current sample is not usable:

1. A native sample for a given point is not present in the image plane of next frame,
2. A native sample for a given point is occulted by the other object,
3. A light path or the shadow rays connecting the light path and the eye path with current point are changed.

When shooting the eye and light paths for the native estimate, the visibility query is used to determine the hit points composing the path and simultaneously check if the corresponding rays are also valid for the other frames within. Whenever the hit points on the eye path or the light path are located on a moving object, we also can reuse the corresponding contributions if they are still accessible. In cases, when

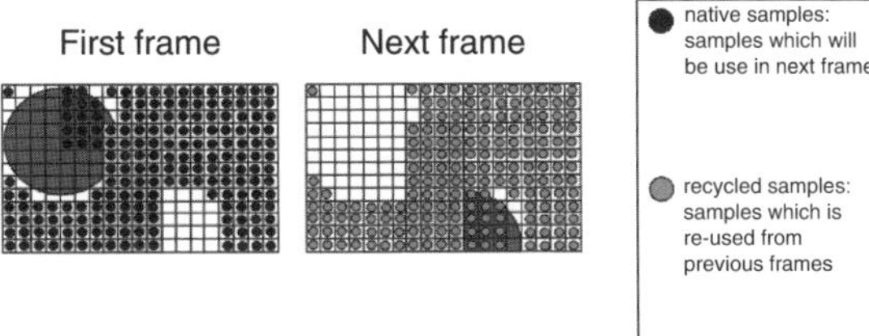

Fig. 1 Idea of native and recycled samples

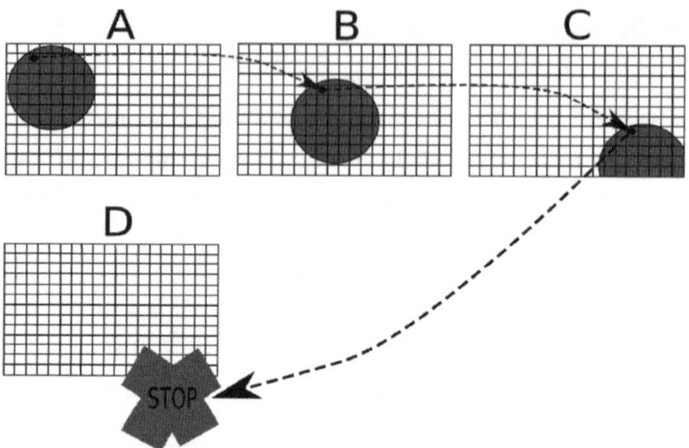

Fig. 2 Simple example for stop recycling process

in order to avoid biasing the solution, for the frames for which these contributions have been invalidated, it is necessary to re-compute a new path starting from the point of camera. Also, for each connection between the eye and light paths, the visibility is simultaneously checked for all next frames. We can unintentionally lose possibilities to recycle the samples in some future frames. We can decide to stop the process because our conditions of non-visibility will be met. But in some next frames they may still be useful. We can assume that image sequence is composed of a number of N images. In each image there is a T calculated samples. Total number of samples are $N_e = N \times T$. In practice it is important to send samples as far as possible. All sent of rays through the pixel is connected with high computations time costs. In next images when we use coherence between frames, we can reduce samples by their effective re-utilization. The goal is to find the way for minimizing the number of new samples N_e. This is what it comes down to calculation of given image it has to use maximum of image samples in next images. Images are noted $I_1, I_2, \ldots, I_N$ and we name μ_k the number of samples calculated in image I_k. We have so on N_e total samples to calculate (3).

$$N_e = \sum_{k=1}^{N} \mu_k. \tag{3}$$

When we minimize this value we will get great computation time advantage. When we use samples of an image I_k to calculate an image I_{k+p} we say that the image I_k propagates samples towards the image I_{k+p}. Image I_k propagates a certain proportion of its sample towards the close images. We note this proportion p_k. Method where it will be possible to find optimal and maximal p_k in the k images should reduce significantly N_e and following on from this, reduce rendering time. A new method of determining maximal p in the following frames will be possible when we know in advance expected value of p_k in the animation. When we find this expected value, we can make propagation in the sense which is helpful to use maximal p, which determines minimal N_e. During the pre-rendering process we can verify the possibilities of appearing objects and cameras arrangement in the future scene. We can try to analyze the angle and position of the camera to find the most similar position in the scene. The results will of course depend on the kind of animation. We can find animation with non-repeated object position and in this case we waste time for the pre-rendering phase. But when there is coherence in some future frames we can save a lot of rendering time. Results from this test will be stored in 2D table, called coherence matrix. Data in the matrix represents number of shared samples between two frames. For example data set in position $[1,6]$ mean same samples in the frame 5 and 1. Problem with locations and with the ordering of frames are similar to the nodes or cities in traveling salesman problem. The traveling salesman problem [1] is a popular illustration of the limits of normal computational techniques and equally a popular subject for demonstrations of alternate approaches to solving such challenges. This problem is usually described as a planning of route through a number of cities such that each city is only visited once, in my case it will be planning of the order of frames. In the sequence the frame can also appeared only once. The challenge in the traveling salesman problem is to minimize the distance. Indeed the same problems arise in my approach; the tasks need to be frame ordered in an optimal manner. In my solution it concerns to take a maximum of distribution of samples in all sequence p. This seems perfectly straightforward, using traveling salesman problem Algorithms to determine more effective frame ordering. One of the first algorithms used to determine a solution to the traveling salesman problem is the nearest neighbor algorithm. The sequence of the visited vertices is the output of the algorithm. This algorithm is easy to implement and executes, quickly yields a short tour, but usually not the optimal one. A genetic algorithm (GA) can be used to find a more optimal solution. There are a couple of basic steps to solving the traveling salesman problem using a GA. First step is to create a group of many random orders which is called a population. Additionally one of the orders is made by an order from nearest neighbor algorithm. Second step, pick 2 of the better order parents in the population (with maximal fitness value) and combine them to make two new child orders. Hopefully, these children orders will be better than either parent. Fitness is a biggest value of p recuperated from before computed coherence matrix. During the small percentage of the time, the child order is mutated. This is done to prevent all orders looking identical in the population. The new child orders are inserted into the population, replacing two of the orders with small fitness. The size of the population remains the same. New children orders are repeatedly created until the desired goal is reached. As the name implies, genetic algorithms mimic nature and evolution using the principles of Survival of the Fittest. Fitness is calculated as mentioned before on the base of coherence matrix. The recuperated order will be used to rendering animated sequence using coherency between frames. Correct order with highest fitness function gives possibility to the optimal recycling of samples. In the situation when sample is not sufficient, a new ray is traced. There exists possibility that in new pixel, placed in different position, there will be more samples than defined pixel sampling rate. If more samples are calculated in one pixel than in the other pixels, it can cause the error of final image. To prevent such a situation it should be made a condition which checks if the current pixel is full, it means it has sufficient number of samples no bigger than needed pixel sampling rate.

3 Results

In all tests several rendering algorithm based on variant of path tracing are used (i.e., classical ray-tracing, Kajiya's path-tracing, bidirectional path tracing, and so on) [5, 9–12]. In such algorithms, the color of every pixel in an image is computed using several samples of the incoming radiance at several points within that pixel. My approach reuses samples hit points in their stochastic light path tracing technique. To test my methods the "Gallo-roman forum of Bavay" Test Scene was applied. This is very complex and complicated geometry. The environment consists of: 46,194 vertex, 74 texture files in jpeg image format, one point light source with Intensity = 1 for each light component (R,G,B). All the images for this scene have been rendered at 320×320 with 60 samples per pixel on a Dual P4 1,700 MHz Linux-PC. To calculate one image we need to send 6,144,000 samples ($T = 320 \times 320 \times 60$). Test animation for this scene contains 340 frames. The test was made for the different ordering methods. The indices method with ascending order is called in the next part of the paper "normal order" and the method with the order generated by genetic algorithm is named "selected order". Primarily the coherence matrix was created, which was used to find optimal ordering of sequence. Time of computation of the coherence matrix and the time of the computation of the genetic algorithm was so small that it can be omitted. In the case of complicated scene such as this one, the number of computed samples was compared, and not the time of computation.

First rendering was made for the normal ordering with the coherence. The comparison of normal ordering with ordering made by genetic algorithm is shown in Figs. 3 and 4. Figure 3 presents the comparison between the traditional method and the method of selected order of calculated number of samples. Before the 82nd

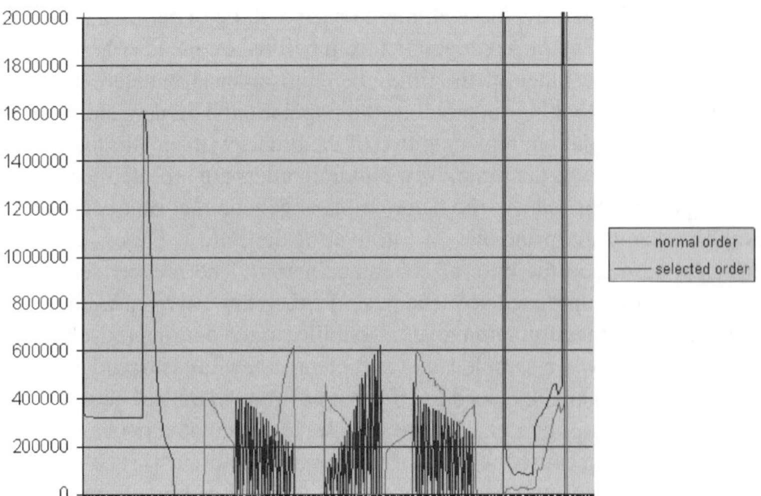

Fig. 3 Normal and selected order number of calculated samples

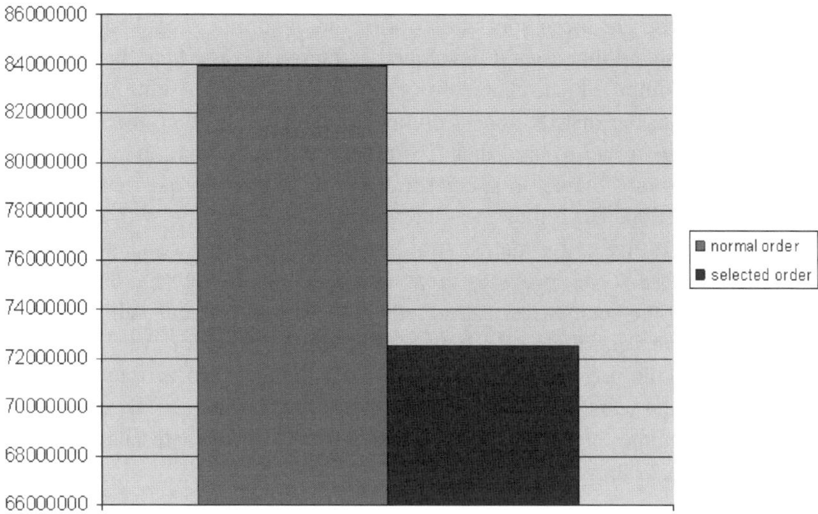

Fig. 4 Normal and selected order number of total calculated samples

frame the results of both methods are identical. Beginning from the 83rd frame the order generated using the genetic algorithm needs smaller number of samples. Thanks to the order selected with the genetic algorithm we do not need to compute any sample until the achievement of the 104th frame. This situation is possible because during the pre-processing phase the genetic algorithm have found the identical frames. Analyzing the results received with the traditional method we can observe that for the frames from 103rd until 123rd the frames are identical. It signifies the static shot of animation. Applying the method called "ordered method", from the 104th to 141st frames we can observe the variation of numbers of the recycling samples. This number changes from 0 to 167,466, but the average value is equal to 167,466. The same results we can observe from the 143rd to 162nd frames. The ordered method for the 182nd to 280th frames gives better results. The total number of samples computed with normal order method is 83,932,775 samples; genetic algorithm needs to compute 72,531,359 samples. In this animation we can remark the advantage of the method of "selecting ordering". In Fig. 4, the comparison between the numbers of simples wanted, using a traditional method and my method is presented. You can observe a significant superiority of the method I proposed. Additionally ordering method clearly increases this advantage. In frame by frame method we need to compute 2,082,816,000 samples. The test scene has typical animation of camera, to move around in a scene to explore across a landscape. This kind of frames similarity we can find when object is moving with static camera and it returns to the same position after some frames. It can be the animation of hand, wings, etc. Images rendered with selected ordering are generated a little fastest. It results from the fact that we can save computational time, when we have to render fewer samples. Certainly there are some wastes of time for computation of order but it is incomparably smallest with total time of rendering. Even though the realistic effects that these algorithms provide are often spectacular, the computational expense is frequently too great for many applications. I developed technique to significantly reduce time by re-utilizing computed samples from one image to others. Additionally I found way to ameliorate the distribution of computed samples to others images without storage great data crowd in hard disk. Even complicated animation can be created in computer memory, thanks to special ordering of images to use optimal way of samples propagation. Usefulness was proved in complex scene with complicated animation shot. In my experiments, stratified sampling, using coherence methods in complete diffuse scene, provides great time reduction. When the determination is realized using the secular effect this reduction will be less but still satisfactory. While the examples and the results may not be representative for every conceivable input scene, they do provide an impression of the value of the various techniques. In future work the tests for more kind of scenes will be made. The optimization algorithm to compute the complicated scenes with all kind of materials and all light prosperity will be elaborated.

4 Conclusions and Future Work

Even though the realistic effects that these algorithms provide are often spectacular, the computational expense is frequently too great for many applications. The developed technique significantly reduces the time of calculations by re-utilizing computed samples in one image to others. Additionally the better distribution of the computed samples to others images without storage the great data crowd in hard disk was made by me. Even complicated animation can be created in computer memory, thanks to special ordering of images to use optimal way of samples propagation. Usefulness was proved in complex scene with complicated animation shot. The method proposed in the paper permits to decrease the time of the individual scenes rendering. The presented method is useful in the visualization connected with the animation of rotating machines and with the animation of the quick changed electromagnetic field distribution. We can remark that during the field simulation inside the transformer the geometry and the texture of this object do not change. The processing of all elements of scene is not necessary. Parts of elements are constant for material as well as for position of devices. The presented method allows in this case to optimize the time of the images rendering and make possible the field visualization even for complex scenes. In my experiments, stratified sampling using coherence methods in complete diffuse scene provides great time reduction.

References

1. Kajiya James T, The Rendering Equation (SIGGRAPH '86 Proceedings), 20(4):143–150, August 1986.
2. Glassner A, An Overview of Ray Tracing in An Introduction to Ray Tracing, Academic Press Limited, 1989.

3. Napieralski P, Optimization Algorithms for Animate Rendering, IEEE ROC&C'2006 ACAPULCO Mexique, 27 October 2006.
4. Havran Vlastimil, An efficient spatio-temporal architecture for animation rendering, EGRW '03: Proceedings of the 14th Eurographics Workshop on Rendering, 2003.
5. Wann Henrik Jensen, Rendering Techniques, Proceedings of the Seventh Eurographics Workshop on Rendering, pages 21–30, 1996.
6. Bekaert Philippe, Mateu Sbert, and John Halton. Accelerating path tracing by re-using paths. In EGRW '02: Proceedings of the 13th Eurographics workshop on Rendering, pages 125–134, Aire-la-Ville, Switzerland, Switzerland, 2002.
7. Besuievsky Gonzalo, A Monte Carlo Approach for Animated Radiosity Environments., PhD Thesis, Barcelona, Spain, 2000.
8. Shelley Kim, Path specification and path coherence, Computer Graphics, 16(3), 1982.
9. Havran Vlastimil, Exploiting temporal coherence in ray-casted walkthroughs, In SCCG '03: Proceedings of the 19th Spring Conference on Computer Graphics, pages 149–155, New York, NY, USA, 2003. ACM Press, 2003.
10. Applegate David L, The traveling salesman problem: a computational study, Princeton Series in Applied Mathematics, Princeton University Press, 2006.
11. Arvo James. Backward ray tracing. In Developments in Ray Tracing, SIGGRAPH'86 Seminar Notes, Vol. 12, 1996.
12. Balazs C, A Review of Monte Carlo Ray Tracing Methods, 1997.

Parallel Computations of Multi-Phase Voltage Forced Electromagnetic Systems

Jakub Kolota, Jakub Smykowski, Slawomir Stepien, and Grzegorz Szymanski

Abstract This research presents a methodology to parallel computations of the finite element models of electromagnetic systems excited from voltage source. In the model an electric circuit equations, field equations and motion equation are sequentially coupled. The circuit equations for each phase are calculated parallel using client – server system.

Simulation results, comparison with conventional sequential system of the computations and computational gain will be presented in this paper.

1 Introduction

Analysis of voltage excited electromagnetic devices with movement is composed of three parts: the analysis of electromagnetic field, analysis of driving circuits and analysis of mechanical motion [1]. The field equations have a current character and to solve the equations an information of the current course in each coil is essential. Thus solution of circuit equations to get coil currents is unavoidable.

Researchers have proposed various methodologies to solve the field-circuit models [1, 2]. They have calculated strong coupled models where circuit equations are coupled with field equations and solved together in one system of equation [2]. Another researchers have proposed sequential models [2], etc.

In this paper a numerical solution methodology of field-circuit-motion models is presented. The circuit, field and motion equations are coupled together. In this approach firstly is solved circuit system by impedance calculation performed for each phase. The impedance calculations are executed in parallel using client – server system with number of nodes equal to number of phases [3]. And next the field equations, force and finally motion equation are solved. An effective approach for parallel computations of multiphase electromagnetic systems is discussed. This paper

Jakub Kolota, Jakub Smykowski, Slawomir Stepien, and Grzegorz Szymanski
Chair of Computer Engineering, Poznan University of Technology, Poznan, Poland

J. Kolota et al.: *Parallel Computations of Multi-Phase Voltage Forced Electromagnetic Systems*, Studies in Computational Intelligence (SCI) **119**, 197–205 (2008)
www.springerlink.com

presents a validity of the socket interface method in the message passing modes and reduction of the time consumption.

Discussed algorithm has been examined in examples of 3D FEM of three phase induction motor and three phase transformer.

2 Model Description

Using the magnetic vector potential **A** and electric scalar potential **V** as electromagnetic field variables, the electric field intensity **E** in conducting region (Ω_C) and magnetic flux density **B** in conducting and non-conducting region ($\Omega_C \cup \Omega_N$) is defined as [1]:

$$\mathbf{B} = \nabla \times \mathbf{A} \text{ in } \Omega_C \cup \Omega_N, \tag{1}$$

$$\mathbf{E} = -\frac{\partial \mathbf{A}}{\partial t} - \nabla V + \mathbf{v} \times (\nabla \times \mathbf{A}) \text{ in } \Omega_C. \tag{2}$$

In this case, the governing equations of an electromagnetic field with the boundary value problem in terms of potentials is expressed as follows:

$$\nabla \times \left(\frac{1}{\mu} \nabla \times \mathbf{A}\right) + \sigma \left(\frac{\partial \mathbf{A}}{\partial t} + \nabla V\right) - \sigma (\mathbf{v} \times (\nabla \times \mathbf{A})) = \mathbf{j}(t) \text{ in } \Omega_C \cup \Omega_N, \tag{3}$$

$$-\nabla \cdot \sigma \left(\frac{\partial \mathbf{A}}{\partial t} + \nabla V - \mathbf{v} \times (\nabla \times \mathbf{A})\right) = 0 \text{ in } \Omega_C, \tag{4}$$

where μ is a permeability, σ represents conductivity, **v** represents velocity of movable armature and **j**(t) current density of the thin coil. If voltage excitation is given, the electric circuit system of equations expressed in term of magnetic vector potential must be considered as [1]:

$$\begin{aligned} &\frac{d}{dt} \oint_{l_1} \mathbf{A}(t)\mathbf{dl} + R_1 i_1(t) = u_1(t) \\ &\vdots \\ &\frac{d}{dt} \oint_{l_n} \mathbf{A}(t)\mathbf{dl} + R_n i_n(t) = u_n(t) \end{aligned} \quad . \tag{5}$$

The above equations are expressed by impedance matrix. For n-phase voltage forced system these equations are expressed by matrix of dynamic impedances. Discrete form of (5) can be defined as bellow:

$$\begin{bmatrix} Z_{11} & \ldots & Z_{1n} \\ \vdots & \ddots & \vdots \\ Z_{n1} & \ldots & Z_{nn} \end{bmatrix} \begin{bmatrix} I_1 \\ \vdots \\ I_n \end{bmatrix} = \begin{bmatrix} U_1 \\ \vdots \\ U_n \end{bmatrix}, \tag{6}$$

where:

$$Z_{jj} = R_j + \frac{1}{I_j}\frac{d}{dt}\oint\limits_{l_j} \mathbf{Adl}_j, \qquad Z_{jk} = \frac{1}{I_j}\frac{d}{dt}\oint\limits_{l_j} \mathbf{Adl}_k. \tag{7}$$

Global impedance matrix includes own impedances Z_{ii} and mutual impedance $Z_{jk} = Z_{kj}$. For $I_j = 1A$ $(j = 1, 2, \ldots, n)$ and for $I_{k\neq j} = 0$ $(k = 1, 2, \ldots, n)$ are performed calculations for matrix of impedance **Z**, next the current vector **I** in both coils are calculated from formulation:

$$\mathbf{I} = \mathbf{Z}^{-1}\mathbf{U}, \tag{8}$$

when a voltage vector **U** is known. Calculation of the force is performed using the Maxwell's stress tensor method. The force density is given by following formula:

$$\mathbf{f} = \nabla \cdot \mathbf{T}, \tag{9}$$

where **T** denotes modified Maxwell's stress tensor [4] proposed as follows:

$$\mathbf{T} = \begin{bmatrix} \frac{1}{2\mu_0}B_x^2 - \frac{1}{2}\mu_0\left(H_y^2 + H_z^2\right) & H_yB_x & H_zB_x \\ H_xB_y & \frac{1}{2\mu_0}B_y^2 - \frac{1}{2}\mu_0 & \left(H_x^2 + H_z^2\right)H_zB_y \\ H_xB_z & H_yB_z & \frac{1}{2\mu_0}B_z^2 - \frac{1}{2}\mu_0\left(H_x^2 + H_y^2\right) \end{bmatrix} \tag{10}$$

Then total force is defined:

$$\mathbf{F} = \int\limits_{\Omega} \mathbf{f} d\Omega. \tag{11}$$

Motion problem is solved by sequentially coupled model with time step verification according to the fixed grid distance in motion direction [4]. Choosing the displacement and velocity as state of 1-DOF mechanical motion, the equation is solved by recurrence Euler's algorithm in state space form as a system of first order differential equations [1]:

$$\frac{d}{dt}\begin{bmatrix} x \\ v \end{bmatrix} = \begin{bmatrix} 0 & 1 \\ -\frac{K}{M} & -\frac{B}{M} \end{bmatrix}\begin{bmatrix} x \\ v \end{bmatrix} + \begin{bmatrix} 0 \\ \frac{1}{M} \end{bmatrix} F, \tag{12}$$

where x represents displacement and v represents velocity of movable armature used in boundary value equations (3)–(4).

3 Numerical Realization

In second step the boundary equation (3)–(4) in integral form are solved by a finite element method with linear shape functions of the potentials. The global matrix system iterative procedure of the system becomes [1]:

$$\begin{bmatrix} \left(\mathbf{C}+\dfrac{\mathbf{D}}{\Delta t_i}\right) & \mathbf{E} \\ \mathbf{F}+\mathbf{H} & \mathbf{G}\Delta t_i \end{bmatrix} \times \begin{bmatrix} \mathbf{z}_{i+1} \\ \mathbf{y}_{i+1} \end{bmatrix} = \begin{bmatrix} \mathbf{r}_{i+1}+\dfrac{\mathbf{D}}{\Delta t_i}\mathbf{z}_i \\ \mathbf{F}\mathbf{z}_i \end{bmatrix}. \tag{13}$$

In above equations **z** represents vector of unknown of magnetic vector potential **A**, **y** represents vector of unknown electric scalar potential **V** and r is a vector of current density **j**(t). Matrices **C, D, E, F, G** and **H** are obtained as result of discretisation of the following formulations: $\int_\Omega \nabla \times \left(\frac{1}{\mu}\nabla\times\mathbf{A}\right) d\Omega - \int_\Omega \sigma\left(\mathbf{v}\times(\nabla\times\mathbf{A})\right) d\Omega$, $\int_\Omega \sigma\frac{\partial \mathbf{A}}{\partial t} d\Omega$, $\int_\Omega \sigma\nabla V d\Omega$, $\oint_S \sigma\frac{\partial \mathbf{A}}{\partial t} dS$, $\oint_S \sigma\nabla V dS$, $-\oint_S \sigma\left(\mathbf{v}\times(\nabla\times\mathbf{A})\right) dS$, Next **r** as result of discretisation $\int_\Omega \mathbf{j}(t) d\Omega$. Obtained system of equation is non-symmetric. Matrix **C** is non-symmetric and diagonally dominant, **D** is a diagonal matrix. Matrix **G** is symmetric. Coefficients of **C, D, E, F, G, H** depends on materials description and type of discretisation. Included in formulations (5) and (6) terms $\int_\Omega \sigma\left(\mathbf{v}\times(\nabla\times\mathbf{A})\right) d\Omega$, and $-\oint_S \sigma\left(\mathbf{v}\times(\nabla\times\mathbf{A})\right) dS$ produces non-symmetric coefficients in global matrix structure. The magnetic vector potential **A** and electric scalar potential **V** are calculated by step by step computation process of equation system (13). For this type of computation the conjugate gradient method with preconditioner called BiCG is worked up [1].

Total force influencing on movable armature is obtained as a combination of calculated potentials **A, V**. The method is widely described in [4]. Then is solving the discrete state space equation of mechanical system to obtain velocity and displacement.

The differential equation of a motion in discrete form

$$\begin{bmatrix} x_{i+1} \\ v_{i+1} \end{bmatrix} = \begin{bmatrix} 1 & \Delta t_i \\ -\frac{K}{M}\Delta t_i & 1-\frac{B}{M}\Delta t_i \end{bmatrix} \begin{bmatrix} x_i \\ v_i \end{bmatrix} + \begin{bmatrix} 0 \\ \frac{\Delta t_i}{M} \end{bmatrix} F_i \tag{14}$$

is successively solved in each iterative step Δt_i to get the armature displacement x.

4 Model of the Parallel System

There are two major methodologies that can be adopted to solve the problem: data parallelism and message passing [5–7]. Decomposition of the problem determined client-server architecture, and message passing with socket interface architecture has been applied. In this model data have a private character for each computing nodes and their exchange follows through communication channels [5].

Linux includes BSD (Berkeley Software Distribution) socket interface, which is the standard network communication protocol. TCP/IP connections as well as communication between processors in Unix domain are used. It's require to define transfer moments using dedicated function to sending and receiving parallel data.

Cluster system using Fast Ethernet base on LAN (100 Mbps Ethernet) was made with Linux operating system. Hardware cluster consists of: one Intel Pentium IV

1.8 GHz with 1.5 GB RAM system memory running as client application and two Intel Pentium IV 1.8 GHz 256 MB as server applications.

Let's describe matrix $\begin{bmatrix} \left(\mathbf{C}+\frac{\mathbf{D}}{\Delta t_i}\right) & \mathbf{E} \\ \mathbf{F}+\mathbf{H} & \mathbf{G}\Delta t_i \end{bmatrix}$ from (16) as **M** and right side of (16) as vector **b**. The idea of parallel calculation is to cut matrix **M** and vector **b** for three fragments simultaneously keeping non-zero extortion in all new part of vector **b**. Calculation of each own impedance requires computations for unknown magnetic vector potential **A**. Mutual impedances are derived from calculated potential for the phase.

For the sake of individuality of each solving problem and to optimise load balancing between cluster nodes the expert's knowledge was used by giving cut dimensions. The total space Ω is split into three calculating subspace ($\Omega \in \Omega_1 \cup \Omega_2 \cup \Omega_3$) taking consideration boundary conditions on division area into account (Fig. 1). Algorithm of solving field issue is divided on parallel field calculation in each from three domains.

In this case we get three systems of equations type $M_N x_N = b_N$ solving by PCG or BiCG algorithm on N node. Continuity of field functions at cutting area has to be kept and internal boundary condition at data divide and integrate should be retained. Then the potentials of next cutted walls in calculate subspace have to accept the same values (Fig. 2). On $\Pi_N \subset \Omega_N$ and $\Pi_{N+1} \subset \Omega_{N+1}$ areas must occur following relation:

$$\underset{M}{\forall} x_{N,M} = x_{N+1,M} \text{ where } x_N \in \Pi_N \text{ and } x_{N+1} \in \Pi_{N+1}. \tag{15}$$

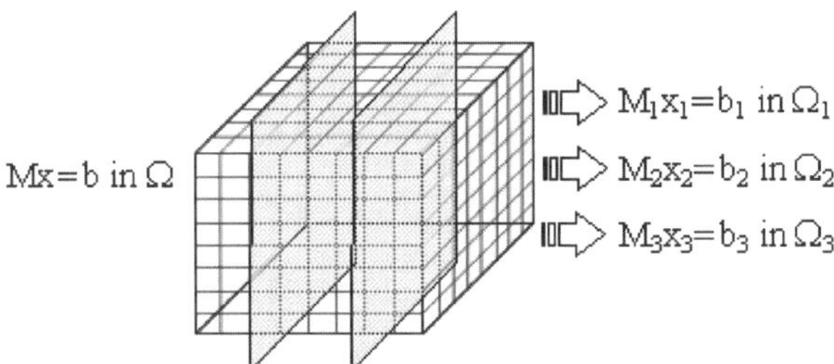

Fig. 1 The model space Ω split into three parts

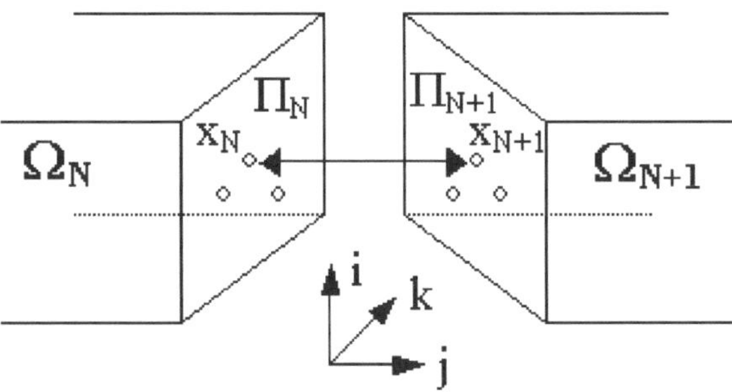

Fig. 2 The way of two subspace merge

Message passing quite often has application in matrix calculus, where individual processors execute computation on selected part of main matrix [7]. In the discussed approach N nodes solve Ω_N part of space Ω. If stopping criteria in the iterative algorithm are achieved on each servers then the results are sent to a client application. Next they are integrated. The whole algorithm is recurred till the maximal number of iteration or stopping criteria is reached.

To realise the communication in the cluster architecture the authors used sockets service. It is implemented at kernel in systems type Unix and execution the operations is enable using system's functions similarly to files. Stream sockets (SOCK_STREAK) use TCP protocol, which main virtues are: sequencing, error control and connection-orient. The method of communication is divided on two levels and the methodology is presented bellow. First data transfer is made once in the whole calculation cycle and its aim is to send the size of matrix M and auxiliary values on each servers.

```
int connection(int* NRP,int* NRM,int* NRP1,int* NRP2,
               int* NRPA, int* NVE, int* INNMAX)
{...
memset(& clin1, 0, sizeof clin1);
clin1.sin_family = PF_INET;
clin1.sin_port = htons(2001);
clin1.sin_addr.s_addr= inet_addr("192.168.10.1");
sock1 = socket(PF_INET, SOCK_STREAM, 0);
gnz1=connect(sock1,(structsockaddr*)& clin1,sizeof clin1);
send(sock1,NRP,sizeof(NRP),0);
...}
```

Data are received by predefined function *recv()* which parameters describe where and how pick data stream up from communication channels.

```
int receive(int* NRPS,double* X1, double* X2)
{
recv(sock1,X1,*NRPS* sizeof(double),MSG_WAITALL);
recv(sock2,X2,*NRPS* sizeof(double),MSG_WAITALL);
...}
```

5 Numerical Examples and Time Consumption

As numerical example 3D models of three phase transformer and three phase induction motor are presented (Figs. 3 and 4). Transformer consists of core made from non-linear material. In parallel implementation the authors have studied linear and non-linear cases of presented device because if non-linear material is taken into account, non-zeros matrix elements are depended from magnetic permeability. Presented transform has six windings – for three on each side (three for primary and

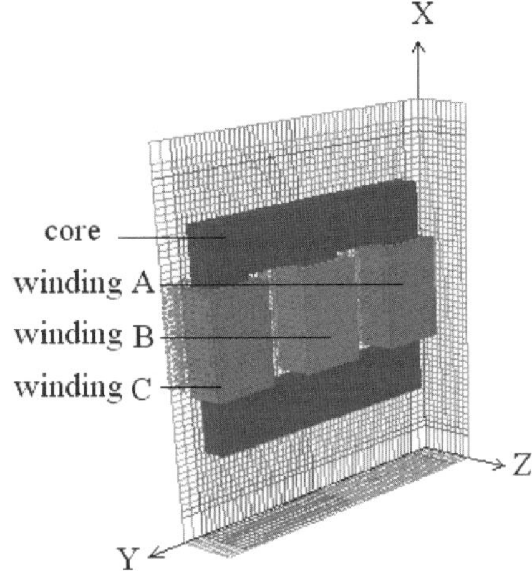

Fig. 3 Model of three phase transformer used as an example of PCG parallel computations

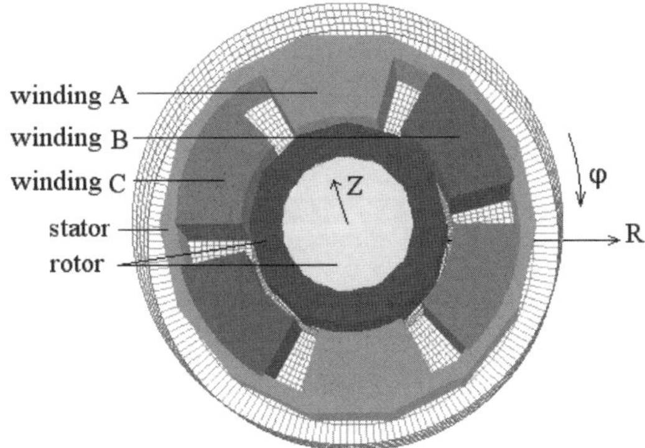

Fig. 4 Model of three phase induction motor used as an example of BiCG parallel computations

three for secondary). Each phase is excited at 230 V voltage amplitude. The transformer core is made from steel with conductivity $\sigma = 7.2 \times 10^6\,\mathrm{V\,m^{-1}}$ and relative permeability $\mu_r = 199$.

As a motion problem example the three phase induction motor with a 45° winding spread per phase is presented. The current density is maintained constant at $310\,\mathrm{A\,cm^{-2}}$. The three phase winding is excited at 60 Hz. The stator steel is laminated and has a conductivity $\sigma = 0\,\mathrm{V\,m^{-1}}$. The central part of rotor is made from aluminum with conductivity $\sigma = 3.72 \times 10^7\,\mathrm{V\,m^{-1}}$. The outside part of rotor is made from steel with conductivity $\sigma = 1.6 \times 10^6\,\mathrm{V\,m^{-1}}$. Both the rotor and stator steel has a relative permeability $\mu_r = 30$.

Discussed models are discretised in three dimensions system. The variable quantities like magnetic vector potential or scalar electric potential are displaced in x-y-z for the transform and r-φ-z for the motor model. It is obvious that number of variables that must be solved is strictly depended from dimensional discretisation. Analyzed phenomena and models have also impact on matrix structure [8–10]. To solve

Table 1 The results of calculations using parallel system for linear and non-linear transform models

Transformer model	Stopping criteria error	Linear model	Size of matrix M	Time of computation	Gain
Only client application	1D-4	YES	162779	Real: 643m 12.03s	—
Parallel system:client+3 servers	1D-4	YES	162779	Real: 424m 21.23s	34.03%
Only client application	1D-5	NO	162779	Real: 177m 22.11s	—
Parallel system:client+3 servers	1D-5	NO	162779	Real: 102m 12.57s	42.37%
Parallel system:client+6 servers	1D-5	NO	162779	Real: 72m 49.41s	58.94%

Table 2 The results of calculations using parallel system for non-linear induction motor model

Induction motor model	Stopping criteria error	Linear model	Size of matrix M	Time of computation	Gain
Only client application	1D-4	NO	412865	Real: 1412m 23.32s	—
Parallel system:client+3 servers	1D-4	NO	412865	Real: 946m 11.42s	33.01%
Only client application	1D-5	NO	412865	Real: 1623m 19.04s	—
Parallel system:client+3 servers	1D-5	NO	412865	Real: 947m 38.24s	41,62%

large sparse system of equations PCG and BiCG (for motion problem) methods are used.

Tables 1 and 2 present time results for parallel calculations models discussed above.

6 Conclusions

Accelerating calculations using more and more sophisticated algorithms when complex models are considered cannot give satisfaction of solving time. In case of fast computer hardware development many researchers reaches for calculations parallelisation. Time of solving problem can be much faster but parallelisation process must be implemented into solving algorithm and software.

In this paper an effective approach for parallel computations of multiphase electromagnetic system in examples of three phase model of transformer and induction motor have been presented. Successful implementation of applied parallelisation method has been shown.

Presented results show that proposed method gives good results especially if number of servers increasing. Massage passing used as technology of parallel computation gives shorter times of solving and can take particular place in multiphase electromagnetic device model calculations.

References

1. S. Stępień and A. Patecki, "Modeling and position control of voltage forced electromechanical actuator", *COMPEL: The International Journal for Computation and Mathematics in Electrical and Electronic Engineering*, Vol. 25, No. 2, pp. 412–426, 2006
2. K. Hameyer, "Field-circuit coupled models in electromagnetic simulation", *Journal of Computational and Applied Mathematics*, Vol. 168, pp. 125–133, 2004
3. J. Kołota, S. Stępień, G. Szymański and J. Wencel, "Parallel computations of 3D models of an electromagnetic device", *Sixth International Conference on Computational Electromagnetics CEM 2006*, 4–6 April 2006, Aachen, Germany, pp. 29–31, 2006
4. W. Demski and G. Szymański, "Comparison of the force computation using vector and scalar potential for 3D", *IEEE Transactions on Magnetics*, Vol. 33, No. 2, pp. 1231–1234, March 1997
5. M.A. Palis, L. Jing-Chiou and D.S.L. Wei, "Task clustering and scheduling for distributed memory parallel architectures", *IEEE Transaction on Parallel and Distributed Systems*, Vol. 7, No. 1, pp. 46–55, 1996
6. R. Janssen, M. Dracopoulos, K. Parrott, E. Slessor, P. Alotto, P. Molfino, M. Nervi and J. Simkin, "Parallelisation of electromagnetic simulation codes", *IEEE Transactions on Magnetics*, Vol. 34, No. 5, pp. 3423–3426, Compumag 1997
7. S. Chingchit, M. Kumar and L.N. Bhuyan, "A flexible clustering and scheduling scheme for efficient parallel computation", *13th International and 10th Symposium on Parallel and Distributed Processing*, pp. 500–505, 1999
8. K. Hamayer and R. Bellmans, *Numerical Modelling and Design of Electrical Machines and Devices*, WIT Press, Southampton, 1999
9. D.A. Lowther, "Automating the design of low frequency electromagnetic devices – a sensitive issue", *COMPEL*, Vol. 22, No. 3, pp. 630–642, 2003
10. T. Nakata, N. Takahashi, K. Fujiwara, K. Muramatsu and Z. Cheng, Comparison of various methods for 3-D eddy current analysis, *IEEE Transactions on Magnetics*, Vol. 24, No. 6, pp. 3159–3162, November 1988

Part III
Applications of Computer Methods

Field – Circuit Coupling with the Time Domain Finite Difference Method for low Frequency Problems

Theodoros I. Kosmanis

Abstract A combined field and circuit approach in time domain is presented in this paper for the realistic study of power frequency devices. The field analysis is performed by an explicit, conditionally stable, finite difference scheme that allows the whole procedure to take place in time domain avoiding large, sparse matrix inversions, a significant shortcoming of implicit methods. The circuit equations are also considered in time and are easily coupled to the field analysis technique via branch currents and conductor voltage drops. The preliminary results by corresponding 2-D problems reveal the algorithm's ability to accurately model transient phenomena.

1 Introduction

The direct connection of electromagnetic parts of power frequency devices to external circuits such as sources, control circuits or loads and the non-linear nature of ferromagnetic material, necessitates the coupling of each electromagnetic field oriented numerical method to circuit equations [1–4] and the inclusion of special techniques for the treatment of non-linearities [1]. Since most numerical methods for eddy-current analysis are implicit, such a combination results in large system matrices which are in general non-symmetrical, thus worsening the great disadvantage of frequency domain methods for quasi-static problems.

An alternative to this shortcoming is the adaptation of a time domain approach for the field solution which would not require matrix inversion but would be based in a time advancing procedure. A class of very efficient time domain schemes for the analysis of eddy-current problems that meets this requirement has recently been

Theodoros I. Kosmanis
Department of Mechanical and Industrial Engineers, University of Thessaly, Pedion Areos, 38334 Volos, Greece
thkosman@uth.gr

T.I. Kosmanis: *Field – Circuit Coupling with the Time Domain Finite Difference Method for low Frequency Problems*, Studies in Computational Intelligence (SCI) **119**, 209–216 (2008)
www.springerlink.com

presented [5]. Based on an algorithmic convergence condition (stability and consistency), a non-standard, finite difference time domain approach of the diffusion equation (part of each low-frequency mathematical model) is derived. The field part of the coupled methodology presented in this paper lies on this time domain technique. As will be verified by the numerical results, it also allows the treatment of non-linear material.

2 Coupled Field-Circuit Analysis

In common with all other techniques, the proposed algorithm starts from the mathematical model of the problem consisting of a set of three equations:

(a) The classic diffusion equation inside conductors or ferromagnets with the magnetic vector potential (MVP), $\boldsymbol{A}$, as main variable

$$\nabla \times \mu_r^{-1} \nabla \times \boldsymbol{A} + \mu_0 \sigma \partial_t \boldsymbol{A} = \mu_0 \boldsymbol{J_S}, \tag{1}$$

where $\boldsymbol{J_S}$ is the imposed current density due to the external circuit,

(b) The equation of total current flowing through conductors

$$\mathbf{I} = -\int_{Sc} \sigma \left(\partial_t \boldsymbol{A} + \Delta\varphi\right) dS. \tag{2}$$

(c) The circuit equations which depend on external circuit structure and are connected to (1) and (2) by means of conductor total currents, $\mathbf{I}$, and voltage drops, $\Delta\varphi$.

The cases encountered, as far as the type of the conductors used concerns, are the solid and filament conductors. In the first one, current carrying conductors are treated as diffusion regions of the field problem since their finite width enforces the development of diffusion phenomena which must be taken care of. Their current density as a function of voltage drops across solid conductors, $\Delta\varphi$, is given by

$$\boldsymbol{J_S} = \sigma \Delta\varphi. \tag{3}$$

On the other hand, filament conductors are too thin to be assumed as diffusion regions, so they are treated as current density sources of value

$$\boldsymbol{J_S} = \sigma \Delta\varphi = l \frac{NI}{S}, \tag{4}$$

where S, l, N are the conductor's cross-section area, turn length and number of turns, respectively.

2.1 Field Analysis: The Explicit Finite Difference Scheme

In the field part of the methodology, all devices requiring electromagnetic analysis, such as conductors or ferromagnetic materials, are treated by means of an explicit finite difference scheme. An appropriate orthogonal grid is used for the discretization of the domain of interest whereas the diffusion equation (1) is discretized according to the general form

$$A_{i,j}^{n+1} = a_{0,0}A_{i,j}^{n} + a_{x_{1,0}}(A_{i-1,j}^{n} + A_{i+1,j}^{n}) + a_{y_{1,0}}(A_{i,j-1}^{n} + A_{i,j+1}^{n}) \\ + a_{0,1}A_{i,j}^{n-1} + a_{x_{1,1}}(A_{i-1,j}^{n-1} + A_{i+1,j}^{n-1}) + a_{y_{1,1}}(A_{i,j-1}^{n-1} + A_{i,j+1}^{n-1}) - a_S J_{i,j}^{n}, \quad (5)$$

the MVP components, $A_{i,j}$, on each grid node $(i,\ j)$ at the *nth* time step, being the degrees of freedom of the computational problem. The values assigned to coefficients $a_{x1,0}$, $a_{y1,0}$, $a_{0,1}$, $a_{x1,1}$, $a_{y1,1}$ and a_S are obtained via a general procedure which is based on the consistency of the scheme with stability requirements. Among the non-standard schemes that have been proposed [5], the most simple, but without lack of efficiency, Dufort–Frankel finite difference one has been selected for the simulations of this paper, where the only non-zero coefficients are

$$a_{0,1} = \frac{1-2\lambda_x-2\lambda_y}{1+2\lambda_x+2\lambda_y} \quad a_{x_{1,0}} = \frac{2\lambda_x}{1+2\lambda_x+2\lambda_y} \quad a_{y_{1,0}} = \frac{2\lambda_y}{1+2\lambda_x+2\lambda_y} \quad a_S = \frac{2\delta t/\sigma}{1+2\lambda_x+2\lambda_y}. \quad (6)$$

Assuming that δt and δx, δy denote the time and space discretization steps, respectively, and μ, σ are the material's permeability and conductivity, coefficients λ_x, λ_y in (6) are given by

$$\lambda_x = \delta t/\mu\sigma\delta x^2 \qquad \lambda_y = \delta t/\mu\sigma\delta y^2. \quad (7)$$

The stability condition of the algorithm requires that [5]

$$\delta t \leq \frac{\mu\sigma\sqrt{\delta x^2+\delta y^2}}{2}. \quad (8)$$

It is important to mention herein that the explicit finite difference scheme can be implemented only in regions of conductive or ferromagnetic material that are physically characterized by the diffusion phenomenon. However, in air which in most cases surrounds electromagnetic devices, the governing equation is the non-homogeneous, elliptic type one

$$\nabla \times \mu_0^{-1} \nabla \times \boldsymbol{A} = \boldsymbol{J}_{S0}, \quad (9)$$

that cannot be treated by the above explicit procedure. In this case, we make the assumption that due to the low frequency, the problem outside diffusion regions is static in each time step and can be solved either by a finite difference static approach [5] or by a finite element static one [6], leading, thus, to a hybrid scheme.

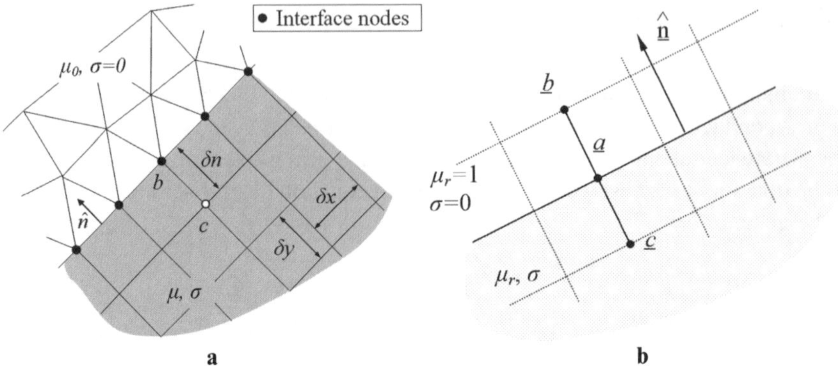

Fig. 1 Discretization of the computational domain according to the hybrid scheme and boundary coupling **a** Dufort–Franklel/finite element and **b** Dufort–Franklel/finite difference

In either case, a unified, fully time domain procedure arises, that is based on the following steps:

(a) The problem is solved in air, treating the interface between air and diffusion regions as additional degrees of freedom (Neumann conditions).
(b) The Dufort–Frankel method is implemented inside diffusion regions having the interface MVP components as boundary values (Dirichlet conditions).
(c) Appropriate interface conditions based on the continuity of the magnetic field along the interface allow the update of the interface values

$$\left.\frac{\partial A}{\partial n}\right|_{\partial\Omega_{air}} = \frac{1}{\mu_r}\left.\frac{\partial A}{\partial n}\right|_{\partial\Omega_{diffusion}} \cong \frac{1}{\mu_r}\frac{A_b - A_c}{\delta n}, \tag{10}$$

where indices a and b refer to the computational grids arising from the aforementioned hybrid approaches as depicted in Fig. 1a,b.
(d) The algorithm advances to the next time step.

2.2 Circuit Time Domain Analysis

The coupling of the time domain field analysis algorithm to circuit equations is performed via the calculation of macroscopic circuit physical dimensions (voltage drop, current) by electromagnetic ones and vice versa. These transitions can be embodied in the numerical procedure of the previous section, thus producing the overall field-circuit coupled algorithm which is fully time-advancing. All dimensions are computed by means of values of the previous time steps.

Therefore, referring to a single time step, after the completion of the field part of the analysis, the surface integral of equation (2) is computed over the cross-section of each current carrying conductor that is connected to both field and circuit part of the analysis. Hence, computation of currents and overall voltage drops along current carrying conductors is possible.

At this point of the algorithm the circuit equations can be solved with the field region replaced by an n-port element that communicates with the rest of the circuit by means of its input and output currents and the voltage drops across ports. The solution of circuit equations, which in their general form are differential ones with respect to time, provides the updates for the currents inserted as inputs for the field analysis of the next time step.

For the circuit-to-field transition, the theoretical separation of solid and filament conductors is also followed. Solid conductors that are closely related to the diffusion phenomenon must be analyzed by the Dufort–Frankel finite difference scheme. They are considered as non-magnetic ($\mu = \mu_0$), conductive ($\sigma \neq 0$) areas of non-zero current density. On the other hand, filament conductors appear in the field computational region as coils of N turns. They are not characterized by diffusion since they are too thin. Therefore, they are inserted in the numerical problem as, static for every time step, current density sources in (9).

The algorithm then advances to the next time step and is repeated as much as required to complete the simulation.

3 Discussion

The key point of the proposed methodology is that the field and circuit analysis is performed completely in the time domain, thus allowing straightforward and accurate transient and multifrequency simulation. Furthermore, the combination of an orthogonal (finite differences) or an adaptive triangular grid (finite elements) offers significant geometrical flexibility.

The repetitive procedure starts with all variables equal to zero, the only excitation term being the voltage or current sources of the circuit equations. After the voltage drops, $\Delta\varphi$, have been computed by circuit equations, the MVP inside all diffusion and non-diffusion regions (solid or filament conductors and ferromagnets) is easily derived by (5) and (9). Finally, the branch currents, $\mathbf{I}$, are computed by (2). The algorithm advances to the next time step and the above procedure is repeated until steady-state is reached.

4 Numerical Results

The proposed methodology has been verified by its implementation in two model field-circuit problems, allowing us to test the abilities of the proposed time domain methodology in various indicative low frequency cases.

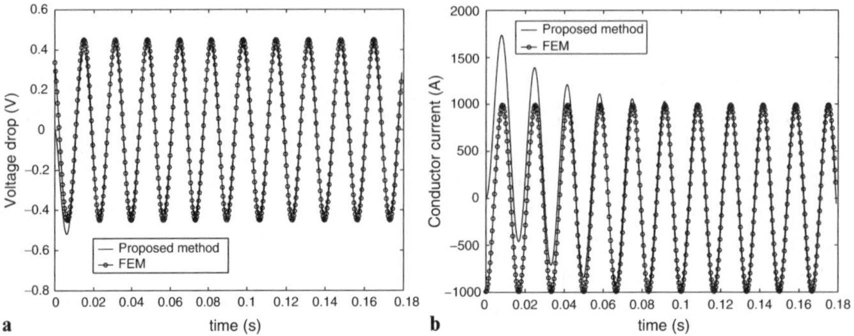

Fig. 2 **a** Voltage drop across the solid conductor. **b** Circuit current

4.1 Solid Conductor in Series with a Sinusoidal Voltage Source

The first coupled field-circuit model problem consists of a solid bar conductor of rectangular cross section $(0.02 \times 0.1\,\text{m}^2)$ placed inside a ferromagnetic slot. The conductor is connected to an external circuit consisting of a resistance $(0.01\ \Omega)$, an inductor (0.265 mH) and a 100 V/60 Hz sinusoidal voltage source connected in series [7]. The necessary circuit equation that completes the system of equations (1)–(3) is given by Kirchhoff's voltage law along the single loop composed, that connects the voltage drop along the solid conductor to the circuit current, that is

$$Ld_t i(t) + Ri(t) + \Delta\varphi = V_S(t). \tag{11}$$

The above equation is easily discretized with respect to time and solved for $\Delta\varphi$ in each time step. The field problem is then fully defined and the MVP across the bar's cross-section is computed. Finally, the circuit current is calculated by (2).

The electric circuit current and the voltage drop across the solid conductor are illustrated in Fig. 2a,b as a function of time. The simulation of the transient phenomenon together with the steady-state is obvious in both figures, promising analogous results with non-sinusoidal stimulations. On the other hand, the standard finite element procedure is unable to model the transient part, approximating only the steady-state due to its monochromatic nature.

4.2 LC Circuit with a Linear/Non-Linear Core Inductor

In the second problem, the series LC circuit, that is embedded in Fig. 3a is considered [8]. The capacitor of 100 nF is powered by a 5,000 V dc voltage step source. The inductor is composed of a ferromagnetic core one edge of which is surrounded by filament wires (Fig. 3a). The material of the core is initially assumed to be linear with $\mu_r = 1,500$.

The necessary circuit equations are again the outcome of Kirchhoff's voltage law along the single loop, which connects the voltage drop along the filament conductor to the circuit current, that is

$$V_C(t) + \Delta\varphi = V_S(t), \tag{12}$$

and the $v - i$ characteristic of the capacitor

$$i(t) = C d_t V_C(t). \tag{13}$$

The above equations are easily discretized with respect to time and solved for $\Delta\varphi$ in each time step. The field problem is then fully defined and the MVP components are computed.

Figure 3b illustrates the magnetic field lines in the core as computed by the combination of the finite difference and finite element methods while the electric circuit current as a function of time is depicted for the linear (Fig. 4a) and the non-linear (Fig. 4b) core. The $B-H$ curve for the latter case is shown in the embedded Fig. 4b.

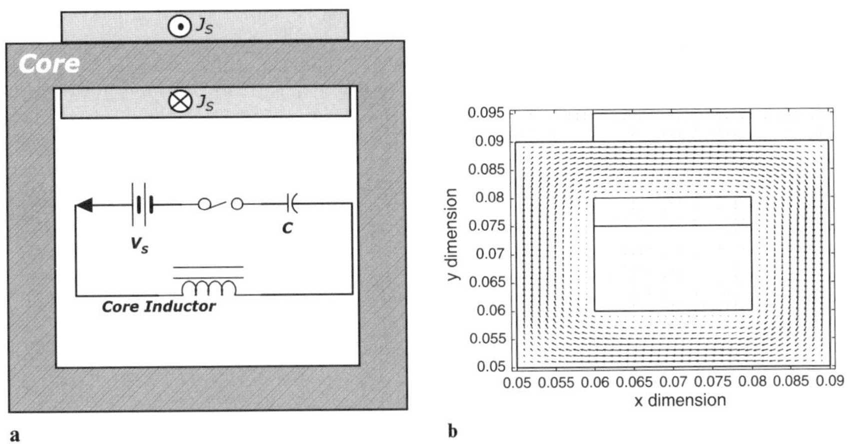

Fig. 3 **a** LC circuit with a core inductor. **b** Magnetic field lines inside the ferromagnetic core

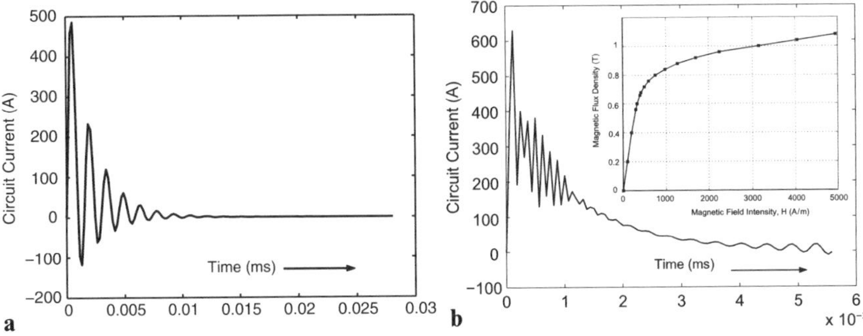

Fig. 4 **a** Circuit current for (a) the linear core and **b** the nonlinear core

The ability of the procedure to simulate the transient phenomenon together with the steady-state is again obvious. It must be mentioned that the time domain analysis as performed by the proposed methodology allows an extensive post-processing procedure for the computation of various circuit and field problem parameters in the frequency domain via a simple fast Fourier transform.

References

1. K. Hamayer and R. Belmans, Numerical Modelling and Design of Electrical Machines and Devices, WIT Press, Bath, UK, 1999.
2. I. A. Tsukerman, A. Konrad, G. Meunier and J. C. Sabonnadiere, Coupled Field-Circuit Problems: Trends and Accomplishments, IEEE Trans. Magn., Vol. 29, pp. 1701–1704, 1993.
3. I. Tsukerman, A Stability Paradox for Time-Stepping Schemes in Coupled Field-Circuit Problems, IEEE Trans. Magn., Vol. 31, pp. 1857–1860, 1995.
4. J. Vaananen, Circuit Theoretical Approach to Couple Two-Dimensional Finite Element Models with External Circuit Equations, IEEE Trans. Magn., Vol. 32, pp. 400–410, 1996.
5. T. V. Yioultsis, K. S. Charitou and C. S. Antonopoulos, The Finite-Difference Time-Domain Technique with Perfectly Matched Layers for Transient Nonlinear Eddy-Current Problems, IEEE Trans. Magn., Vol. 38, pp. 621–624, 2002.
6. T. I. Kosmanis, T. V. Yioultsis and T. D. Tsiboukis, Computational Analysis of Power Frequency Devices by a Novel Hybrid Quasi-Static Finite Difference – FEM Technique, ICEM'06 (XVII International Conference on Electrical Machines), Chania, Greece, 2006, p. 507.
7. S. J. Salon, Finite Element Analysis of Electrical Machines, first ed., Kluwer, Boston, 1995, pp. 63–72.
8. A. Nicolet, F. Delince, N. Bamps, A. Genon and W. Legros, A Coupling Between Electric Circuits and 2D Magnetic Field Modeling, IEEE Trans. Magn., Vol. 29, pp. 1697–1700, 1993.

Rotor Shape Optimisation of a Switched Reluctance Motor

Marian Łukaniszyn, Krzysztof Tomczewski, A. Witkowski, Krzysztof Wróbel, and Mariusz Jagieła

Abstract In the paper the method of torque pulsations reduction of a switched reluctance motor by means of rotor shape optimization is presented. The optimization procedure is constructed on a basis of a genetic algorithm. The objective function is evaluated using variations of the electromagnetic torque predicted by two-dimensional magnetostatic field analysis. To account for torque pulsations due to current switching a circuit model is used that utilizes the electromagnetic quantities predicted by the finite element analysis. To reduce the time of computations an environment for distributed computing was developed.

1 Purpose

Basic structure and the specific commutation algorithms of switched reluctance motors (SRMs) are two fundamental contributions to electromagnetic torque pulsations in these machines [1, 2]. Torque pulsations can be reduced by appropriate shaping of the current waveforms. This, however results in rise of electric power loss due to increased switching frequency. It is also possible to optimise the magnetic circuit, particularly the rotor shape, to reach minimum of torque pulsations. Such an optimization can be realized in two different ways. First of them (first method) is to carry out the magnetostatic field analysis under assumption of constant currents in the stator winding. Such an approach does not allow for accounting the torque pulsations due to current switching. Therefore, a second method is developed where linear circuit-based dynamic analysis is realized at the given operating point in a step subsequent to the analysis of the magnetic field. Such an approach makes the design of significantly improved constructions of SRMs with respect to those designed using only the magnetostatic field analysis possible.

M. Łukaniszyn, K. Tomczewski, A. Witkowski, K. Wróbel, and M. Jagieła
Opole University of Technology, ul. Luboszycka 7, 45-036 Opole, Poland
m.lukaniszyn@po.opole.pl, k.tomczewski@po.opole.pl, a.witkowski@po.opole.pl, k.wrobel@po.opole.pl, m.jagiela@po.opole.pl

M. Łukaniszyn et al.: *Rotor Shape Optimisation of a Switched Reluctance Motor*, Studies in Computational Intelligence (SCI) **119**, 217–221 (2008)
www.springerlink.com

2 Method of Computations

The optimization procedure developed, that is based on a genetic algorithm is linked with the two-dimensional magnetostatic field solver (program FEMM [3]). Only the rotor shape is optimized. In a genetic algorithm an initial randomly generated population, that represents the decision variables, corresponds with coordinates that define the curvature of the rotor surface. The objective function takes into account several values of torque at different positions of the rotor. The finite element model of the motor must be thus created several times during each call of the objective function by the optimization procedure. Moreover, in case of second method the variations of electromagnetic quantities such as linkage fluxes and torque versus angular position of the rotor must be determined with desired accuracy to create appropriate look-up tables. Since more than 100 different rotor positions must be taken into account, at this stage of the study it is impossible to account for material nonlinearity.

Normally, such an analysis would require vast amount of time to reach the solution as the number of nodes in the finite element mesh usually exceeds 15,000. To reduce the time of computations the Authors developed an environment for distributed computing, which is based partially on the publicly available applications, and partially on the programs developed by the authors [4]. The system is developed in such a way that the main program, which implements the optimization procedure, works on a single computer. The calculations of the magnetic field are realized on distributed workstations. The suite of programs for distributed computing is managed by system Condor [5]. The whole environment works under Windows XP system.

3 Results of Analysis

The analysis presented in this work was carried out for a small low-voltage (24 V), two-phase SRM having maximum torque of 0.5 Nm at rated rotor speed. The model, due to its specific structure is characterized by high level of torque pulsations.

The optimization was done so that the average value of torque was kept constant and equal to nominal torque developed by the motor. Only the rotor shape is optimized by subdivision of the contour describing the rotor surface into 15 arcs as shown in Fig. 1. The decision variables were radii of the arcs R_k and their angular length α which is the same for all arcs. The total number of decision variables was 16.

The calculations have been realised for the population including 42 individuals. The optimization has been stopped after 200 generations. The average time of calculations required for each individual was approximately 6 min using only the magnetostatic analysis. In case of additional calculations involving the circuit analysis the time of calculations increased to approximately 7 min. The calculations were

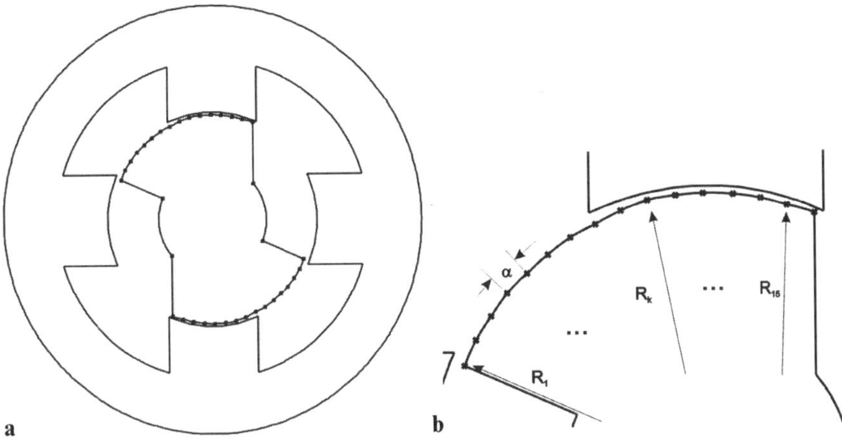

Fig. 1 Motor cross-section **a** and the rotor configuration **b** showing subdivision of the rotor surface into arcs

carried out on distributed workstations equipped with Sempron 3100+ processors and 256 Megabytes of memory.

The objective function is defined in such a way as to minimize the torque pulsations. The following form of the objective function was used:

$$f = \frac{T_{eAV}}{s \cdot \frac{(T_{eMAX} - T_{eMIN})}{T_{eAV}}}, \tag{1}$$

where T_{eAV} is an average value of torque, s is a standard deviation of torque from its average value, T_{eMAX} is the maximum value of torque, and T_{eMIN} is the minimum value of torque.

The optimization was primarily realised using only the magnetostatic field analysis under constant current (first method). For the rotor structure found by the optimisation procedure the dynamic linear circuit-based analysis was carried out at the given operating point. In this case the torque pulsations coefficient TRF (2) was equal to 0.73 Nm Nm^{-1}.

$$TRF = \frac{T_{eMAX} - T_{eMIN}}{T_{eAV}}. \tag{2}$$

In the second case (second method) the optimisation was carried out under previously described system for distributed computing. The basic difference with reference to the first method is that the goal function is determined on a basis of the dynamic circuit-based analysis, where the current switching processes are accounted for. In this case the torque pulsations factor was reduced to 0.091 Nm Nm^{-1}.

To compare the torque waveforms developed by the two optimized motors the circuit analysis was performed and the waveforms that were obtained are shown plotted in Fig. 2.

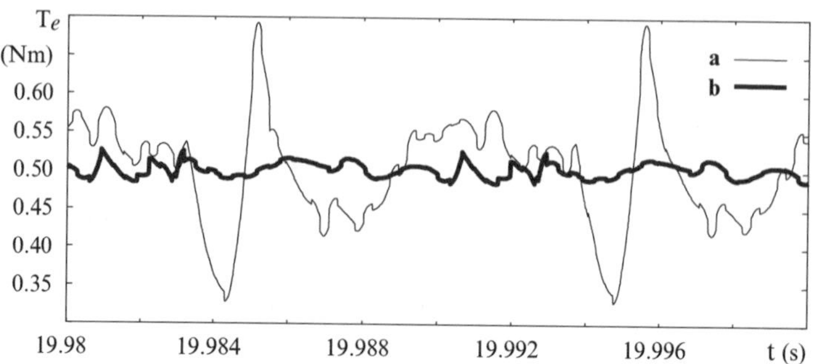

Fig. 2 Electromagnetic torque waveforms for the optimized motor versions using first method **a** and second method **b**

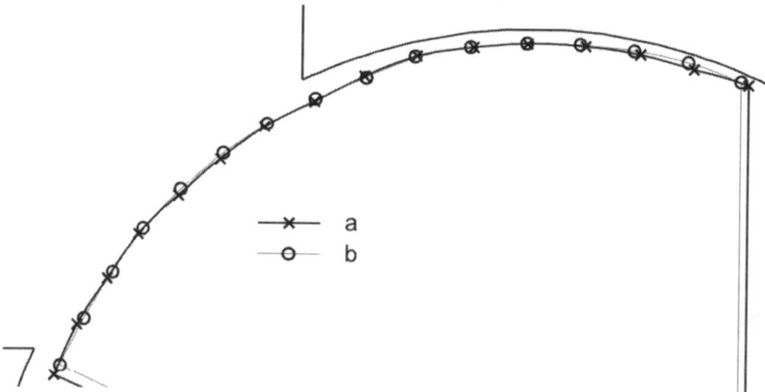

Fig. 3 Optimal rotor shapes obtained using first method **a** and second method **b**

Table 1 Minimum values of torque pulsations factor for the given operating conditions

Using optimisation procedure based only on the magnetostatic field analysis (first method)			Using optimisation procedure based on instantaneous torque waveforms obtained from circuit model (second method)		
U	T_L	TRF	U	T_L	TRF
V	Nm	Nm Nm^{-1}	V	Nm	Nm Nm^{-1}
24	0.5	0.73	24	0.5	0.091
24	0.2	0.802	24	0.2	0.238
12	0.5	0.698	12	0.5	0.276
12	0.2	0.696	12	0.2	0.164

Figure 3 compares the optimal rotor shapes obtained.

The analysis presented was carried out under nominal value of supply voltage. In a second stage the optimal motors were analyzed in dynamic conditions. The circuit analysis was performed where the load torque and the supply voltage were varied in order to compare the torque pulsations produced by both motor versions. The results of the analysis are summarized in Table 1. It is shown that the optimal model being the result of the second method on an optimization has always lower torque pulsations factor.

4 Conclusions

It was shown in the paper that using the rotor shape, found by the genetic optimization algorithm coupled with the magnetostatic field analysis under constant current, leads to the substantial reduction of torque pulsations. Much better results are obtained if the objective function takes into account dynamic torque variations due to current switching. Although useful, such an algorithm requires plenty of time to reach solution. For that algorithm to be effective the software environment for distributed computing must be used.

In further investigations the Authors will try to enable accounting for material nonlinearity in the optimization method based on instantaneous torque waveforms (second method). It will be made possible by increasing the number of distributed workstations.

Acknowledgements This work was realized under Grant No. KBN N510 011 31/0782 of Polish Ministry of Scientific Research and Information Technology – the State Committee for Scientific Research.

References

1. Miller T.J.E., Switched Reluctance Motors and Their Control, Oxford University Press, 1993.
2. Takashi K., Nobuyuki M., Yo-Ichi T., Hideo D., Some considerations on torque ripple suppression in reluctance motors, Electr. Eng. Jpn, 130(1): 118–128, 2000.
3. Meeker D., Femm Version 4.1, available at: http://www.foster-miller.net.
4. Witkowski A., Wróbel K., Implementation of a distributed computing environment for massive computations using the finite elements method, In Proceedings of Conference on Computers Applications in Electrical Engineering, ZKwE 2006, Poznan, str. 129–130 (in Polish).
5. Condor Project, available at: http://www.cs.wisc.edu/condor.

Numerical Model of Optimization Lead-Acid Accumulator Grids

Eva Kroutilova, I. Behunek, and Pavel Fiala

Abstract This article deals with chemical processes during charging and discharging of the lead-acid accumulator. There are presented one shape of electrode grids in this work but we researched more variants. We prepared numerical models based on combined finite element method (FEM) and finite volume method (FVM) of those variants and computed current density distribution on the surface of electrodes. The model joins magnetic, electric and current field, flow field and chemical nonlinear ion model. Results were obtained by means of FEM/FVM as a main application in ANSYS software.

1 Introduction

The lead-acid battery uses lead dioxide as the active material of the positive electrode and metallic lead [1], in a high-surface-area porous structure, as the negative active material. Typically, a charged positive electrode contains both α-PbO_2 (orthorhombic) and β-PbO_2 (tetragonal). The equilibrium potential of the α-PbO2 is more positive than that of β-PbO_2 by 0.01 V The α form also has a larger, more compact crystal morphology which is less active electrochemically and slightly lower in capacity per unit weight; it does, however, promote longer cycle life. Neither of the two forms is fully stoichiometric. Their composition can be represented by PbO_x, with x varying between 1.85 and 2.05. The introduction of antimony, even at low concentrations, in the preparation or cycling of these species leads to a considerable increase in their performance. The preparation of the active material precursor consists of a series of mixing and curing operations using leady lead oxide ($PbO + Pb$), sulfuric acid, and water. The ratios of the reactants

E. Kroutilova, I. Behunek, and P. Fiala
Department of Theoretical and Experimental Electrical Engineering, Brno University of Technology, Kolejni 2906/4, 612 00 Brno, Czech Republic
kadleca@feec.vutbr.cz, behunek@feec.vutbr.cz, fialap@feec.vutbr.cz

E. Kroutilova et al.: *Numerical Model of Optimization Lead-Acid Accumulator Grids*, Studies in Computational Intelligence (SCI) **119**, 223–230 (2008)
www.springerlink.com

and curing conditions (temperature, humidity, and time) affect the development of crystallinity and pore structure. The cured plate consists of lead sulfate, lead oxide, and some residual lead ($<5\%$). The positive active material, which is formed electrochemically from the cured plate, is a major factor influencing the performance and life of the lead-acid battery. In general the negative, or lead, electrode controls cold-temperature performance (such as engine starting). The electrolyte is a sulfuric acid solution, typically about 1.28 specific gravity or 37% acid by weight in a fully charged condition. On the grids there are following reactions.

Negative electrode

$$\mathrm{Pb} \underset{\text{charging}}{\overset{\text{discharging}}{\Leftrightarrow}} \mathrm{Pb}^{2+} + 2\mathrm{e}^-, \tag{1}$$

$$\mathrm{Pb}^{2+} + \mathrm{SO_2}^{2-} \underset{\text{charging}}{\overset{\text{discharging}}{\Leftrightarrow}} \mathrm{PbSO_4}, \tag{2}$$

Positive electrode

$$\mathrm{PbO_2} + 4\mathrm{H}^+ + 2\mathrm{e}^- \underset{\text{charging}}{\overset{\text{discharging}}{\Leftrightarrow}} \mathrm{Pb}^{2+} + 2\mathrm{H_2O}, \tag{3}$$

$$\mathrm{Pb}^{2+} + \mathrm{SO_4}^{2-} \underset{\text{charging}}{\overset{\text{discharging}}{\Leftrightarrow}} \mathrm{PbSO_4}, \tag{4}$$

Overall reactions

$$\mathrm{Pb} + \mathrm{PbO_2} + 2\mathrm{H_2SO_4} \underset{\text{charging}}{\overset{\text{discharging}}{\Leftrightarrow}} 2\mathrm{PbSO_4} + 2\mathrm{H_2O}. \tag{5}$$

2 Mathematical Model

Electromagnetic part is derived from reduced Maxwell equations

$$\operatorname{rot} \boldsymbol{H} = 0, \tag{6}$$

$$\operatorname{div} \boldsymbol{B} = 0, \tag{7}$$

where $\boldsymbol{H}$ is the vector of magnetic field intensity, $\boldsymbol{B}$ is magnetic field induction, $\boldsymbol{J}$ is vector of current density.

$$\operatorname{rot} \boldsymbol{E} = 0, \tag{8}$$

$$\operatorname{div} \boldsymbol{J} = 0, \tag{9}$$

where $\boldsymbol{E}$ is the vector of electric field intensity. Vector functions of electric, magnetic field are expressed by means of a scalar potentials ϕ_e, ϕ_m, Final current density from (9) $\boldsymbol{J}$ is influenced by velocity $\boldsymbol{v}$ of the flowing ion solution and outer magnetic field

$$\boldsymbol{J} = \gamma(\boldsymbol{E} + \boldsymbol{v} \times \boldsymbol{B}). \tag{10}$$

If electrodes E_1 and E_2 have different electrical potentials then the current density $\boldsymbol{J}$ is created in the Ω area according to (10) and the current I_L flows in the ion solution

$$I_L = \iint\limits_{S_e} \boldsymbol{J} \cdot d\,\boldsymbol{S} = \iint\limits_{S_e} \gamma(\boldsymbol{E} + \boldsymbol{v} \times \boldsymbol{B}) \cdot d\,\boldsymbol{S}, \tag{11}$$

where $\boldsymbol{S}_e$ is a directed area of electrodes E_1 and E_2 into space Ω. In (11) there is the electric field intensity $\boldsymbol{E}$ for ion solution much smaller than product of $\boldsymbol{v} \times \boldsymbol{B}$, so we ignore the influence of electric field intensity. The specific force $\boldsymbol{f}$ affects the moving charge q and the force $\boldsymbol{F}$ in whole Ω area is

$$\boldsymbol{F} = \iiint\limits_{\Omega} \boldsymbol{J} \times \boldsymbol{B} dV. \tag{12}$$

We obtain voltage between electrodes E_1, E_2 from

$$U_L = \int\limits_{E_1}^{E_2} \frac{\boldsymbol{F}}{q} \cdot d\ell. \tag{13}$$

where the electric field intensity is derived from the force $\boldsymbol{F}$ which affects a charge q. After modification voltage on electrodes is

$$U_L = \iiint\limits_{\Omega} \left(\frac{\boldsymbol{J}(\boldsymbol{v})}{I_L} \times \boldsymbol{B} \right) \cdot (\boldsymbol{v})\, dV. \tag{14}$$

The current density $\boldsymbol{J}(\boldsymbol{v})$ depends on immediate ion velocity between E_1 and E_2. *The magnetic field*, which is expressed in (14) by the induction B, we gain from Biot–Savart law by means of different scalar magnetic potentials (DSP)

$$T = \frac{1}{4\pi} \int\limits_{\Omega} \frac{\boldsymbol{J}_c \times \boldsymbol{R}}{|\boldsymbol{R}|^3} dV, \tag{15}$$

where $\boldsymbol{R}$ is a distance between a point where we look for the magnetic field intensity $\boldsymbol{T}$ and a point where magnetic field source is with the current density $\boldsymbol{J}_c$. We can write the magnetic field intensity $\boldsymbol{H}$ in the area as

$$\boldsymbol{H} = \boldsymbol{T} - \text{grad}\ \phi_m, \tag{16}$$

where $\boldsymbol{T}$ is the previous or estimated magnetic field intensity. Boundary conditions are

$$\boldsymbol{n} \cdot \mu_0 \mu_r (\boldsymbol{T} - \text{grad}\ \phi_m) = 0 \text{ on the boundary } \Gamma_{Pb-0}, \tag{17}$$

where $\boldsymbol{n}$ is the normal vector. $\Gamma_{\text{Pb-0}}$ is the interface between areas Ω_{Pb} and Ω. The continuity of tangential components of magnetic field intensity on the area interface with a ferromagnetic material is

$$\boldsymbol{n} \times (\boldsymbol{T} - \text{grad}\ \phi_m) = 0 \text{ on the boundary } \Gamma_{\text{Pb}-0}. \tag{18}$$

By the help of relations (6), (7) we get

$$div\mu_0\mu_r\boldsymbol{T}-div\mu_0\mu_r grad\ \phi_{\mathrm{m}}=0. \tag{19}$$

We get discretization of (19) by means of an approximation of the scalar magnetic potential

$$\phi_m=\sum_{k=1}^{N_\phi}\varphi_{mk}W_k\left(x,y,z\right),\quad \forall\left(x,y,z\right)\subset\Omega, \tag{20}$$

where φ_m is the nodal value of the scalar magnetic potential, W is base function, N_φ is number of mesh nodes. We obtain semi-discrete solution by means of the approximation (20) in the relation (19) and Gallerkin method

$$\sum_{j=1}^{N_\phi}-\int_\Omega \mu\ t_j\cdot gradW_i+\mu\ grad\ \varphi_{mj}\cdot gradW_i d\Omega=0, i=1,\ldots,N_\varphi, \tag{21}$$

where $\boldsymbol{t}_{\mathrm{j}}$ is the nodal value of known magnetic field intensity. Equations (20) is possible to write briefly

$$-\left[k_{Tij}\right]+\left[k_{ij}\right]\{\varphi\}=0\ i,\ j=1,\ldots,N_\varphi. \tag{22}$$

Coefficients for (22) are written as

$$\begin{aligned} k_{Tij}^{em}&=-\int_{\Omega^e}\mu^e t_j\cdot grad\ W_i d\Omega\ i,\ j=1,\ldots,N_e,\\ k_{ij}^{em}&=-\int_{\Omega^e}\mu^e\ grad\ \varphi_j\cdot grad\ W_i\ d\Omega \end{aligned} \tag{23}$$

where Ω^{e} is the area of the selected element of mesh, μ^{e} is the material permeability of the selected element, N_{e} is the number of elements of mesh. The equation system changes into relation

$$-\left[k_{Tij}^{em}\right]+\left[k_{ij}^{em}\right]\{\varphi\}=0\ e=1,\ldots,N_e. \tag{24}$$

We can solve the equation system (24) by means of standard algorithms. The solution by means of DSP consists of two parts. Firstly, we express the distribution of magnetic field intensity $\boldsymbol{T}$ from current sources according to (16) with the respect to boundary conditions (17) and (18) in the area Ω_{Pb}. Secondly, we have to find out the solution of the magnetic intensity $\boldsymbol{H}$ distribution according to (16) from the previous step. The *model of electrical or current* field is formulated from previous equations

$$\gamma\, div\ grad\ \phi_{\mathrm{e}}=0. \tag{25}$$

On the interface there are conditions

$$\boldsymbol{n}\cdot\gamma(grad\ \phi_e)=0\ \text{on the boundary}\ \Gamma_{\mathrm{E-k}}, \tag{26}$$

where $\boldsymbol{n}$ in the normal vector to the surface of the electrode. $\Gamma_{\mathrm{E-k}}$ is the interface between liquid and an electrode. The continuity of tangential components of electrical field intensity on the interface is

$$\boldsymbol{n} \times (\mathrm{grad}\ \phi_e) = 0 \text{ on the boundary } \Gamma_{\mathrm{E-k}}. \tag{27}$$

We can do the approximation of the scalar electric potential in the similarly way like in the relation (20) and by means of (20), (25) and Gallerkin method we get semi-discrete solution

$$\sum_{j=1}^{N_\phi} - \int_\Omega \gamma\ \mathrm{grad}\ \varphi_{ej} \cdot \mathrm{grad} W_i d\Omega = 0,\ i = 1, \ldots, N_\varphi, \tag{28}$$

where φ_{ej} is the nodal value of the scalar electric potential. We can rewrite the equation system (28) by the help of

$$\left[k_{ij}^J\right]\{\varphi\} = 0\ i,\ j = 1, \ldots, N_\varphi. \tag{29}$$

$$k_{ij}^{Je} = -\int_{\Omega^e} \gamma^e\ \mathrm{grad}\varphi_{ej} \cdot \mathrm{grad} W_i\ d\Omega, \tag{30}$$

where Ω^{e} is the area of the selected element of mesh, γ^{e} is the specific conductivity of liquid in the static state of the selected element, N_{e} is the number of elements. The relation for a voltage drop during discharging $U(t) = U_0 - \Delta u$ is

$$\Delta u = \sqrt{Z\left(\int_\Omega \frac{\boldsymbol{J}_e^- q_e^- \cdot \boldsymbol{v}_e^-}{\Delta V_e \gamma_e}\right) d\Omega + Z\int_\Omega (\boldsymbol{J}_e \times \boldsymbol{B}_e) \cdot \boldsymbol{v}_e^- d\Omega + Z\left(\int_\Omega \frac{\boldsymbol{J}_e^+ q_e^+ \cdot \boldsymbol{v}_e^+}{\Delta V_e \gamma_e}\right) d\Omega + Z\int_\Omega (\boldsymbol{J}_e \times \boldsymbol{B}_e) \cdot \boldsymbol{v}_e^+ d\Omega}, \tag{31}$$

where

$$v_e^+ = \frac{E_e \gamma}{F_c \Delta V_e \sum_{k=1}^{N_{ion+}} c_k^+ N_{\mathrm{k}}^{+ion}}, \quad v_e^- = \frac{E_e \gamma}{F_c \Delta V_e \sum_{k=1}^{N_{ion-}} c_k^- N_{\mathrm{k}}^{-ion}},$$

$$q_e^+ = F_c \Delta V_e \sum_{k=1}^{N_{ion+}} c_k^+ N_{\mathrm{k}}^{+ion}, \quad q_e^- = F_c \Delta V_e \sum_{k=1}^{N_{ion-}} c_k^- N_{\mathrm{k}}^{-ion}, \tag{32}$$

where F_{c} is the Faraday constant, $F_{\mathrm{c}} = 96{,}484\,\mathrm{C\ mol^{-1}}$, $\boldsymbol{E}_{\mathrm{e}}$ the electric field intensity in direction of ions motion in an element of mesh, c^+ the positive ions concentration, c^- the negative ions concentration, ΔV_{e} is the element volume, $N_{\mathrm{k}}^{+\mathrm{ion}}$ is the integer multiple of electron charge for specific positive ion, $N_{\mathrm{k}}^{-\mathrm{ion}}$ is the integer multiple of electron charge for specific negative ion, q_{e}^- is the whole charge of negative ions in one element, q_{e}^+ is the total charge of positive ions in one element, $N_{\mathrm{ion+}}$ is the number of different positive charge carriers (elements, compounds), $N_{\mathrm{ion-}}$ is the number of negative charge carriers. There are concentrations c_{k} in fluid

Positive ions		*Negative ions*
H^+	$\ldots 9{,}9193 \times 10^2\,\mathrm{mol\,m^{-3}}$	$SO_4^{2-} \ldots 4{,}9597 \times 10^3\,\mathrm{mol\,m^{-3}}$
Pb^{4+}	$\ldots 2{,}2028 \times 10^4\, mol\, m^{-3}$	$OH^- \ldots 8{,}8110 \times 10^4\, mol\, m^{-3}$

3 Numerical Solution FEM/FVM

The numerical model was prepared by means of ANSYS tools [2–6] and the main FEM/FVM solution was solved with APDL program over ANSYS system. In Fig. 1 we can see the geometrical model and in Fig. 2 the distribution of current density

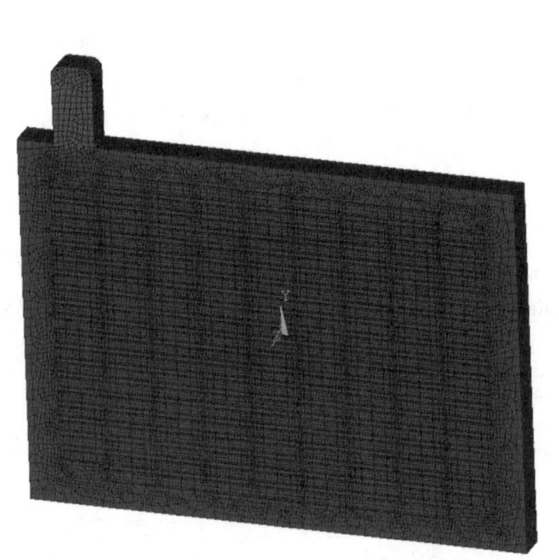

Fig. 1 Geometry of the numerical model

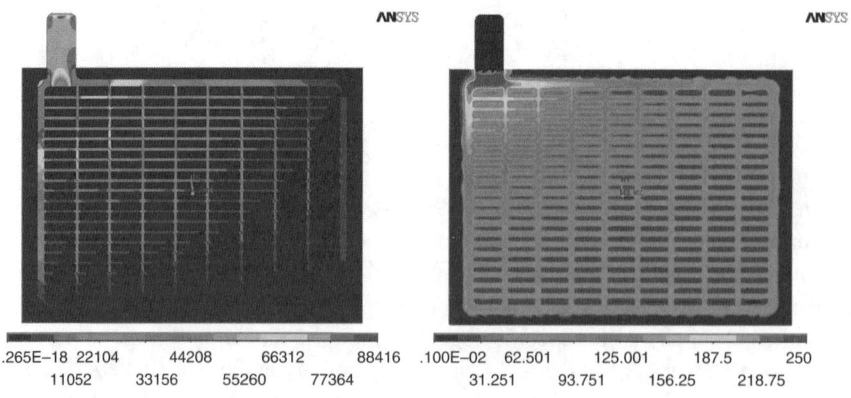

Fig. 2 Current density distribution on the grid surface (*left*) and in the electrolyte (*right*), charged state

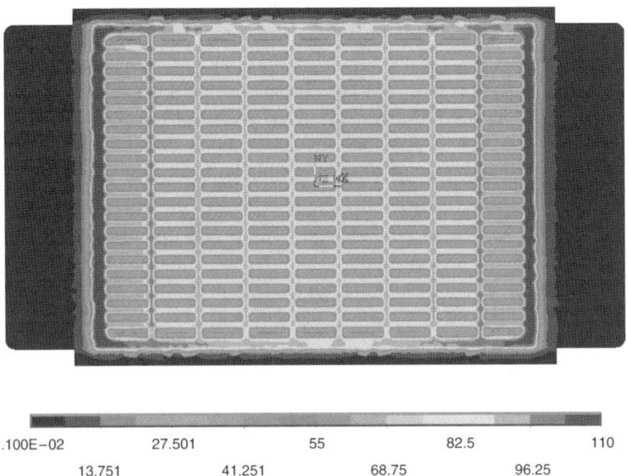

Fig. 3 Current density distribution in the electrolyte, charged state

module J on the grid surface and in the electrolyte between the positive and the negative electrodes. The computation was carried out under the following conditions: initial voltage 2.3 V, minimal voltage of discharged accumulator, supposed discharging time t_s, and time period between the steps of computation. The best solution of grid design (optimal) we can see in Fig. 3. The optimal solution is presented to the intent that the lead grid has regular contour.

4 Conclusion

This work deals with chemical processes on the electrodes and in the electrolyte of the lead-acid accumulator. There is the mathematical and the numerical description in the article and the results of current density distribution in the electrodes area. Such analysis could be used for an optimization of grid geometry. The best shape is obtained if the module of current density is constant on the electrode surface.

Acknowledgements The research described in the paper were financially supported by FRVŠ by research plan No. MSM 0021630513 ELCOM, No. MSM 0021630516.

References

1. Salkind, A.J., Cannone, A.G., Trumbure, F.A., *Handbook of Batteries*. Boca Raton (Florida, USA): CRC Press, Inc., 2001.

2. *Ansys User's Manual*. Huston (USA): Svanson Analysys System, Inc., 2006.
3. Vojtek, T., Skoupil, T., Fiala, P., Bartušek, K. *Accuracy of Air Ion Field Measurement*, In Progress In Electromagnetics Research Symposium 2006 Proceedings. Progress In Electromagnetics Research Symposium. Boston, USA: The Electromagnetics Academy, 2006, ISBN 1-933077-08-05.
4. Fiala, P. *Model of Induction flowmeter DN100*. Research Report no. 2/01, Numerical Modeling and Optimisation Laboratory in Electromechanical Systems FECT VUT BRNO, 21.6.2001, Brno, Czech Republic.
5. Běhunek, I., Fiala, P., Kroutilová, E. *A Heat Acumulator Layer – Properties and Solution*, In Proceedings of the International Workshop ISEP – UTEE 3. – 6. 9. 2006 Paris. International Workshop ISEP – DTEE 3. – 6. 9.2006 Paris. L'Institut Supérieur d'Electronique de Paris, ISBN 80-214-3250-0.
6. Kroutilová, E., Běhunek, I., Fiala, P. *Numerical Model of Optimization of the Lead-Acid Accumulator Grids*. Progress In Electromagnetics, ISSN 1559-9450, 2007.

Parametric Models Based on Sensitivity Analysis for Passive Components

Gabriela Ciuprina, Daniel Ioan, Dragos Niculae, Jorge Fernández Villena, and Luis Miguel Silveira

Abstract Passive components with significant high frequency field effects have to be modeled taking into consideration full wave electromagnetic field equations. Such a field formulation with appropriate electromagnetic circuit element boundary conditions is numerically analyzed in the time domain with the finite integral technique, a sparse state-space representation of the component being obtained. The novelty of the presented approach is the use of model parameterization and the extraction of the model sensitivities needed by parametric model order reduction procedures. The paper investigates the validity of first order Taylor Series expansion with respect to the parameters as approximation for the extracted semi-state space models.

1 Introduction

An extremely important RF design activity is the post-layout verification step. It entails the accurate prediction of the behaviour of the chip after each design iteration, but before actual production. Trying to correct a design through trial and error of silicon implementation is far too costly and far too slow in view of the economics and decreased time to market. Therefore, it is crucial to have a state-of-the-art procedure for the design verification of future RFICs up to 60 GHz. Such procedure needs accurate models for passive components, able to describe all relevant electromagnetic field effects inside the devices. Moreover, in order to allow the connection of these models to the rest of the circuit, robust and efficient methods are needed for extraction of parameterized reduced order models able to take into account the

Gabriela Ciuprina, Daniel Ioan, and Dragos Niculae
"Politehnica" University, Spl Independentei 313, 060042 Bucharest, Romania
lmn@lmn.pub.ro

Jorge Fernández Villena and L. Miguel Silveira
INESC-ID/IST – Technical University of Lisbon, Rua Alves Redol, 9, 1000-029 Lisboa, Portugal
jorge@algos.inesc-id.pt, lms@algos.inesc-id.pt

G. Ciuprina et al.: *Parametric Models Based on Sensitivity Analysis for Passive Components*, Studies in Computational Intelligence (SCI) **119**, 231–239 (2008)
www.springerlink.com

variability induced by lithography or process variations as well as changing operating conditions. This is one of the issues addressed by the FP6/CHAMELEON-RF project [1].

Efficient methodologies to extract compact parametric models of passive components valid for parameter variations were developed based on the effective adjoint field technique (AFT) and adjoint circuit technique (ACT) [2, 3]. An alternative to the adjoint techniques is to compute the sensitivities of the semi-state space matrices directly from the discretized model, to use Taylor Series (TS) expansion w.r.t. the variable parameters, and to apply appropriate model order reduction procedures in the time-domain [4]. This paper investigates the validity of first order Taylor Series expansion with respect to the parameter as approximation for the extracted semi-state space models.

2 Parametric Full Wave Models with FIT

The electromagnetic field effects at high frequency are quantified by Maxwell equations of the electromagnetic field in full wave regime. Therefore, at the first level of approximation, the model of a passive device is defined by the EM field problem correctly formulated, i.e. with appropriate boundary and initial conditions. This EM field problem defines a consistent I/O system which has a unique response, described by the output signals, for any input signal applied as terminals excitation. The most appropriate formulation for passive devices with distributed parameters compatible with external electrical circuits is the electric circuit element (ECE) [5]. The next level of approximation in the modelling process is obtained by applying the finite integration technique (FIT) method to discretize the continuous model defined above [6]:

$$C\frac{dx}{dt} + Gx = Bu, \tag{1}$$

$$y = Lx, \tag{2}$$

where $\mathrm{x} = [u_e,\ u_m]^{\mathrm{T}}$ is the state space vector, consisting of electric voltages u_e defined on the electric grid and magnetic voltages u_m defined on the magnetic grid, and u is the vector of input quantities. When all terminals are current excited, the matrices C,G,B and L have the following structure:

$$C(p) = \begin{bmatrix} C_e(p) & 0 \\ 0 & G_m(p) \\ C_{e-Sl}(p) & 0 \\ C_{e-T}(p) & 0 \end{bmatrix}, \qquad G(p) = \begin{bmatrix} G_e(p) & -B'_{E-F} \\ B_{F-E} & 0 \\ G_{e-Sl}(p) & 0 \\ G_{e-T}(p) & 0 \end{bmatrix}, \tag{3}$$

$$B = \begin{bmatrix} 0 \\ 0 \\ 0 \\ I \end{bmatrix}, \qquad L = \begin{bmatrix} S_{T-G}\ 0 \end{bmatrix}, \tag{4}$$

The first group of equations is obtained by writing Ampere's law for all magnetic loops. Thus, C_e and G_e are diagonal matrices holding, respectively, the electric capacitances and conductances that correspond to the edges of the electric grid, and B'_{E-F} is a topological matrix, namely the edges to faces incidence matrix defined on the electric grid, having four nonzero entries on each column.

The second group of equations is obtained from Faraday's law for all electric loops. Here G_m is a diagonal matrix, holding the inverse of the magnetic reluctances that correspond to the edges of the magnetic grid and B_{F-E} is a topological matrix, namely the faces to edges incidence matrix defined on the electric grid, having four nonzero entries on each row.

The third group of equations represents the current conservation for nodes on the boundary that are not on terminals, C_{e-Sl} and G_{e-Sl} being capacitances and conductances of the edges that are connected to these nodes.

The last group of equations represents the current conservation law for terminals, C_{e-T} and G_{e-T} being, respectively, the capacitances and conductances of the edges that touch the terminals.

As to the output equation, since the voltage of a terminal is an algebraic sum of electric voltages (which are included among the dofs), then the matrix L is topological, including a selection matrix S_{T-G}, showing those paths that link each terminal to the reference terminal.

The possible variation of parameters does not affect all entries of the matrices. The affected blocks are marked with (p) in the expressions above. In the case of voltage-excited terminals, the corresponding output quantities (the current through the terminals) are included in the vector of dofs, and the output equation is merely a selection of quantities from the dofs.

3 First Order Sensitivity of State Space Matrices

The simplest way to analyse the parameter variability is to compute first order sensitivities. These are the derivative of device characteristics with respect to the design parameters. Let us consider $p_1, p_2, \ldots, p_n$ n real numbers which represent the design parameters, either geometric data or material "constants". Any real quantity used in the correct formulation of the field problem may be considered as a design parameter. Let F be one of the device characteristics, represented by a real number which depends on the design parameters $(p_1, p_2, \ldots, p_n)$. F may be for instance the real or imaginary part of the device impedance or admittance at a given frequency. The parameter variability is completely described by the real function $F : \mathrm{R}^n \rightarrow \mathrm{R}$, defined over the parameter space Σ part of R^n. Considering $\mathbf{p}_0 = (p_{01}, p_{02}, \ldots, p_{0n})$ the nominal values of the design parameters, if F is smooth enough, its truncated Taylor series is the best polynomial approximation in the vicinity of the expansion point $\mathbf{p}_0$. The first order truncation of the Taylor series expansion is the affine function:

$$F_1(p_1, p_2, \ldots, p_n) = F(p_{01}, p_{02}, \ldots, p_{0n}) + S_{p1}(p_1 - p_{01}) \\ + S_{p2}(p_2 - p_{02}) + \ldots + S_{pn}(p_n - p_{0n}), \quad (5)$$

where $S_{pk} = \partial F / \partial p_k$ are first order sensitivities, defined as partial derivatives of device characteristics w.r.t. the design parameters, computed for the nominal values of parameters. This definition is valid not only for real characteristics, but also when F is a complex number, a vector or a matrix.

The first order sensitivities of the matrices defined in the previous section are essential for the analysis of the parameter variability in the time domain. Considering as design parameters the geometrical variables and material constants, it can be noted that the integer (topological) elements of these matrices do not depend on these parameters. Only the elements of the Hodge matrices C_e, G_e, G_m can be influenced by these design parameters. The derivatives of topological matrices are zero regardless the design parameter p_k, so the sensitivities of state representation matrices are:

$$\frac{\partial C(p_k)}{\partial p_k} = \begin{bmatrix} \frac{\partial C_e(p_k)}{\partial p_k} & 0 \\ 0 & \frac{\partial G_m(p)}{\partial p_k} \\ \frac{\partial C_{e-SI}(p_k)}{\partial p_k} & 0 \\ \frac{\partial C_{e-T}(p)}{\partial p_k} & 0 \end{bmatrix} \qquad \frac{\partial G(p_k)}{\partial p_k} = \begin{bmatrix} \frac{\partial G_e(p_k)}{\partial p_k} & 0 \\ 0 & 0 \\ \frac{\partial G_{e-SI}(p)}{\partial p_k} & 0 \\ \frac{\partial G_{e-T}(p_k)}{\partial p_k} & 0 \end{bmatrix} \quad (6)$$

The derivative w.r.t. to the parameter of the electric capacitance of edge number k:

$$C_{ek} = \left(\sum_{j=1}^{4} \varepsilon_j A_j \right) / l_k$$

is nonzero only if the parameter affects either the material of the four cells that include that branch, or if the parameter is a geometrical one, affecting a cell that touches the interface parameterized by means of that parameter. The computation of derivatives is straightforward. However, a sign factor has to be taken into account for cells that shrink (-1) and for cells that enlarge $(+1)$.

To validate the computation of the first order sensitivities of the matrices, the output quantity will be computed for a given frequency range, in three ways:

- Using simulation results given by FIT (a new discretized model is generated for each set of parameters)

$$y(p) = L(j\,\omega\, C(p) + G(p))^{-1}\, B\, u. \quad (7)$$

- Using the Taylor series for state space matrices:

$$C_{TS} = C_0 + \frac{\partial\, C}{\partial\, p}(p - p_0), \quad (8)$$

$$G_{TS} = G_0 + \frac{\partial\, G}{\partial\, p}(p - p_0), \quad (9)$$

$$y(p) = L\,(j\,\omega\,C_{TS} + G_{TS})^{-1}\;B\,u, \tag{10}$$

where $C_0 = C(p_0)$, $G_0 = G(p_0)$ are the matrices computed for the nominal values of the parameters

- Using the Taylor series for the output quantity

$$y(p) = y(p_0) + \frac{\partial\,y}{\partial\,p}\,(p - p_0)\,, \tag{11}$$

where the sensitivity of the output quantity can be computed from the sensitivities of the state space matrices as follows:

$$\frac{\partial\,y}{\partial\,p} = L\frac{\partial\,x}{\partial\,p}, \text{where} \tag{12}$$

$$\frac{\partial\,x}{\partial\,p} = -\,(j\,\omega\,C + G)^{-1}\left[\left(j\,\omega\frac{\partial\,C}{\partial\,p} + \frac{\partial\,G}{\partial\,p}\right)x\right], \tag{13}$$

$$x = (j\,\omega\,C + G)^{-1}\;B\,u. \tag{14}$$

4 Validity Range of First Order Taylor Series Expansion

Let us consider only one varying design parameter p. Then, the output quantity can be expressed as

$$y(p) = y(p_0) + \frac{\partial y}{\partial p}(p_0)\,(p - p_0) + \frac{\partial^2 y}{\partial p^2}(\xi)\,(p - p_0)^2. \tag{15}$$

If we denote by

$$S_p(y) = \frac{p_0}{y_0}\frac{\partial y}{\partial p}(p_0), \tag{16}$$

the relative first order sensitivity of the output quantity w.r.t. the parameter p, where $y_0 = y(p_0)$, then it follows that the relative error of the output quantity $\varepsilon_y = \frac{y(p) - y_0}{y_0}$ can be expressed as

$$\varepsilon_y = S_p(y)\varepsilon_p + \varepsilon_1, \tag{17}$$

where $\varepsilon_p = \frac{p - p_0}{p_0}$ is the relative variation of the parameter p, and

$$\varepsilon_1 = \frac{p_0^2}{2\,y_0}\frac{\partial^2 y}{\partial p^2}(\xi_1)\varepsilon_y^2 \tag{18}$$

depends on the second order derivative of the output quantity. Thus, to ensure a relative validity range of the first order approximation of the output quantity less a given threshold t_1, the absolute variation of the parameter must be less than $\sqrt{\frac{2 y_0 t_1}{D_2}}$, where D_2 is an upper limit of the second order derivative of the output quantity y with respect to the parameter p.

The validity range can be increased in some cases if the Taylor Series expansion is used for the quantity $1/y(p)$. In this case, to obtain the same validity range of the first order approximation for the "reversed" output quantity, the variation of the parameter has to be less than $\sqrt{\frac{2t_1}{y_0 D_2{}^r}}$, where D_2^r is an upper limit of the second order derivative of the reversed output quantity.

5 Numerical Example

This example refers to an L-shape parameterized conductor. The varying parameters are p_2 and r_2 (Figs. 1 and 2). Figure 3 shows the parameter impact on the answer at 3 GHz. First order TS approximation is accurate enough at this frequency, even for a quite large variation range of the parameter. Figure 4 represents the relative variation of the answer with respect to the relative variation of the parameter for the system reconstructed using TS, but also for the reduced order models obtained with the

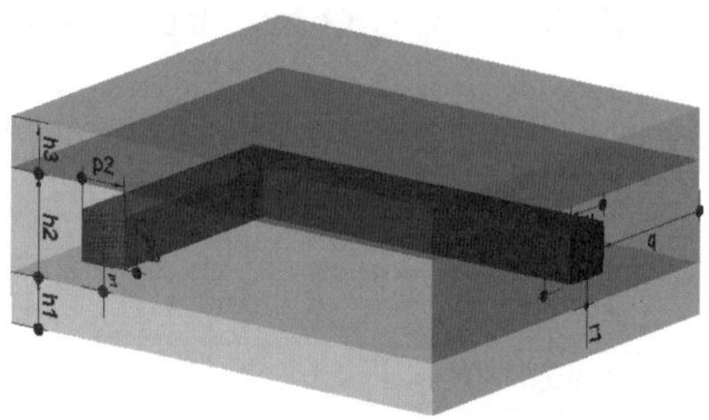

Fig. 1 L-shape parameterized conductor

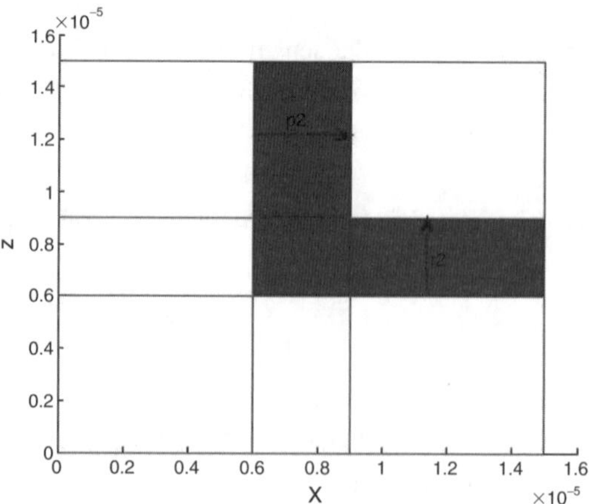

Fig. 2 Parameters that vary: p_2 and r_2

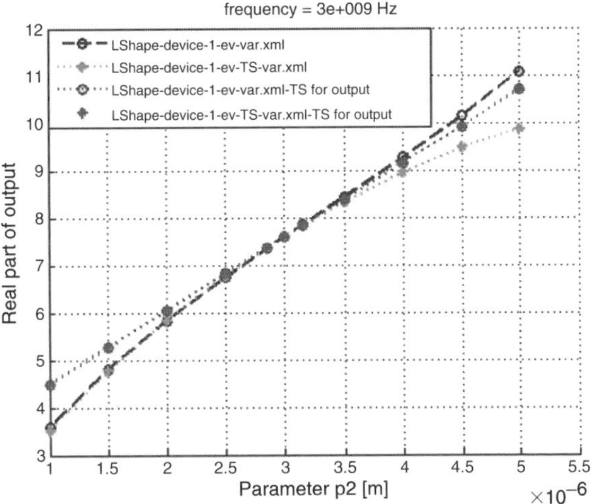

Fig. 3 Reconstruction of the answer at 3 GHz, from TS first order expansion

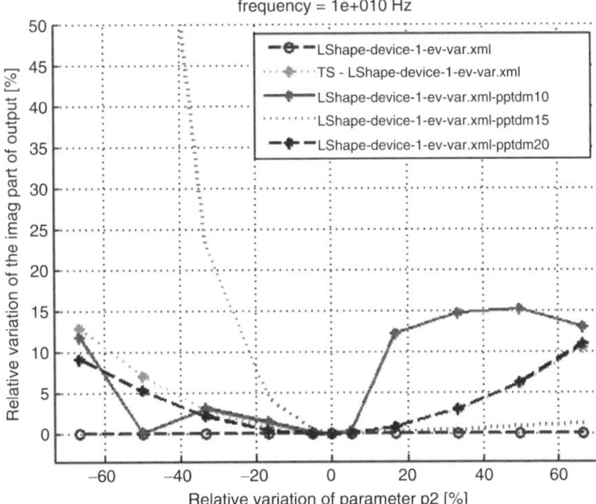

Fig. 4 Relative error of the answer vs. the parameter relative variation

PPTDM method [4] and different values for the reduced order. The reference values are the ones obtained using FIT for each sample individually. For an appropriate set of parameters of the PPTDM method, the accuracy of the parameterized reduced order model is up to the accuracy of the reconstruction using the TS expansion for the initial, nonreduced model.

Figures 5 and 6 illustrate how the validity range of the Taylor Series expansion can be increased by inverting the transfer function, which is equivalent to changing the terminals excitation type, from voltage to current excitation in this case. For example, with the parameter r_2 kept to its nominal value, to obtain a validity range of the output quantity less than 5%, in the case of voltage excitation the relative variation of the parameter p_2 has to be less than 8%, whereas in the case of current excitation, the relative variation of the same parameter is less than 40%.

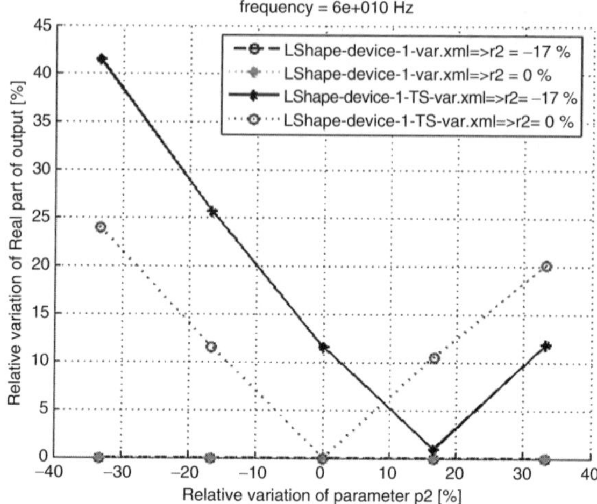

Fig. 5 Voltage excitation – output is admittance

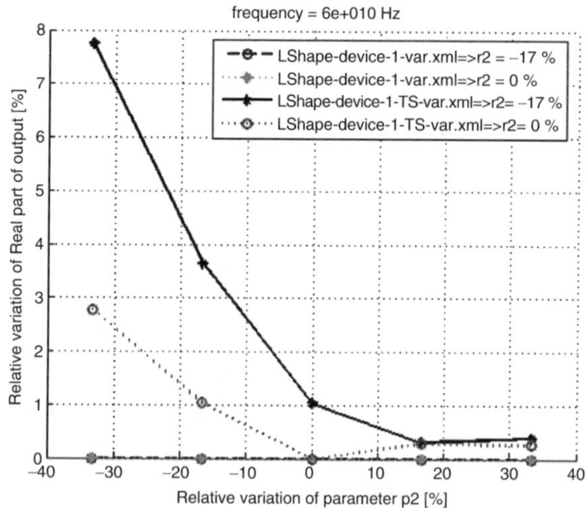

Fig. 6 Current excitation – output is impedance

6 Conclusions

The model parameterization and the extraction of state space matrices sensitivities can be easily included in a FIT discretization scheme for electromagnetic circuit element formulation. The first order sensitivities of the state space matrices can be assembled simultaneously with the matrices themselves, by using a direct differentiation technique. First order Taylor series approximation can be successfully used, the validity range of the expansion depending on the second order derivative of the output quantity. If the model allows several excitations, the one that ensures the largest validity range has to be chosen, and thus the use of higher order terms in the Taylor expansion can be avoided.

References

1. CHAMELEON-RF website: http://www.chameleon-rf.org
2. Ioan, D., I. Munteanu, G. Ciuprina, Adjoint field technique applied in optimal design of a nonlinear inductor, IEEE Transactions on Magnetics, Vol. 34, No. 5, pp. 2849–2852, 1998.
3. Nikolova, N.K., J.W. Bandler, M.H. Bakr, Adjoint techniques for sensitivity analysis in high-frequency structure, IEEE Transactions on MTT, Vol. 52, No. 1, pp. 403–413, 2004.
4. Gunupudi, P.K., R. Khazaka, M.S. Nakhla, T. Smy, D. Celo, Passive parameterized time-domain macromodels for high-speed transmission line networks, IEEE Transactions on Microwave Theory and Techniques, Vol. 51, No. 12, pp. 2347–2354, 2003.
5. Ioan, D., I. Munteanu, Missing link rediscovered: The electromagnetic circuit element concept, JSAEM Studies in Applied Electromagnetics and Mechanics, Vol. 8, pp. 302–320, Oct. 1999.
6. Weiland, T., A discretization method for the solution of Maxwell's equations for 6 component fields, AEÜ, Electronics and Communication, Vol. 31, pp. 116–120, 1977.

Optimisation of a Drive System and Its Epicyclic Gear Set

Nicolas Bellegarde, Philippe Dessante, Pierre Vidal, and Jean-Claude Vannier

Abstract This paper describes the design of a drive consisting of a DC motor, a speed reducer, a lead screw transformation system, a power converter and its associated DC source. The objective is to reduce the mass of the system. Indeed, the volume and weight optimisation of an electrical drive is an important issue for embedded applications. Here, we present an analytical model of the system in a specific application and afterwards an optimisation of the motor and speed reducer main dimensions and the battery voltage in order to reduce the weight.

1 Introduction

The system studied in this paper is a linear electrical drive system realized with a Ni-MH battery bank, a DC/DC converter, a DC motor, a speed reducer and a lead-screw device. The aim of the system is to move a load along a linear displacement. Regarding the load, we can define mainly two specifications. Firstly, it has to apply a rather high static force to overcome some static friction force. This has to be done at constant speed or at standstill. Secondly, it has to be driven from one point to another in a given time. This specification implies a dynamic force, a given acceleration and a maximum speed depending on the kind of displacement.

In order to optimise the weight of the system, and mainly the battery, DC motor and speed reducer weights, geometrical and physical relations have to be written for each component. These relations are then linked with the others to make a global optimisation of the system. The constraints are based upon the load specifications. The mathematical optimisation is performed with the help of various numerical methods like genetic algorithm, random search, differential evolution and Nelder Mead.

Nicolas Bellegarde, Philippe Dessante, Pierre Vidal, and Jean-Claude Vannier
Département Énergie – Supélec, 1 rue Joliot-Curie 91192 Gif-sur-Yvette, France
nicolas.bellegarde@supelec.fr

N. Bellegarde et al.: *Optimisation of a Drive System and Its Epicyclic Gear Set*, Studies in Computational Intelligence (SCI) **119**, 241–248 (2008)
www.springerlink.com

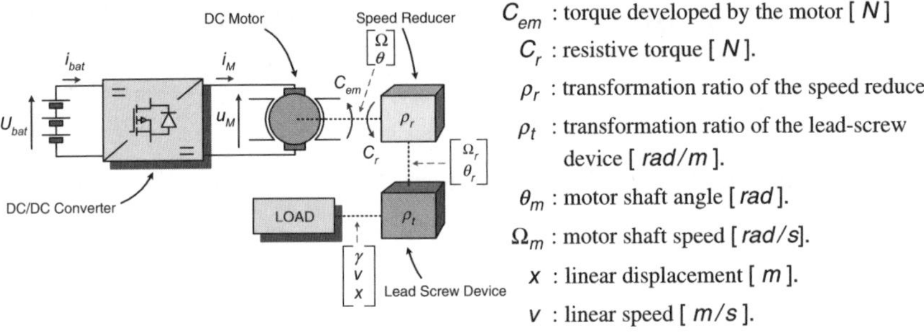

Fig. 1 Power conversion system

2 System Model

2.1 Power Conversion System

A general representation of the system is given in Fig. 1.

Concerning the mechanical part, the lead-screw is represented by its transformation ratio deduced from the screw pitch while the speed reduction system introduces a speed transformation ratio.

The relation between the motor shaft angle θ_m and the linear displacement x can be written as follows:

$$\theta_m = \rho_r \rho_t \cdot x = \rho \cdot x, \tag{1}$$

Where ρ is the global transformation ratio.

2.2 Speed Reducer Volume

A representation of the speed reducer is given in Fig. 2.

The speed reducer is constitued by two epicyclic gears (characterized by R_1, R_2) and one cylindrical gear with straght outer teeth (characterized by R_3, R_4). We can express the volume V_r of this speed reducer as follows:

$$V_r = \pi \cdot b \cdot \left[[R_1 + 2 \cdot R_2]^2 \cdot [\alpha_1 + \alpha_2] + R_3{}^2 \cdot \alpha_3 + R_4{}^2 \cdot \alpha_4 \right]. \tag{2}$$

2.3 Load Specifications

Here, all load specifications are expressed on the motor shaft. In the considered application, the motor has to generate two sorts of torques, imposed by the load. A static torque which is necessary to reach the breakaway force on the load just before

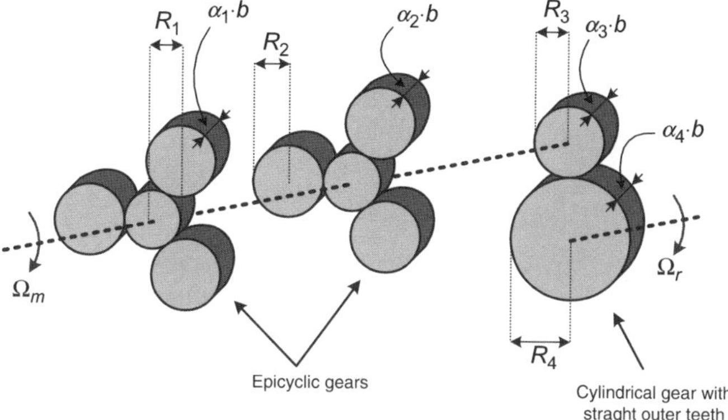

Fig. 2 Speed reducer structure

it starts to move or to maintain the speed at a constant value. The static torque C_{em_s} is given by

$$C_{em_s} = K_1 \cdot \frac{R_3}{\left[1 + \frac{R_2}{R_1}\right]^2 \cdot R_4}, \tag{3}$$

Where K_1 is a constant depending on the speed, the lead-screw transformation ratio, the different resistive forces due to the load.

The motor must also generate a dynamic torque which is required when the different resistive forces are at their maximum values.

The dynamic torque C_{em_d} can be expressed as follows:

$$C_{em_d} = K_2 \cdot \frac{R_3}{\left[1 + \frac{R_2}{R_1}\right]^2 \cdot R_4}, \tag{4}$$

Where K_2 is a constant depending on the same parameters as K_1.

An other torque can be also considered. Indeed, the RMS torque $C_{em_{RMS}}$ is interesting because it is directly linked to the copper losses in the motor. Its expression is given by

$$C_{em_{RMS}}{}^2 = K_3 \cdot J^2 \cdot \frac{\left[1 + \frac{R_2}{R_1}\right]^4 \cdot R_4{}^2}{R_3{}^2} + K_4 \cdot \frac{R_3{}^2}{\left[1 + \frac{R_2}{R_1}\right]^4 \cdot R_4{}^2} + K_5 \cdot J \tag{5}$$

J is the total inertia of the system. The constants K_3, K_4, K_5 are functions of the average and the rms values of the load speed and acceleration, the lead-screw transformation ratio and the average and the rms values of the different resistive forces.

2.4 Motor Specifications

Among the limits concerning the motor, there is a maximal rotor speed Ω_M, function of the maximal linear speed v_M, defined by

$$\Omega_M = \rho_r \rho_t \cdot v_M. \tag{6}$$

The maximum peak power P_M given by the motor to the load has also to be considered, where K_6 is a constant depending on the different resistive forces:

$$P_M = K_6 \cdot v_M. \tag{7}$$

For the motor's design, it is possible to define the main dimensions by the peak torque C_{em_M}, the nominal torque C_{em_n} and the rotor inertia J_m as seen in previous papers [1–6]:

$$\begin{cases} C_{em_n} = 2\pi \cdot A_L \cdot B \cdot \gamma_p \cdot R^2 \cdot L \\ C_{em_M} = 4 \cdot p \cdot H_M \cdot B \cdot R \cdot L \cdot E \\ J_m = \frac{\pi}{2} \cdot \mu_m \cdot R^4 \cdot L \end{cases}. \tag{8}$$

These three relationships introduce three main dimensions parameters for the design: the rotor radius R, the rotor length L and the permanent magnet thickness E. The remaining parameters are constants and have the following significations:

- $p = 1$: pole's number
- $H_M = 80\,kA/m$: magnet's peak magnetic field
- $B = 0,4\,T$: air gap flux density
- $A_L = 7850\,A/m$: rms current linear density
- $\gamma_p = 0,75$: pole's overlapping factor
- $\mu_m = 5070\,kg/m^3$: motor mass density.

3 System Optimisation

The established relationships are used to define the constraints during the optimisation procedure. Two types of constraints are considered: the physical constraints which permit to ensure that the motor can supply load requirements and the geometrical constraints which permit to define a feasible motor and speed reducer (see [7] for futher details).

Concerning the physical constraints, the motor peak torque has to be greater than the static and dynamic torques. The nominal torque has also to be greater than the required rms torque.

$$\begin{cases} C_{em_M} > C_{em_d} \\ C_{em_M} > C_{em_s} \\ C_{em_n} > C_{em_{RMS}} \end{cases}. \tag{9}$$

The maximum power supplied by the battery bank has to be greater than the maximum power consumption.

$$[U_{bat} \cdot i_{bat}]_M > P_M. \tag{10}$$

The maximum speed required by the load has to be kept lower than the nominal speed of the motor Ω_n.

$$\Omega_n > \rho_r \rho_t \cdot v_M. \tag{11}$$

Concerning the geometrical constraints, the following will be used in order to have feasible motor and speed reducer:

$$\begin{cases} R \geq 4mm \\ R \leq L \leq 5 \cdot R \\ R \geq E \geq 3mm \\ R_2 \geq R_1 \geq 4mm \\ R_4 \geq R_3 \geq 6mm \end{cases} \tag{12}$$

Depending on the application, different cost functions can be minimized. Here, we present an optimisation where the weight of the system is the cost function. We sum the motor, speed reducer and battery bank masses:

$$M = M_{bat} + M_m + M_r. \tag{13}$$

The motor and speed reducer masses are defined as follows:

$$M_m = \mu_m \cdot \pi \cdot R^2 \cdot L. \tag{14}$$

$$M_r = \pi \cdot \mu_r \cdot b \cdot \left[[R_1 + 2 \cdot R_2]^2 \cdot [\alpha_1 + \alpha_2] + {R_3}^2 \cdot \alpha_3 + {R_4}^2 \cdot \alpha_4 \right]. \tag{15}$$

The battery bank mass is a function of the voltage if we take a constant capacity.

$$M_{bat} = f(U_{bat}). \tag{16}$$

The total weight can be expressed by the following expression:

$$M = f(U_{bat}) + \mu_m \cdot \pi \cdot R^2 \cdot L + \pi \cdot \mu_r \cdot b \cdot \left[[R_1 + 2 \cdot R_2]^2 \cdot [\alpha_1 + \alpha_2] + {R_3}^2 \cdot \alpha_3 + {R_4}^2 \cdot \alpha_4 \right]. \tag{17}$$

The seven optimisation parameters are: $R, L, E, R_1, R_2, R_3, R_4$.

4 Results

We present here the results concerning the definition of the battery, the speed reducer and the motor whose characteristics are optimised for a given load. In this example, the load characteristics are presented in the Table 1.

Table 1 Load characteristics

$\gamma_{eff} = 30,49mm/s^2$	$v_{eff} = 12,42mm/s$
$\gamma_M = 60mm/s^2$	$v_M = 15mm/s$
$K_1 = 0,23N$	$K_2 = 4,47N$
$K_3 = 23496N^2/kg^2 \cdot m^4$	$K_4 = 3,46N^2$
$K_5 = -140,4N^2/kg \cdot m^2$	$K_6 = 14995W \cdot s/m$

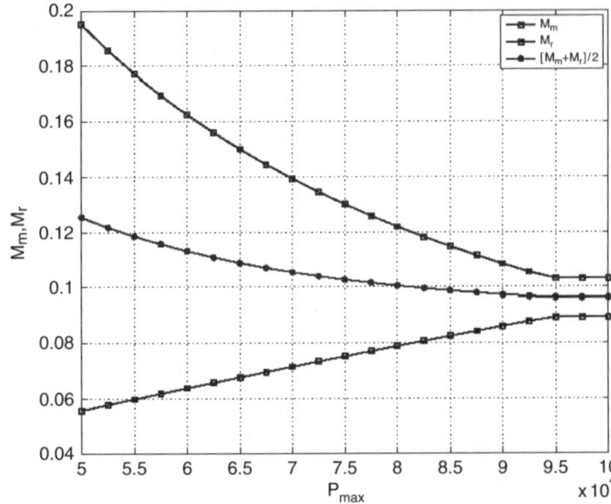

Fig. 3 Motor and speed reducer weights vs.ρ_{max}

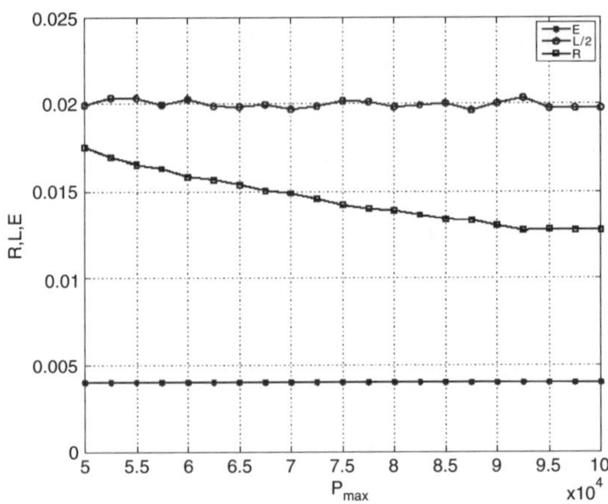

Fig. 4 Motor dimensions vs. ρ_{max}

Before optimisation, the battery voltage is $14V$ and the weight M is equal to $1180g$. The optimisation procedures uses the constraints (9–12) and searches a set of values for $R, L, E, R_1, R_2, R_3, R_4$ and U_{bat} which minimises the total mass of the system in equation (17).

The optimisation is made with Mathematica and verfied with Matlab and an home made genetic algorithm.

Figures 3–5 show the results obtained after the optimisation procedure. The battery voltage is equal to $12V$ and its weight is $640g$.

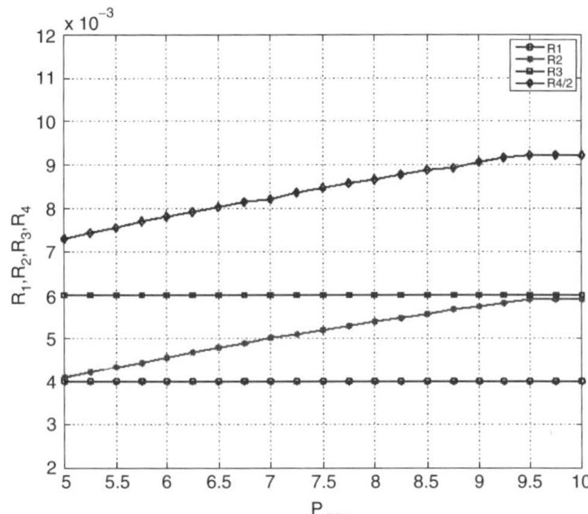

Fig. 5 Speed reducer dimensions vs. $\rho_{\max}$

As seen in a previous paper [4,5], boundaries are used to limit the variation of the parameters to feasible values. It appears that the results are controlled by the upper boundary value for the transformation ratio $\rho = \rho_r \rho_t$.

Figure 3 shows the motor and speed reducer weights of the system in function of the constraint $\rho_{\max}$. The system mass decreases until $\rho_{\max} = 95rad/mm$, where it remains constant. Before the previous value, the minimum is obtained for $\rho = \rho_{\max}$. There is no need to have a transformation ratio greater than $95rad/mm$ because the system weight will not decrease anymore. At this point, the decrease of the motor weight is compensated by the increase of the speed reducer weight. The resulting active mass is $830g$.

Figures 4 and 5 show the dimensions of the motor and the speed reducer vs. the upper constraint $\rho_{\max}$. Concerning the motor, we can see in Fig. 4 that the radius decreases until $\rho_{\max} = 95rad/mm$ and the two others parameters are independant of $\rho_{\max}$. In Fig. 5, we can note that two radius increase until $\rho_{\max} = 95rad/mm$ while the two others stay constant.

5 Conclusion

In this paper, the weight of an electromechanical conversion system has been optimised. Firstly, a model of the motor and the speed reducer has been done. This model links the motor and speed reducer main dimensions to their performances. Then, the battery bank, motor and speed reducer weights have been written in function of the optimisation parameters. Secondly, an optimisation procedure was executed in order to minimise the objective function which is the weight of the previous components. Different numerical optimisation methods were used to valid the methodology. All gave the same results. The initial total weight was 1,180 g and the obtained total weight after optimisation is equal to 830 g. We can note that the weight is reduced by 30%.

Acknowledgements The authors would like to thank Virax company, and particularly Mr. Frédéric Bernier for his help.

References

1. Macua E., Ripoll C., Vannier J.-C., *Design and Simulation of a Linear Actuator for Direct Drive*, PCIM2001, pp. 317–322, June 19–21, 2001, Nürnberg, Germany.
2. Macua E., Ripoll C., Vannier J.-C., *Design, Simulation and Testing of a PM Linear Actuator for a Variable Load*, PCIM2002, pp. 55–60, May 14–16, 2002, Nürnberg, Germany.
3. Macua E., Ripoll C., Vannier J.-C., *Optimization of a Brushless DC Motor Load Association*, EPE2003, Sept. 02–04, 2003, Toulouse, France.
4. Dessante, Ph., Vannier J.-C., Ripoll Ch., *Optimisation of a Linear Brushless DC Motor Drive*, ICEM 2004, Sept. 2004.
5. Dessante Ph., Vannier J.-C., Vidal P., *Optimisation of a Linear Brushless DC Motor Drive and the Associated Power Supply*, AES 2005 Civil or Military All-Electric Ship Conference, 13–14 Oct. 2005, Paris-Versailles, France.
6. Rioux C., *Théorie Générale Comparative des Machines Electriques Etablie à Partir des Equations du Champ Electromagnétique*, Revue générale de l'Electricité (RGE), mai 1970.
7. Nurdin M., Poloujadoff M., Faure A., Synthesis of Squirrel Cage Motor: A Key to Optimisation, *IEEE Transactions on Energy Conversion*, Feb. 1994.

Torque Ripple Reduction Using Evolution Optimization Method and Varying Supply Voltage

Andrej Stermecki, Peter Kitak, Igor Tičar, Oszkar Biró, and Kurt Preis

Abstract The purpose of this research was to reach torque ripple minimization using a relatively inexpensive manufacturing method – shifting of the rotor's permanent magnets and varying supply voltage. In the performed analyses, the differential evolution optimization method combined with FEM analysis was used to find the optimal shifting angles of the magnetic poles. Furthermore the procedure of varied supply voltage was simulated using the transient FE model, in order to study the feasibility of commutation torque ripple minimization at the optimized motor model.

1 Introduction

The object of the analysis was the 12 pole brushless PM motor with a nominal power of 800 W. The permanent magnets used were standard ferrites with a remanent flux-density of $\mathrm{Br} = 0.39\,\mathrm{T}$ and a coercive force of $\mathrm{Hc} = 310{,}000\,\mathrm{A\,m^{-1}}$. The motor winding was a standard three-phase winding, with star-connected phase windings. These properties and the characteristics of the power supply system correspond to square-wave PM brushless motor drive. Detailed characteristics of these types of brushless motors can be found in literature [1] so they are not explicitly elucidated here.

The designing of brushless PM machines requires discontinuances in the stator and rotor structures. These lead to dissimilar reluctance values when observed in different angular directions. This is mainly caused by openings in the stator slots towards the air gap area resulting from technological restrictions during the winding process. A tangential component of magnetic attraction between the rotor-mounted

Andrej Stermecki, Peter Kitak, and Igor Tičar
Faculty of Electrical Engineering and Computer Science, Smetanova 17, 2000 Maribor, Slovenia
andrej.stermecki@uni-mb.si

Oszkar Biró and Kurt Preis
Technical University of Graz, IGTE, Koppernikusgasse 24, A-8010 Graz, Austria

A. Stermecki et al.: *Torque Ripple Reduction Using Evolution Optimization Method and Varying Supply Voltage*, Studies in Computational Intelligence (SCI) **119**, 249–257 (2008)
www.springerlink.com

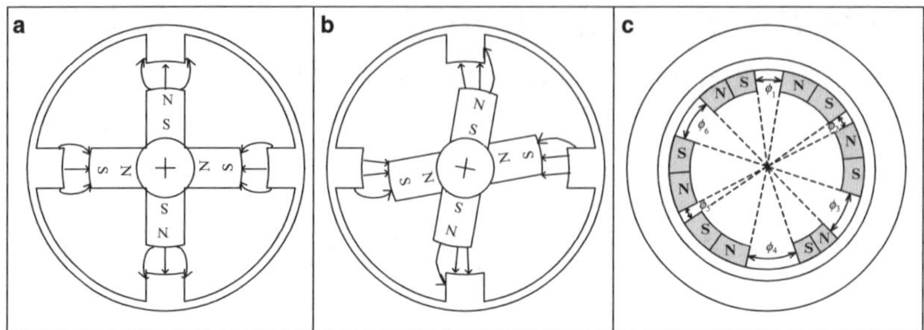

Fig. 1 **a** Schematic presentation of the basic magnetic poles position; **b** schematic presentation of the shifted magnetic poles position; **c** schematic presentation of the shifting angles

permanent magnets and the stator teeth is produced due to dissimilar reluctance values. This attraction, the cogging torque, is an undesired effect that causes additional irregularities in the torque characteristic of the brushless PM machine. Such irregularities can cause problems when starting the motor, difficulties in motor speed/torque control, additional noise, vibrations, etc. The cogging torque value can be minimized in one way by reducing the magnetic flux in the machine (varying the magnets' strength). As this is unavoidably followed by a reduction in the resulting torque of the motor, it is not the most appropriate method and is, therefore, not discussed in details. Other methods are based on reducing the rate of reluctance change with respect to rotor position, which can be achieved in several ways: changing the stator tooth geometry, varying the magnet arc length, shifting the magnetic poles and skewing the magnetizing direction of the permanent magnets.

In our previous research work [2], shifting of the magnetic poles proved to be a highly efficient and economical method for cogging torque reduction. It is possible to shift the magnetic poles in such a way, that cogging torque producing tangential forces are partially mutually canceled (Fig. 1). As shown in our previous research, a 50% reduction of cogging torque was achieved compared to the basic model [2]. As holds for all cogging torque minimization techniques this procedure is also very delicate – namely, by shifting the magnetic poles, the total torque can be substantially reduced. To obtain the desired results, a shifting angle should be very carefully chosen and the search for its optimal value should be approached by numerical analysis possibly supported by an automated optimization algorithm. The basic idea of shifting magnetic poles is evident from a simplified schematic presentation of a four-pole motor (Fig. 1a,b).

2 Optimization Procedure

In the presented paper the differential evolution method (DE) was chosen for the optimization strategy. It is a relatively new optimization algorithm based on the origins of new populations in the evolution of mankind. This method was first presented by Price and Storn [3].

Since the goal of optimization, in our case, was to reduce the cogging torque (T_{cog}) without lowering the total torque (T), it was necessary to consider two objective functions. The total torque value is dependent on the total magnetic arcs' lengths (Φ_{ARC}); therefore, the *minimum/maximum* objective $u_{ARC}(\Phi_{\text{ARC}})$ was used to describe this objective function taking the total torque into account. A similar *minimum/maximum* objective $u_{Tcog}(T_{\text{cog}})$ was used to compose the objective function describing the FEM calculated cogging torque. Both objective functions were analytically given by bell-shaped fuzzy sets and merged into a unified objective function $u(\Phi_{\text{ARC}},\ T_{\text{cog}})$ that was used in the optimization algorithm (1):

$$u(\Phi_{\text{ARC}}, T_{\text{cog}}) = w_{\text{ARC}} u_{\text{ARC}}(\Phi_{\text{ARC}}) + w_{Tcog} u_{Tcog}(T_{\text{cog}}), \tag{1}$$

where w_{ARC} and $w_{T\text{cog}}$ are the weights of individual quantities. Now we can describe the optimization algorithm as searching for the minimum of the objective function:

$$\min\left(u(\Phi_{\text{ARC}}, T_{cog}), x_i\right); i = 1 \ldots 6, \tag{2}$$

where $u(u_{\text{ARC}},\ u_{\text{Tcog}})$ is the function describing mutual dependence on FEM-calculated cogging torque values, and an additional function describing the magnets' arcs, while x_{i} represents six shifting angles, limited by the lower limit l_{i} and the upper limit h_{i}. In this way, the optimal cogging torque reduction should be accomplished with the largest magnetic arcs as possible, thus fulfilling the conditions for high overall electromagnetic torque. Six shifting angles, that are presenting the optimization parameters, are shown in the schematic presentation of the rotor's magnets (Fig. 1c).

Cogging torque values were calculated using a series of static FEM calculations of the motor at different rotor positions. The FEM analysis was performed in the Ansys 7.1 computer program [4] and the optimization algorithm was programmed in Matlab 6.5. Both programs were linked together using file input/output exchange.

3 Analysis of the Torque Ripple

The electromagnetic torque is additionally affected by factors that are difficult to predict when using the static FEM analyses performed during the optimization procedure. For studying types of brushless motor, the results of the ideal commutation process controlled by power supply electronics are square-wave phase currents where only two phases are conducting at a time [1]. In practice, the current waveforms deviate from the ideal trapezoidal form due to the effects of inductance, non-ideal (square-wave) back EMF waveform, induced voltages by commutation process, impact of PWM modulation and other parasitic influences. The resulting non-rectangular currents produce considerable torque ripple also known as *commutation torque ripple* [5]. The total torque ripple of the final optimized motor model was, therefore, verified by transient numerical analysis with the motion of the rotor also taken into account. Such action is required, since it is possible that the

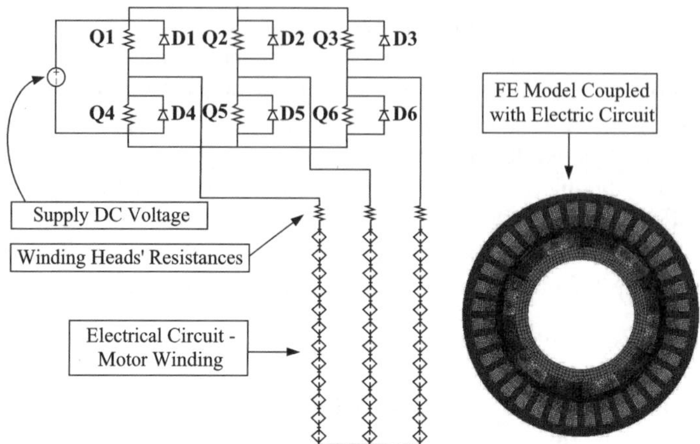

Fig. 2 FE model coupled with a three-phase bridge circuit

change in motor's geometry in order to reduce the cogging torque, additionally produces increases in the commutation torque ripple. The final goal of this verification procedure was to propose a motor design with low torque ripple and high overall electromagnetic torque.

This study used a numerical formulation based on a magnetic vector potential $\boldsymbol{A}$. To solve the non-steady problem, time discretization was applied, using the Hughes generalized trapezoidal rule [4]. In order to include the rotor movement into the analysis, the coupling of the neighboring nodes in the air gap was performed by constraint equations. Constraint equations tie together two layers of non-coincident nodes with respect to the element-shape function. Therefore, the final result is virtually disconnected mesh, but the layers of slip surface are tied-up by the mentioned equations. The FE model was coupled to an electrical circuit, in order to methodically include simulation of the square-wave brushless motor's commutation process into the transient FEM analysis. A three-phase bridge circuit with six transistors and freewheeling diodes was, therefore, modeled, as presented in Fig. 2.

4 Results

The graph in Fig. 3a shows the convergence course of the objective function towards the optimal result. For each iteration step, the objective function had as many values as there were population numbers. To plot the presented course, the average objective function values within one iteration were considered.

The cogging torque waveforms of basic and optimized models are presented in Fig. 3b. Maximum cogging torque was reduced from 0.7 Nm to only 0.02 Nm, which represented a reduction of more than 95% compared to the basic model. This result

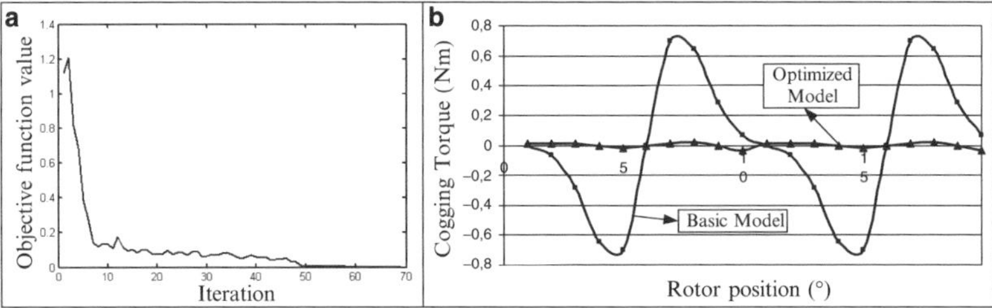

Fig. 3 **a** Convergence of the differential evolution optimization algorithm; **b** Cogging torque waveforms

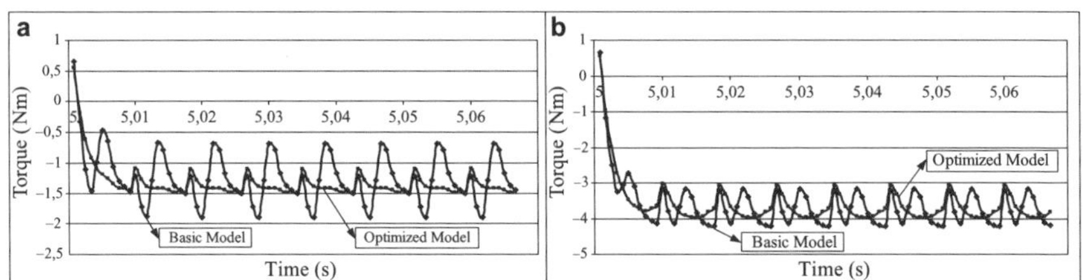

Fig. 4 **a** Torque ripple at the lower supply voltage; **b** torque ripple at the higher supply voltage

was achieved when the total magnets' arc was decreased from 354° to 332.5° which correspond to 6%. All optimization parameters were set to 1 at the basic model ($\phi_1 = \phi_2 = \phi_3 = \phi_4 = \phi_5 = \phi_6 = 1$) and the optimized values were $\phi_1 = 2.0$, $\phi_2 = 5.9$, $\phi_3 = 1.0$, $\phi_4 = 6.5$, $\phi_5 = 5.4$ and $\phi_6 = 6.7$.

Cogging torque represents only one contribution to the final torque ripple of the motor. The modifications in the geometry of the motor as a result of the optimization process can be reflected in the changes of other factors influencing torque ripple. Transient analysis was performed where the movement of the rotor was also taken into consideration to verify the newly proposed motor model.

Simulation was performed under the presumption of a constant rotor speed of $n = 200\,\text{RPM}$. It can be seen from Fig. 4a,b that, with any increase in the supply voltage and, consequently the supply current, the torque ripple also becomes more distinctive. In Fig. 4a the torque ripple of the optimized model is only 32% compared to the original model. In Fig. 4b this difference is not as advantageous, since the ripple of the optimized model is 87% compared to the original model.

Increase in torque ripple at a higher supply current (voltage) is the result of commutation torque ripple. Commutation torque ripple is provoked by the non-ideal current waveform. The current ripples produced by the commutation process are, of course, more noticeable at higher supply voltages. Torque ripple follows the same outline as corresponding current ripple, as is clearly seen in Fig. 5a. A typical anomaly in the current waveform is seen from the simulated result (Fig. 5a), as well as from the measurement results (Fig. 5b).

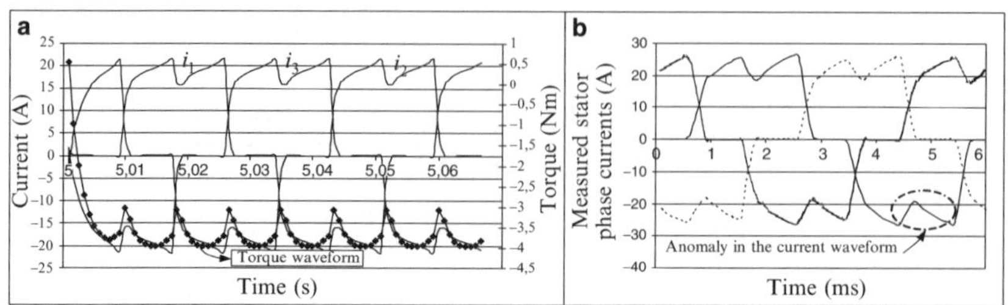

Fig. 5 **a** Stator current waveforms and corresponding torque waveform; **b** measured current waveforms

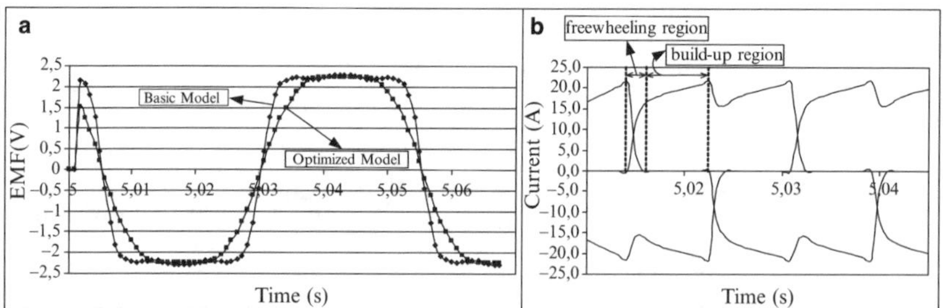

Fig. 6 **a** EMF waveforms of basic and optimized models; **b** current waveform of the brushless motor

5 Reduction of the Comutation Torque Ripple

The commutation torque can also be partially reduced using some of the available techniques. These strategies are mostly carried out when presuming an ideal shape (trapezoidal) of the open circuit back-EMF. After the optimization procedure the EMF waveform of the studied motor was noticeably changed and strongly deviate from the ideal trapezoidal shape (Fig. 6a).

Transient FEM analysis was again used to verify whether the impact of transformed EMF waveform was too high to reduce the commutation torque ripple, using the procedure of varied input voltage [6]. The main idea behind this strategy [6] is to overcome the anomalies in the current waveform, which originate in the commutation process, by changing the input voltage at the time of commutation [6].

The commutation process is divided into two intervals: *freewheeling* region and *build-up* region (Fig. 6b). The freewheeling region denotes the start of the commutation and lasts until the commutated current falls to zero. The falling of the commutated current occurs through those freewheeling diodes presented in Fig. 2. The build-up region represents the remaining interval until the rising current achieves its nominal value (Fig. 6b). The transient current waveforms for freewheeling region can be described by the following equations:

$$U = i_a R + L\frac{di_a}{dt} + E + E + i_b R + L\frac{di_b}{dt};\ 0 = i_c R + L\frac{di_c}{dt} + E' + E + i_b R + L\frac{di_b}{dt};\ i_b = i_a + i_c. \tag{3}$$

And for the build-up region:

$$U = 2i_a R + 2E + 2L\frac{di_a}{dt}, \tag{4}$$

where U stands for the supply voltage, E for the back-EMF in the build-up region, E' for the back-EMF at the start of commutation, L for winding inductance, R for winding resistance, and i_a, i_b and i_c for phase currents. The time instant of the commutation starting was chosen in such a way, that commutation occurred in phases A and C (i_a represents the rising and i_c the falling current in the commutation process).

The system of presented differential equations can be solved using Laplace transformation, as suggested in [6]. In this way we obtain an analytical expression for the stator current waveforms, taking the commutation process into account. Thus we obtain for the freewheeling region:

$$i_a = \frac{2U-3E-E'}{3R}\left(1-e^{-(R/L)t}\right);\ i_b = -\left(I_0 e^{-(R/L)t} + \frac{U-3E-E'}{3R}\left(1-e^{-(R/L)t}\right)\right)$$

$$i_c = I_0 e^{-(R/L)t} - \frac{U-2/3E'}{3R}\left(1-e^{-(R/L)t}\right). \tag{5}$$

And for build-up region:

$$i_a = -i_b = \frac{U}{2R} - \frac{E}{R} - \frac{U-2RI_0-2E}{2R}e^{-(R/L)t}, \tag{6}$$

where I_0 denotes the initial current of the required current for obtaining the torque. As it can be seen from (5), the current of phase C is falling faster than the current of phase A is rising. The current of phase B, that is conducted continuously, is represented by the sum of the currents of phases A and C. Therefore, in the freewheeling region, a ripple in the current of phase B is produced. This anomaly in the current is seen in the transient FEM simulation, as well as in the measurement results (Fig. 5a,b).

The torque ripple can be reduced by supplying the various input voltage in order to compensate for the difference between the falling and rising currents in (5). This is achieved by requiring a constant current ($i_\text{b} = I_0$) in the freewheeling region. In this way we obtain the voltage that is required to make the current constant in the freewheeling and build-up region:

$$U_{free} = 3RI_0 + 3E + E';\quad U_{build} = 2(RI_0 + E). \tag{7}$$

Transient FEM analyses of basic and optimized models were performed, while applying the varied supply voltages (7) to the motor models. As pointed out previously, the goal of this research was to verify if the modifications in geometry, resulting from the optimization procedure, affected the applicability of this method for commutation torque reduction. The analysis was performed under the presumption of a constant rotor speed of $n = 200\,\text{RPM}$.

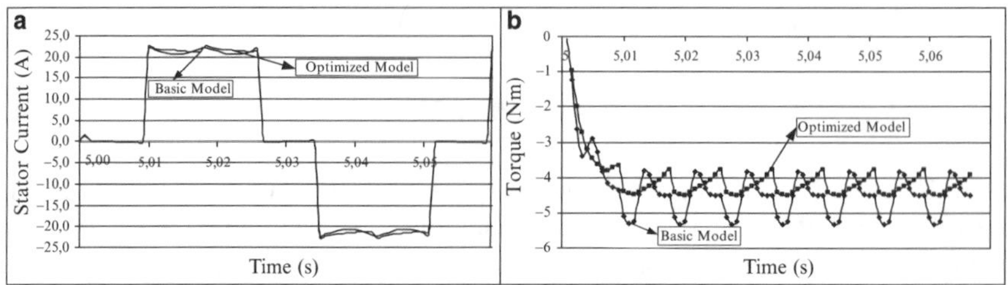

Fig. 7 **a** Current waveforms of the basic and optimized models; **b** torque waveforms of the basic and optimized models

Based on the simulation results (Fig. 7a) the conclusion can be drawn, that it is possible to reduce the current ripple by varying the input voltage for both studied models – basic and optimized models. In this way the final torque ripple of the optimized model is efficiently reduced (Fig. 7b). The torque ripple of the optimized model fed by the varying input voltage is reduced by 52% compared to a basic model with the same voltage supply.

6 Conclusion

Efforts were made to reach cogging torque minimization using a relatively inexpensive manufacturing method. In the performed analyses, the differential evolution optimization method was used to find the optimal shifting angles of the magnetic poles. In this way a cogging torque was successfully reduced by over 95%. In our previous work [2] similar calculations were performed, but without the automated optimization algorithm, using a simple "*cut and try*" approach. This way the cogging torque had been reduced by only 50%. The comparison between these two results shows the advantages of the optimization algorithm when combined with FEM.

Furthermore the total torque ripple of the motor was calculated using the transient FE model coupled with an electric circuit presenting the three-phase bridge circuit. This analysis took in all the transient effects emerging from changing electric values and from rotor movement into account. The findings of this study lead to the conclusion that, at higher supply currents, the commutation torque ripple becomes more expressive and predominates over the cogging torque. Hence, the benefits of cogging torque optimization are not as evident when the total torque ripple is studied.

Finally the procedure of varied supply voltage was simulated using the same transient FE model, in order to study the feasibility of commutation torque ripple minimization at the optimized motor model. Based on these results, the conclusion can be drawn, that the presented procedure is efficient in both studied motor models. Using an optimization algorithm and by varying supply voltage, the final torque ripple was reduced by 52%.

References

1. J. R. Hendershot Jr and T. J. E Miller, Design of Brushless Permanent-Magnet Motors, Oxford: Magna Physics Publishing and Clarendon Press, 1994.
2. I. Tičar, A. Stermecki and I. Zagradišnik, Numerical analysis of the brushless motor magnetic field and torque, in ISEF 2003, vol. 2, A. Krawczyk, and M. Trlep (Eds). Maribor: FERI, 2003, pp. 847–852.
3. R. Storn and K. Price, Differential evolution: a simple and efficient adaptive scheme for global optimization over continuous spaces, J. Glob. Optim., 11:341–359, 1997.
4. ANSYS 7.0 Documentation. Canonsburg: SAS IP, 1998.
5. S. Chen, C. Namuduri and S. Mir, Controller induced parasitic torque ripples in a PM synchronous motor, IEEE Trans. Ind. Appl., 38(5):1273–1281, 2002.
6. Ki-Yong Nam, Woo-Taik Lee, Choon-Man Lee, and Jung-Pyo Hong, Reducing Torque Ripple of Brushless DC Motor by Varying Input Voltage, IEEE Trans. Magn., vol. 42, no. 4, 2006, pp. 1307–1310.

Combined Electromagnetic and Thermal Analysis of Permanent Magnet Disc Motor

Goga Cvetkovski, Lidija Petkovska, and Sinclair Gair

Abstract The paper presents a methodology for coupling electromagnetic and thermal phenomena in a permanent magnet disc motor performance analysis. Both the electromagnetic and thermal analysis is performed using two-dimensional finite element method (FEM). Due to the complex geometry of the disc motor a proper modelling of the motor is performed. The thermal analysis is performed based on the losses calculated from the electromagnetic FEM analysis, as well as the measured ones. To show the validity of the proposed method, a test bench is realised and temperatures in specific points are measured and afterwards compared with the calculated ones using FEM analysis.

1 Introduction

In the design stage of a permanent magnet disc motor, a crucial problem concerns the determination of the temperature distribution. Since the heating of the machine depends on the electromagnetic losses, there is a coupling between the electromagnetic and the thermal phenomena that results from the temperature dependence of the winding resistivity. In the last decade, the finite element method has been widely used to compute the electromagnetic and thermal behaviours of permanent magnet motors and electrical machines in general. In this paper, we describe a method to achieve a magneto-thermal computation of a permanent magnet disc motor for electric vehicle. Therefore, first the magnetic field calculation has been performed

Goga Cvetkovski and Lidija Petkovska
Ss. Cyril and Methodius University, Faculty of Electrical Engineering and Information Technology, P.O. Box 574, 1000 Skopje, Macedonia
gogacvet@feit.ukim.edu.mk, lidijap@feit.ukim.edu.mk

Sinclair Gair
University of Strathclyde, 204 George street, G1 1XW Glasgow, Scotland, UK
s.gair@eee.strath.ac.uk

G. Cvetkovski et al.: *Combined Electromagnetic and Thermal Analysis of Permanent Magnet Disc Motor*, Studies in Computational Intelligence (SCI) **119**, 259–267 (2008)
www.springerlink.com

in order to gain some information data necessary for the thermal computation and analysis of the motor. In order to be able to get the necessary data for the PM disc motor, a calculation of the magnetic field has to be performed. The 2D analysis is very suitable for this type of geometry and has a lot of advantages over the 3D calculation, such as lower memory storage and reduced time computation.

2 PM Disc Motor Description

The modelled motor is a brushless three phase synchronous permanent magnet disc motor, with rated torque 54 Nm and speed 750 rpm@50 Hz, fed by a pulse width modulated (PWM) inverter and rechargeable batteries or fuel cell. The PMDM is a double sided axial field motor with two laminated stators having 36 slots and a centred rotor with eight skewed neodymium–iron–boron permanent magnets with $B_r = 1.17\,\mathrm{T}$ and $H_c = -883\,\mathrm{kA\,m^{-1}}$. The real view and side view presentation of the prototype permanent magnet disc motor are given in Figs. 1 and 2, respectively.

Fig. 1 Permanent magnet disc motor real view

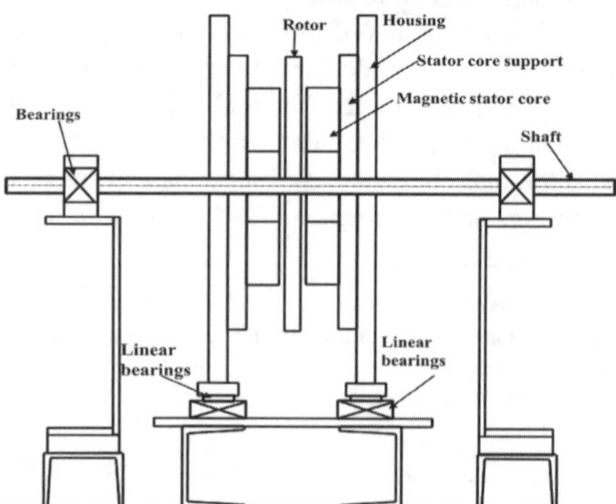

Fig. 2 Permanent magnet disc motor side view

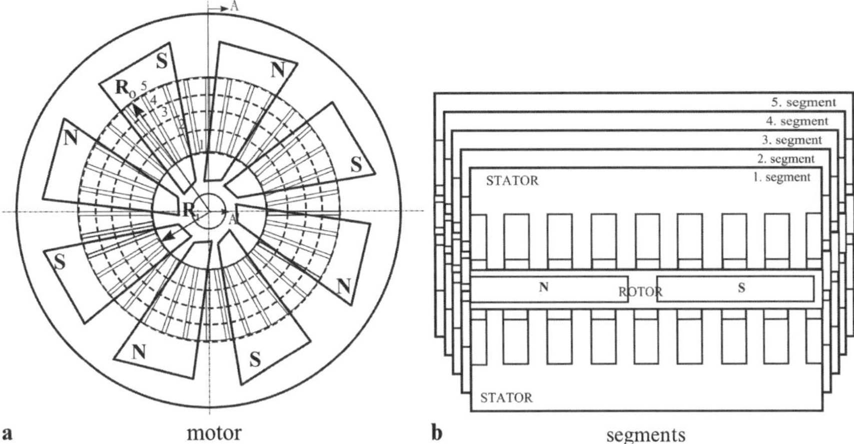

Fig. 3 Radial division of the motor into five segments

3 PM Disc Motor FEM Modelling

In order to be able to get the necessary data for the PM disc motor, a calculation of the magnetic field has to be performed. The 2D analysis is very suitable for this type of geometry and has a lot of advantages over the 3D calculation, such as lower memory storage and reduced time computation. The quasi-3D method [1] which is adopted for this analysis consists of a 2D FEM calculation of the magnetic field in a three-dimensional radial domain of the axial field motor. For this purpose, a notional radial cut through the two stators and one rotor of the disc motor is performed and then opened out into linear form, as shown in Fig. 3. By using this linear quasi three-dimensional model of the disc motor, which is divided into five segments, it is possible to model the skewing of the magnets and also to simulate the vertical displacement and rotation of the rotor. Due to the symmetry of the machine the calculation of the motor is performed only for one quarter of the permanent magnet disc motor or for one pair of permanent magnets.

4 FEM Magnetic Field Analysis

After the proper modelling of the permanent magnet disc motor and the adequate mesh size refinement, especially in the air gap a magnetic field calculation is performed for each segment separately, for different current loads and different rotor displacements. As an example the magnetic field distribution of the motor at no load and one rotor position for the first, third and fifth segment, as specific ones, is presented in Fig. 4a–c, respectively. On the other hand the magnetic field distribution

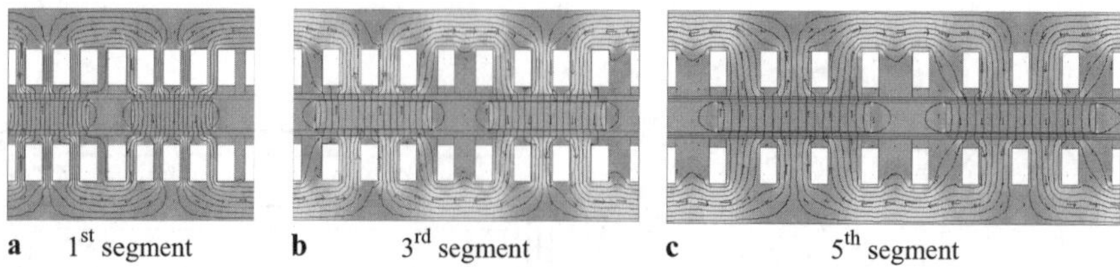

Fig. 4 Magnetic field distribution at no load

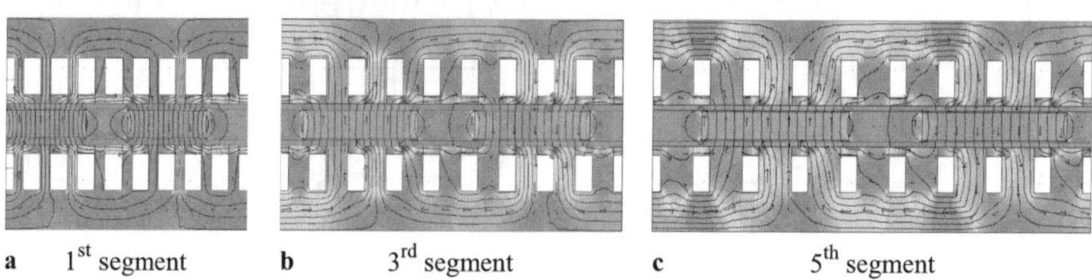

Fig. 5 Magnetic field distribution at rated current load

of the motor at rated current load at the same rotor position, for the first, third and fifth segment, is presented in Fig. 5a–c, respectively.

The calculated data from the FEM magnetic field analysis in the postprocessor mode could also be used to estimate certain magnetic and electric parameters [2] necessary for further investigation and performance evaluation of the permanent magnet disc motor. Due to the complexity of the analysed motor model, such as specific geometry and proper modelling for the FEM calculation, certain modifications of the standard equations for parameters calculation using FEM data are made.

4.1 Air Gap Flux Density Calculation

An interesting and very important parameter that can be used for further motor analysis is the average value of the air gap flux density and its distribution under different current load conditions.

The air gap flux density is calculated by using the results of the FEM magnetic field calculation, applying them in (1) and solving it numerically by the same programme:

$$\mathbf{B} = \text{curl}\,\mathbf{A}. \tag{1}$$

The distribution of the air gap flux density is calculated and presented for no load and for rated load for the first, third and fifth segment in Figs. 6 and 7, respectively.

The finite element analysis enables to evaluate the magnetic field properties in the whole investigated domain of the motor which is very important for the performance analysis of the motor. Using the data from the FEM analysis some motor parameters can be calculated and determined, such as: the previously presented air gap flux

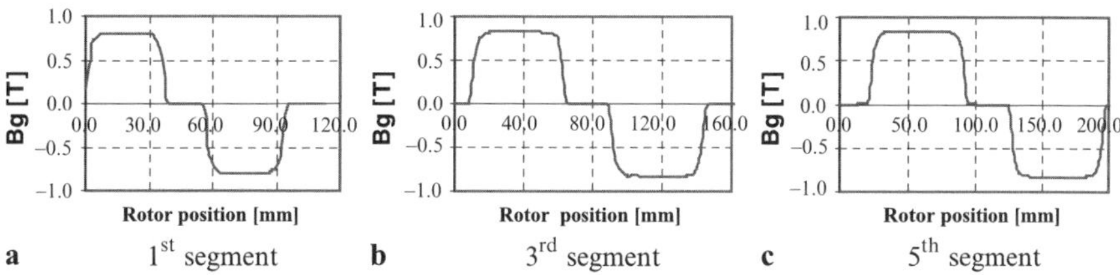

Fig. 6 Air gap flux density distribution at no load

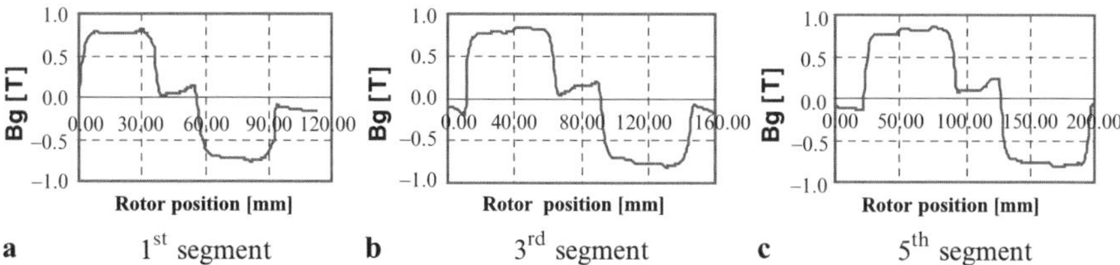

Fig. 7 Air gap flux density distribution at rated current load

density distribution, the reactances along the d, q axes, the electromagnetic torque, the cogging torque, copper and iron losses, which are necessary for the performance analysis of the motor [3], as well as input data for the thermal analysis.

5 FEM Thermal Analysis

An accurate estimation of the thermal behaviour of an electrical machine is important considering the fact that safe operating conditions and overloading capabilities are dependent on the temperature rise. Temperature limits exist for permanent magnets and the winding insulations. Beside, the winding resistances and consequently I^2R losses, the permanent magnet flux is also temperature dependent. Having a lot of variable factors, unknown loss contributions with their complicated three-dimensional distributions, it is obvious that an exact determination of the machine thermal behaviour is impossible.

One way of determining the thermal distribution in the motor is by using a thermal equivalent circuit, which is an analogy of the electric circuit in which the loss sources are represented as current sources, where on the other hand thermal resistances and thermal capacitances are represented as resistances and capacitances, respectively. All machine elements are described by node sources, having an average surface temperature with respect to the ambient temperature and the thermal capacitance. The elements are connected to each other by conduction or convection resistances [4, 5].

Another way of determining the thermal distribution in the motor is by using the finite element method in order to perform a numerical thermal field analysis.

Models based on numerical field solution of the heat flow problem feature strongly in the literature. In some cases it has been performed by using a three-dimensional calculation of the heat flow, where in other cases it has been done by using a two-dimensional calculation. Both approaches have advantages and disadvantages over different maters, such as computational time, quality of the results, and others. The authors of this paper have taken into consideration all the pros and cons, and have decided to use the two-dimensional calculation of the heat flow with a specific modelling of the motor that has been previously described. For the two-dimensional, steady state problem, the governing partial differential equation is:

$$\frac{\partial}{\partial x}\left(k_x \cdot \frac{\partial T}{\partial x}\right) + \frac{\partial}{\partial y}\left(k_y \cdot \frac{\partial T}{\partial y}\right) + \dot{q} = 0, \tag{2}$$

where k_x, k_y are the thermal conductivity coefficients, T the temperature and $\dot{q}$ the strength of the heat source per unit volume [6]. Compared to the magnetostatic or electrostatic case, a more versatile set of boundary conditions has to be taken care of:

- Prescribed temperature T at a node or surface;
- Surface heat flow across the boundary due to convection

$$k_x \frac{\partial T}{\partial x} \cdot l_x + k_y \frac{\partial T}{\partial y} \cdot l_y + h(T - T_o) = 0, \tag{3}$$

  where T_o is the ambient temperature, h is the convection heat coefficient, l_x and l_y are the direction cosines of the outward normal to the boundary surface;
- Surface heat flow across the boundary due to radiation

$$k_x \frac{\partial T}{\partial x} \cdot l_x + k_y \frac{\partial T}{\partial y} \cdot l_y + \sigma F\left(T^4 - T_o^4\right) = 0, \tag{4}$$

  where T_o is the ambient temperature, σ is the Stefan–Boltzmann constant, and F is an emission function;
- Internal heat source $\dot{q}$.

The steady state problem formulation described by (2) and considering all boundary conditions is equivalent to the problem of finding the temperature in the whole predefined domain. After the proper geometrical modelling of the motor that is the same as for the magnetic field calculation, as well as the proper definition of the material properties and boundary conditions is done, the temperature distribution is calculated for all five segments and for different current loads. In this paper only a part of the temperature distribution is presented in Fig. 8 for the first, third and fifth segment for rated current load.

Table 1 contains temperatures extracted from the FEM thermal model as well as temperatures gained from test measurements.

The comparative results presented in Table 1 show fairly good agreement between the FEM calculated data and the measured ones. Since the permanent magnet disc motor is a prototype model and is still under investigation for this analysis the

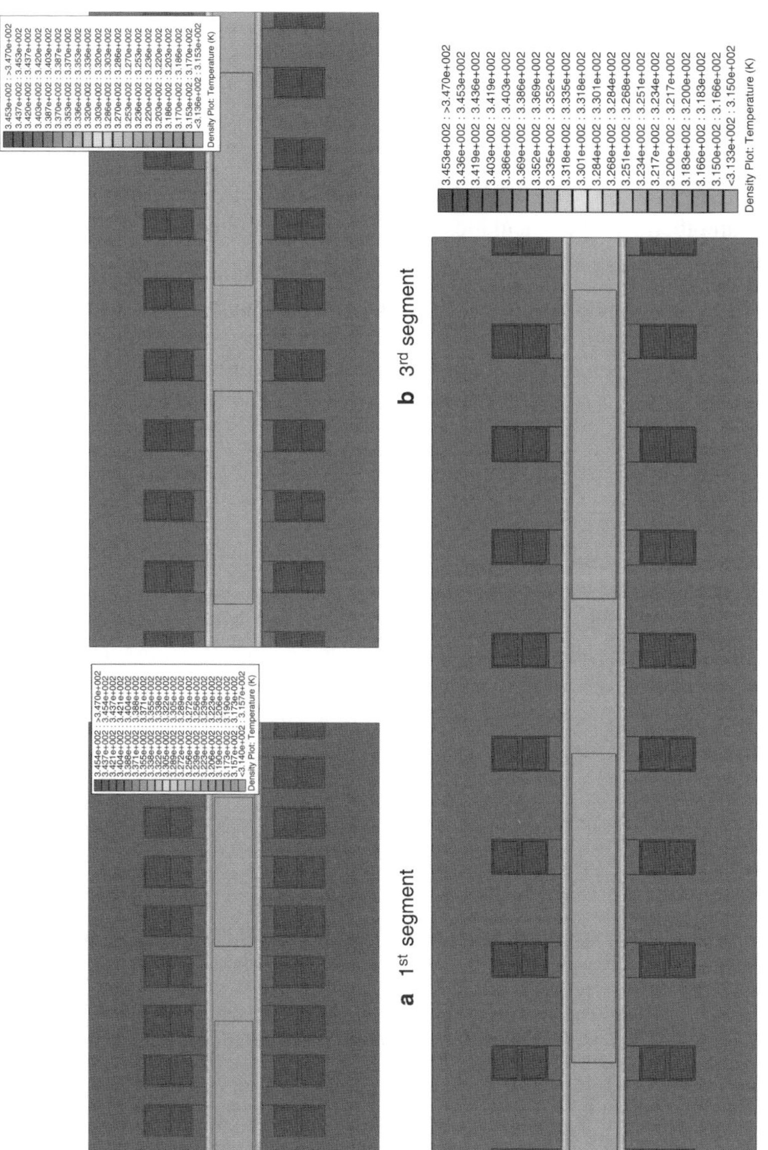

Fig. 8 Temperature distribution at rated current load

Table 1 Comparison of selected temperatures in the FEM thermal model and measurements

Location	FEM [K]	Test measurement [K]
Stator no. 1 outside radius	340.6	338
Stator no. 2 outside radius	340.6	337
Stator inside radius	342.7	336
Stator end winding	/	347
Stator winding	346	/

motor was loaded up to $I = 8.723\,\text{A}$. Based on the results from the magnetic field calculation and thermal field calculation it can be concluded that:

1. Either the motor can be loaded even more that has been done so far without doing any damage to the stator winding insulation as well as the permanent magnets;
2. Or that the motor could be optimised [7] or totally redesigned, using an optimisation tool such as genetic algorithm [8], or any other optimisation method.

6 Conclusion

An efficient method to simulate coupled magnetic-thermal fields in PM disc a motor is presented. This methodology can be also performed on other electrical machines, as well as devices. With this approach an efficient and robust coupled magnetic and thermal field simulation scheme is presented. The gained results from the magneto-thermal FEM analysis were used to present a performance analysis of the motor and suggest future steps that can be taken in the development procedure of the prototype model of the permanent magnet disc motor.

References

1. G. Cvetkovski, L. Petkovska, et al., Quasi 3D FEM in Function of an Optimisation Analysis of a PM Disk Motor, Proceedings of the 14th IEEE International Conference on Electrical Machines-ICEM '2000, Helsinki, Finland, Vol. 4/4, pp. 1871–1875, 2000.
2. G. Cvetkovski, L. Petkovska, et al., PM Disc Motor Parameter Estimation Using FEM Data, Proceedings of the 6th International Symposium on Electric and Magnetic Fields-EMF2003, Aachen, Germany, pp. 191–194, 2003.
3. G. Cvetkovski, L. Petkovska, Performance Evaluation of an Axial Flux PM Motor Based on Finite Element Analysis, Proceedings of the 12th International Symposium on Electromagnetic Fields in Mechatronics, Electrical and Electronic Engineering-ISEF'2005 on CD, Baiona, Spain, pp. 1–6, 2005.
4. P. H. Mellor, D. Roberts, D. R. Turner, Lumped Parameter Thermal Model for Electrical Machines of TEFC Design, IEE Proceedings-B, Vol. 138, No. 5, pp. 205–218, September 1991.
5. F. Sahin, A. J. A. Vandenput, Thermal Modelling and Testing of High-speed Axial-flux Permanent-magnet Machine, Proceedings of 15th IEEE International Conference on Electrical Machines-ICEM '2002, Brugge, Belgium on CD, pp. 1–9, 2002.

6. K. Hamayer, U. Pahner, R. Belmans, H. Hedia, Thermal Computation of Electrical Machines, Proceedings of the International Symposium on Electric and Magnetic Fields–EMF1996, Liege, Belgium, pp. 61–66, 1996.
7. G. Cvetkovski, L. Petkovska, S. Gair, Efficiency Improvement of PM Disc Motor Using Genetic Algorithm, Proceedings of the 9th Spanish Portuguese Congress on Electrical Engineering – 9CHLIE'2005 on CD, Marbella, Spain, pp. 1–7, 2005.
8. G. Cvetkovski, L. Petkovska, et al., GA Approach in Design Optimisation of a PM Disk Motor, COMPEL – The International Journal for Computation and Mathematics in Electrical and Electronic Engineering, Vol. 19, No. 2, pp. 608–614, 2000.

Computer Modelling of Energy-Saving Small Size Induction Electromechanical Converter

Maria Dems, Krzysztof Koméza, and Sławomir Wiak

Abstract In the paper, circuit and field-circuit analyses of the energy-saving small size induction motors are presented. The possibility of determination of the influence of the shape of core and windings data change on motors characteristics and parameters, based on this analysis, will be shown. From the comparison of classical and energy-saving constructions some general remarks about the possible improvements of motor efficiency by core and windings dimension change were stated.

1 Introduction

Higher expectation for the economical aspect of the production of electromechanical converters leads to design new variants of induction motor constructions. The increase of efficiency of electric machines is the main problem of serialised production of induction motors, especially in the range of small motors. Usually, the production of the energy-saving small size induction motors is based on the typical motors constructions. The change of the motor construction would lead to increase of the efficiency of these electric machines; such change would also have an influence on other parameters and load characteristics of the motor. In the paper, two small induction motors of the same rated values: classical and the energy-saving were analysed. Both motors have the same external diameter of the stator core, but different length and different windings.

The subject of this paper is a field-circuit and circuit analyses of the energy-saving small size induction motors [1, 2]. The aim of this analysis is to determine the influence of the shape of core and windings data change on motors curves and parameters.

Maria Dems, Krzysztof Koméza, and Sławomir Wiak
Institute of Mechatronics and Information Systems, Technical University of Lodz,
ul. Stefanowskiego 18/22, 90–924 Lodz, Poland
mdems@p.lodz.pl, komeza@p.lodz.pl, wiakslaw@p.lodz.pl

M. Dems et al.: *Computer Modelling of Energy-Saving Small Size Induction Electromechanical Converter*, Studies in Computational Intelligence (SCI) **119**, 269–276 (2008)
www.springerlink.com

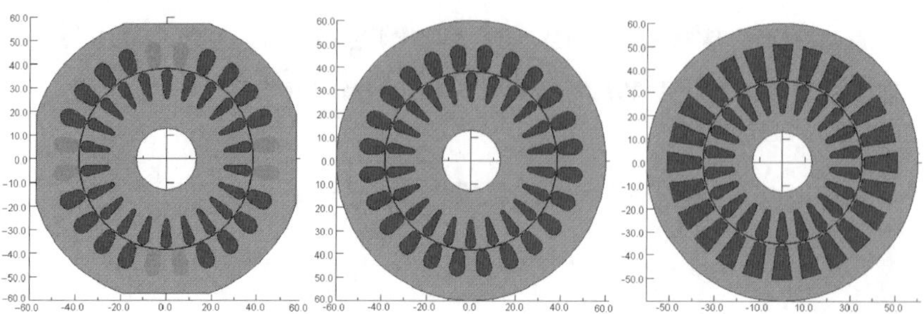

Fig. 1 The shapes and dimensions of the stator and rotor core of the induction motor classical constructions with and without cuts and energy-saving

2 Object of Investigation

An energy-saving construction of the small induction motor was designed basing on the classical structure of the four-pole induction motor model size 80. The output power of this motor is 0.75 kW, stator windings are star connected, and supply voltage is 400 V for the frequency 50 Hz. This motor has stator core shape with cut of parts making stator yoke width not constant; the external maximal diameter of the stator core Dse max = 120 mm, and external minimal diameter Dse min = 114 mm. Based on this construction the energy-saving induction motor of the same rated data has a round shape of the stator core with the external diameter Dse = 120 mm. The stack length of this motor was increased by 42%, the internal stator core diameter was decreased by 8.5% and the number of series turns of stator windings by 15% and the width of the air-gap was decreased from 0.25 to 0.225 mm.

Figure 1 shows stator and rotor core shapes and dimensions of the both motors, respectively.

Two constructions of both motors are possible: with half-closed and closed rotor slots. In the case of the induction motor with closed rotor slots the high saturation effect of the magnetic circuit is caused in the upper part of the rotor core placed over the rotor slot bar. This phenomenon has a strong influence on the field distribution in the rotor core causing that the part of the leakage field is penetrating the rotor slot bar. Therefore, the classical-circuit approach gives insufficient results, in the static states as well as transient states of the motor.

3 Computing of Operating Curves Using Circuit Model

For the energy-saving induction motor the parameters and curves of current, power factor, efficiency and rotor speed vs. output power and also current and electromagnetic torque vs.rotational rotor speed were computed. The results of this calculation for the motor with half-closed and closed rotor slots are shown in Figs. 2 and 3.

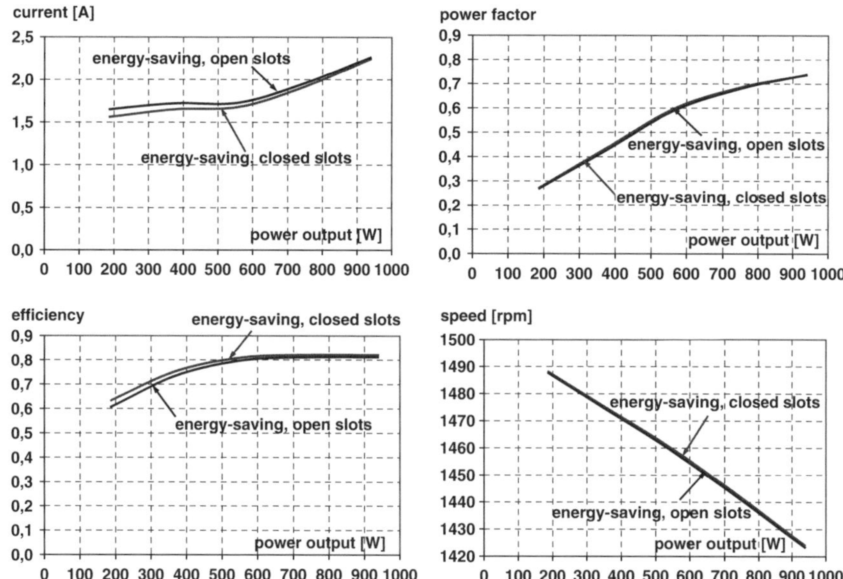

Fig. 2 Current, power factor, efficiency and rotor speed vs. output power for the energy-saving motor

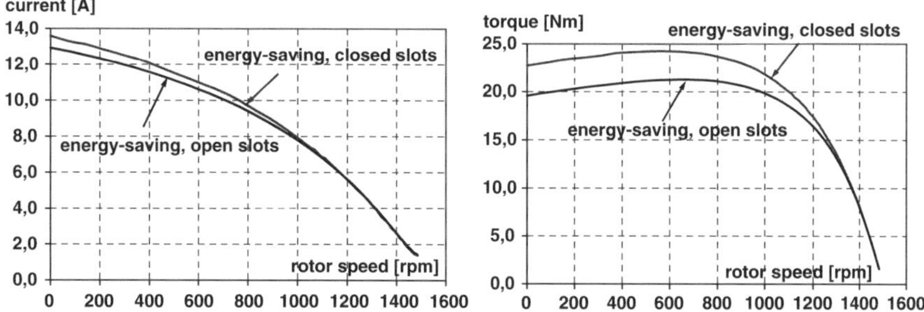

Fig. 3 The current and electromagnetic torque vs. rotational rotor speed for the energy-saving motor

Comparing curves shown in Figs. 2 and 3 we can state that better results can be obtained by using closed rotor slots construction. For the construction with closed rotor slots these curves were compared with the results of the calculation obtained for the classical motor, for the average value of the external stator diameter Dse av = 117 mm, and maximal value Dse max = 120 mm. The results of this comparison are shown in Figs. 4 and 5.

Table 1 presents the values of the magnetic flux density and total power losses in both motors.

As a result of lower magnetising current, a lower total stator current in the energy-saving motor, rather than in the classical motor with external stator diameter

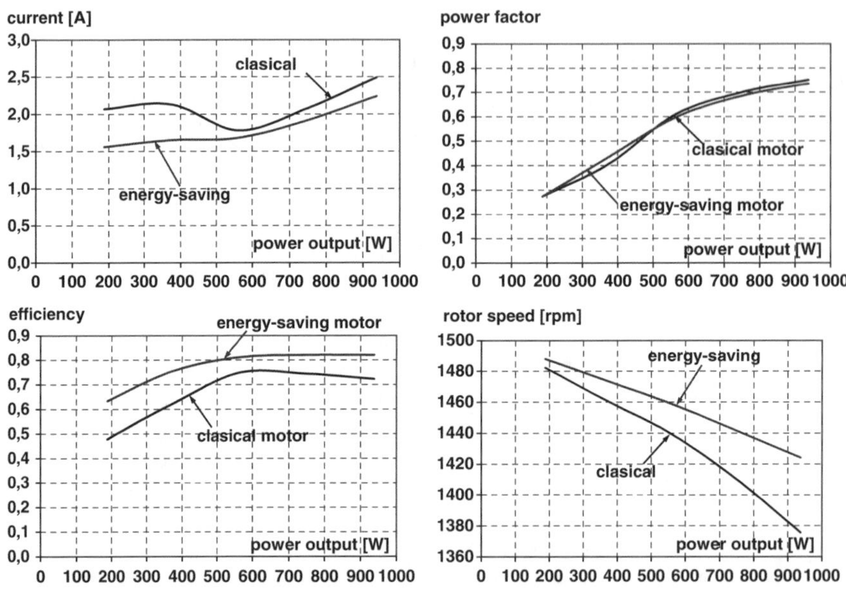

Fig. 4 Current, power factor, efficiency and rotor speed vs. output power for the energy-saving induction motor and classical induction motor (Dse av = 117 mm)

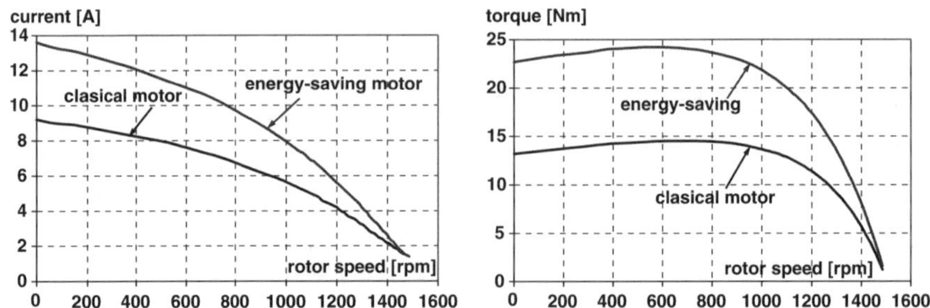

Fig. 5 The current and electromagnetic torque vs. rotational rotor speed for the energy-saving induction motor and classical induction motor

Dse av = 117 mm was obtained. In the energy-saving motor, the characteristic of efficiency vs. output power is higher than in the classical motor, for all range of output power. For the rated power output the difference of efficiency has the value:

$$\Delta\eta = 82.06\% - 74.38\% = 7.68\%. \tag{1}$$

This significant increase in efficiency for the energy-saving motor is obtained thanks to the lower copper losses in the stator and rotor windings. For the energy-saving motor the resistances of stator and rotor windings are lower than for the classical motor because of a smaller number of the series turns of the stator winding and the larger cross section diameter of the stator wire (two paralleled wires), and also

Table 1 Flux density and total losses in the classic and energy-saving motor

	Energy-saving motor	Classical motor Dse = 117 mm	Classical motor Dse = 120 mm
Flux density in the stator yoke [T]	1.714	1.969	1.74
Flux density in the stator tooth	1.552	1.819	1.72
Flux density in the rotor tooth	1.850	1.775	1.73
Magnetizing current [A]	1.424	1.958	1.19
Core losses [W]	58.6	47.1	44.1
Stator winding losses [W]	67.5	158.4	121
Rotor winding losses [W]	29.8	47.1	45.3
Total losses [W]	164.0	258.4	217

a bigger cross-section of the rotor slots. This effect is decreased by the bigger length of the energy-saving motor, which is necessary for the sake of acceptable values of magnetic flux density in the motor core shown in Table 1.

The bigger value of the rotor losses of the classical motor gives the lower values of the rotor speed vs. output power.

The higher value of the rotor losses of the classical motor gives the lower values of the rotor speed vs. output power.

4 Computing of Magnetic Field Distribution

The field-circuit analysis is done for 2-D structure of the motor and for all its constructions with frequency = 50 Hz. Magnetic field distributions were calculated for three structures of the motor with stator core: with cuts, and without cuts, and for energy-saving with different effective stack lengths. Figure 6 shows the vector plot and distribution of the magnitude of magnetic flux density for energy-saving construction.

As seen from Figs. 6–8 the flux density distribution for energy-saving construction is much more regular than for the construction with cuts and also for the construction without cuts. The stator core is less saturated for energy-saving construction but, magnetic flux density in the rotor is higher than in classical round construction. This is consistent with the results of calculations obtained from the circuit method, shown in Table 1. This fact connected with the smaller number of series turns for energy-saving construction leads to a higher value of magnetising current for energy-saving than round classical construction. Figure 9 illustrates this phenomenon. In Fig. 9 the comparison of magnetising currents for motors with half-closed and closed rotor slots are shown. Additionally, the comparison with the measured values for model motor with half-closed slots shows that the used field-circuit model is very adequate. The transient analysis instead of quasi-static analysis was necessary, due to local saturation of the part of the motor core, to achieve proper accuracy.

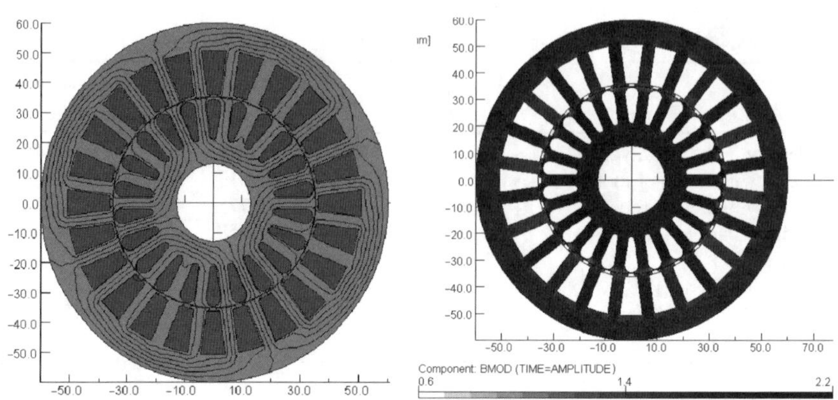

Fig. 6 Vector plot and distribution of magnitude of magnetic flux density for energy-saving construction

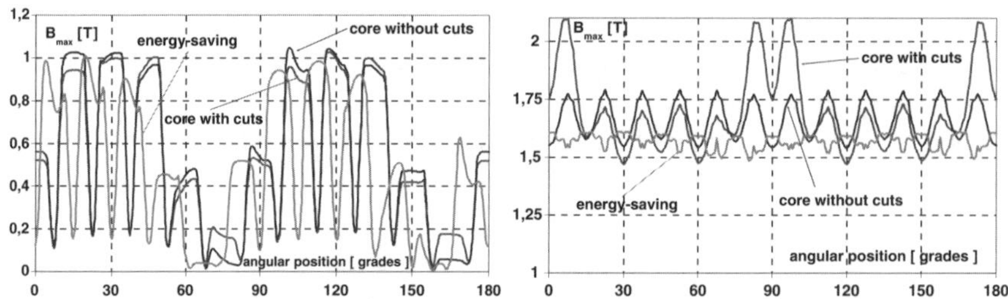

Fig. 7 Distribution of flux density in the middle of the air gap and in the stator yoke vs. angle

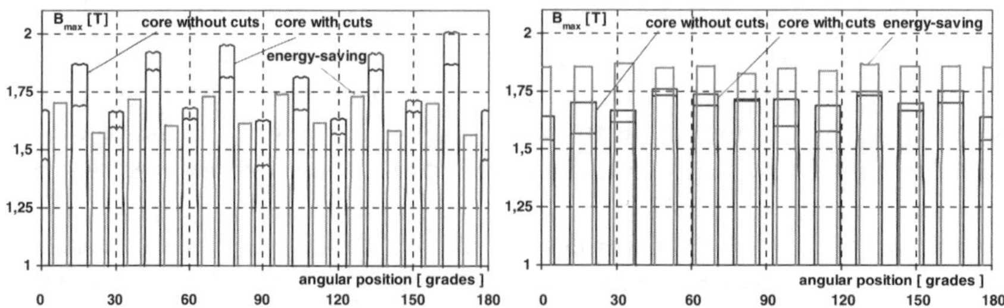

Fig. 8 Distribution of flux density in the stator teeth and in the rotor teeth vs. angle

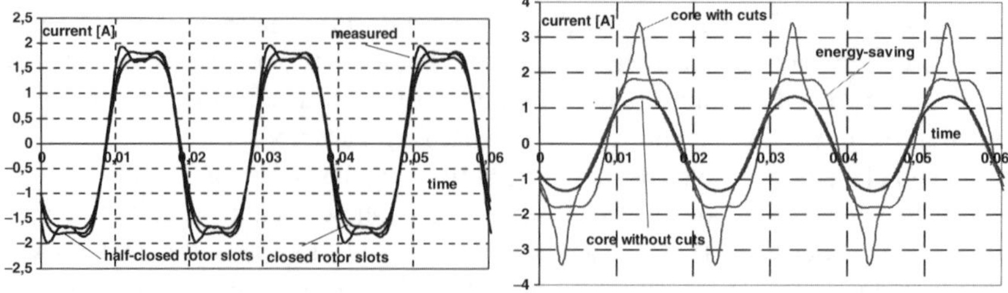

Fig. 9 The magnetizing current vs. time for energy-saving construction with half-closed and closed rotor slots and for different constructions of the motor

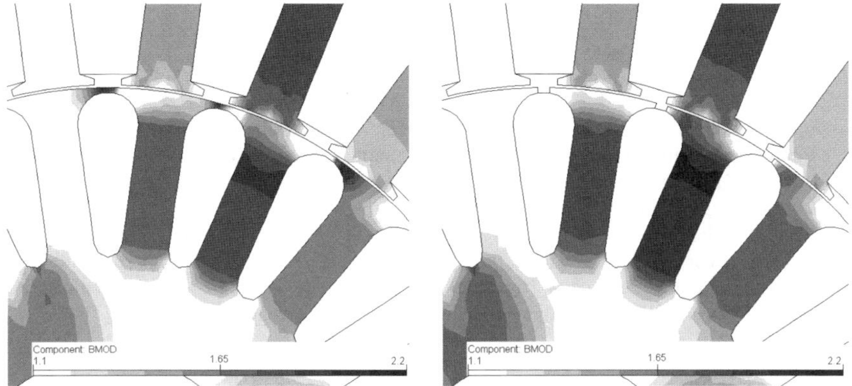

Fig. 10 The flux density distribution for half-closed and closed rotor slots constructions

The reason for small differences in magnetising current values for half-closed and closed rotor slots constructions can be seen from Fig. 10 where the flux density distribution is shown.

5 Conclusion

In small power induction motors, winding losses play a crucial role in the global losses total effect. The core losses are not so important as in high power machines. On the other hand, small machines exhibit a high magnetizing current up to 70% of full load current. Therefore, looking for energy-saving construction it is more important to decrease either a magnetizing current or resistance of stator winding. The second solution was used in the presented construction by increasing the machine length, stator and rotor slots area and simultaneously decreasing the number of the series turns of the stator windings. However, decrease in number of turns leads to increase of magnetizing current which of course slightly reduces the positive effect. Therefore, the best solution will be the use of magnetic materials with better magnetizability like amorphous iron for stator core or alternatively the use of copper squirrel-cage. For both solutions, there are still some technological problems to be solved.

References

1. Wiak S., Koméza K., Dems M., Electromagnetic field and parameters modeling of induction motors by means of FEM, Proceedings 32 Spring. International Conference MOSIS'98, Vol. 3, May 5–7, 1998, Ostrava, Czech Republic, pp. 275–281.
2. PC OPERA-2D – version 11, Software for Electromagnetic Design from VECTOR FIELDS.

3. Dems M., Koméza K., Wiak S., Stec T., The highly efficient three-phase small induction motors with stator cores made from amorphous iron, COMPEL, The International Journal for Computation and Mathematics in Electrical and Electronic Engineering, Vol. 23, No. 3, 2004, pp. 625–632.
4. Dems M., Koméza K., Wiak S., Stec T., Kikosicki M., Application of circuit and field-circuit methods in designing process of small induction motors with stator cores made from amorphous iron, COMPEL, The International Journal for Computation and Mathematics in Electrical and Electronic Engineering, Vol. 25, No. 2, 2006, pp. 283–296.
5. Dems M., Koméza K., Kikosicki M., Wiak S., Stec T., Numerical models and experimental verification of small size asynchronous electromechanical converter with stator core made from amorphous iron, Proceedings of Third Slovenian – Polish Joint Seminar – Computational and Applied Electromagnetics, 6–8 June 2005, Maribor – Slovenia, pp. 21–24.

Integral Parameters of the Magnetic Field in the Permanent Magnet Linear Motor

Tomczuk Bronisław and Waindok Andrzej

Abstract The magnetic field integral parameters have been calculated for the five-phase permanent magnet tubular linear motor (PMTLM). For the field analysis the finite element method has been used. Magnetic force and inductance of the stator winding have been calculated for different values of the supplying current and different mover positions. The physical model of the motor has been tested and a good conformity between the calculated and the measured characteristics has been obtained.

1 Introduction

Linear machines have been known for many years [1]. Owing to modern magnetic materials and permanent magnets usage, the linear motors are used widely in speed and position controlled drive systems [2]. In some applications they is an alternative to pneumatic or hydraulic linear actuators [2]. Due to the exclusion of a mechanical gear, the direct drive linear motor systems gain high-speed and high-accuracy of the mover [3, 4]. In comparison with flat linear motors, the tubular ones have higher efficiency and higher force/weight ratio.

The presented permanent magnet tubular linear motor (PMTLM, Fig. 1a) can be included in a group of actuators [5, 6]. It is a five-phase construction. The stator is divided into five separated sections. Each section forms one phase. The mover is assembled from a permanent magnet and ferromagnetic rings, which are fixed on a nonferromagnetic tube. The permanent magnets are characterized by axial magnetization and they are situated alternately with ferromagnetic rings. The motor could

Tomczuk Bronisław and Waindok Andrzej
Opole University of Technology, Departament of Industrial Electrical Engineering,
ul. Luboszycka 7, 45-036 Opole, Poland
b.tomczuk@po.opole.pl

T. Bronisław and W. Andrzej: *Integral Parameters of the Magnetic Field in the Permanent Magnet Linear Motor*, Studies in Computational Intelligence (SCI) **119**, 277–281 (2008)
www.springerlink.com

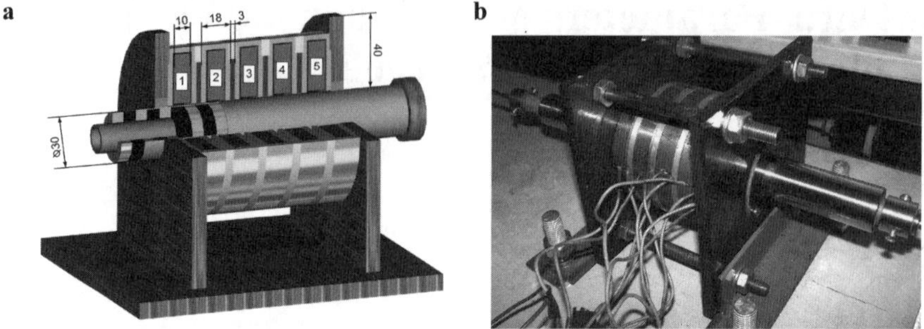

Fig. 1 PMTLM motor: **a** outline; **b** physical model

operate as a synchronous one or as a step motor. In the second case the motor low precise due to the cogging force arising from the permanent magnet field.

The linear motors almost always operate in the transient state. Thus, it is important to consider its transients. One of the analysis method is to construct the field-circuit mathematical model of the linear actuator [7]. The integral parameters of the presented motor (Fig. 1a) magnetic field are included in the equivalent diagram of the model [7, 8]. To calculate the integral parameters, the magnetic field analysis has been performed and the flux distributions have been carried out.

The magnetic force derives from the permanent magnet field or from excitation current of the windings. The winding coils are excited independently in accordance with the supplying system. We studied both operation conditions for supplying of one coil and five-phase sinusoidal current excitation (synchronous work).

2 Magnetic Force Calculation

The axisymmetrical field model was created using the 2D FEMM program [9] and including the nonlinear characteristics of the magnetic materials. The magnetic force was calculated using the Maxwell's stress tensor. Taking into account the axial symmetry of the motor only two components of the field must be considered [7]

$$\overleftrightarrow{\mathbf{T}} = \begin{bmatrix} \mu(\mathrm{B})(H_r^2 - \frac{1}{2}H^2) & \mu(\mathrm{B})H_r H_z \\ \mu(\mathrm{B})H_r H_z & \mu(\mathrm{B})(H_z^2 - \frac{1}{2}H^2) \end{bmatrix}. \tag{1}$$

The magnetic force generated by the motor strongly depends on exciting current i and the runner position z. For the field-circuit model it is important to put the characteristic $F = f(i, z)$ into the circuit diagram for each phase, separately. Using the symmetry of the motor, we can calculate this characteristic for first, second and third coil, respectively. In each case we assumed that the current value varies from -10 to 10 A and the runner position changes from -30 to 45 mm. Due to clarity, only the force for the coil three excitation is presented as a function of current value

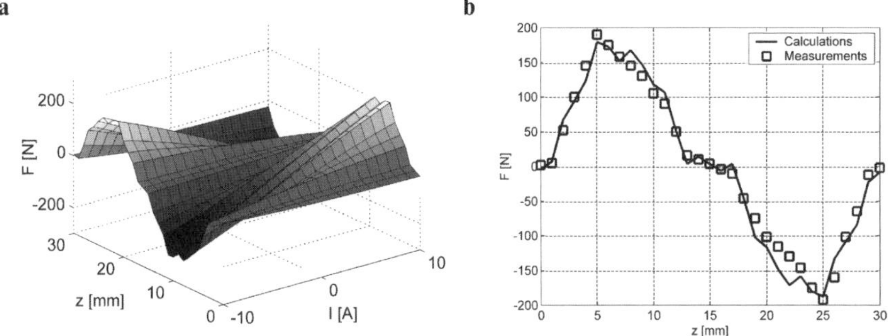

Fig. 2 **a** Force as a function of runner position and current value for coil three excitation, **b** measurement verification for current value $I_3 = 8\,\mathrm{A}$

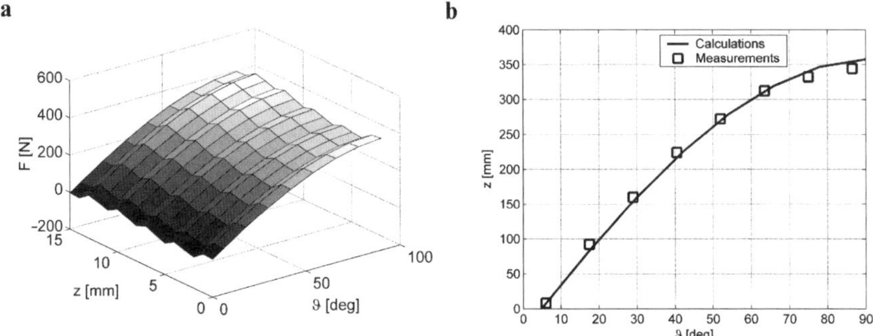

Fig. 3 **a** Force as a function of load angle and mover position, **b** measurement verification for the fixed mover ($z = 2\,\mathrm{mm}$)

and the runner position (Fig. 2a). For the position with the coordinate $z = 0$ and for the net current excitation value, there are no forces acting on the mover.

The calculations results were compared with those obtained from measurements (Fig. 2b). The current value was fixed ($I_3 = 8\,\mathrm{A}$) and the mover position was changed from 0 to 30 mm. Although the measurements are very troublesome, a good agreement has been obtained between both calculations and experiment results. The difference is for the maximum value of the magnetic force. In the calculation results two local maximums are visible, whereas in the measurement wave there is only one maximum point. It could be due to the physical model inaccuracies (there are some static friction forces).

The force calculations have been carried out for the motor supplying with a sinusoidal wave of the current, as well. The characteristic verso mover position and load angle is similarly to mechanical characteristics of the synchronous machines. For the nominal current excitation ($I_m = 8\,\mathrm{A}$) the maximum force value, obtained by the nominal load angle $\vartheta = 90°$, amounts 400 N (Fig. 3a). Because of many calculation points, the measurement verification was investigated only for $z = 2\,\mathrm{mm}$ mover position (Fig. 3b), but for different load angles ($\vartheta = 0 \div 90°$). The measurements confirm calculation results.

3 Static and Dynamic Inductances

The inductance of the stator winding is another important integral parameter of the PMTLM. Its value is especially important for controlling system. In our analysis we have calculated the inductance of all coils within the stator winding. Due to the slotted construction of the stator, each coil is screened against the others. Thus, the magnetic field of one coil is separated from the other ones.

The inductance of each stator coil for different mover positions can be found from the field analysis by magnetic energy calculation or the integration of the vector potential within the coil domain. The self-inductance of each stator coil can be calculated from ratio of the Ψ/i or from the vector potential (in each coil domain) knowledge.

$$L_s = \frac{\int \mathbf{A} \cdot \mathbf{J} \mathrm{d}V}{i^2}, \tag{2}$$

where i is the current intensity in the coil.

In the field-circuit models [8], the static inductance is less important than the dynamic one. The dynamic one performs in the transient equations and combines the field model with the circuit one. The measurement verification of the dynamic inductance is difficult to do. However, it can be simply calculated for each point of the Ψ/i curve as a partial derivative

$$L_{\mathrm{d}} = \frac{\partial \Psi(i,z)}{\partial i}. \tag{3}$$

The comparison between experiment and calculations results of the static inductance is presented in Fig. 4a. The results correspond to the neutral mover position, for which the cogging magnetic force vanishes [5]. For the current value $I_3 = -5\,\mathrm{A}$, the permanent magnet and current fluxes neutralize each other. Thus, the resulted flux vanished and the static inductance is forced to be zero (Fig. 4a). Contrary to the static inductance L_{s}, the current variations do not change the inductance L_{d} significantly (Fig. 4b). However, small variations in the dynamic inductance values influenced the transients of the motor.

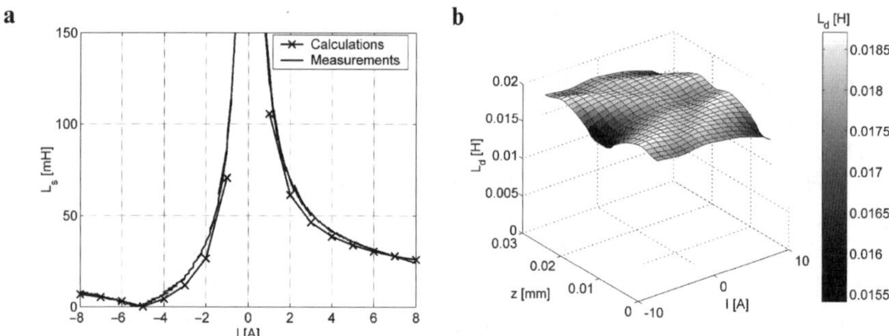

Fig. 4 **a** Static inductance vs. current values for neutral mover position (coil three excitation), **b** dynamic inductance as a function of current and mover position (for coil three excitation)

4 Conclusions

The integral parameters of the magnetic field have been calculated for the PMTLM. They are magnetic force and inductances of the stator coils. The computer simulations certify that their values depend on the dimensions of the stator coils and the magnetic circuit. The magnetic force values for supplying of one coil strongly depend on the mover position, as well. In the case of positioning mode operation, the five-phase current from PWM supplier can be used for excitation of the motor coils. For five-phase excitation, the positioner is more accurate than that with three-phase excitation [3].

The static inductance value depends on permanent magnet position and current value, naturally. Because of permanent magnet field, under the zero current value, the static inductance is undefined. The derivative of the flux linkage with respect to the current intensity has been calculated for each stator coil. It is an important parameter for the transient calculations, where we included the field-circuit model. From the computer simulations and experiments we have learnt that the dynamic inductance changes with the mover position and the current value. Both, forces and inductances, have been measured and a good agreement has been obtained.

References

1. Basak A., Permanent-Magnet DC Linear Motors, Oxford, Clarendon Press, 1996.
2. Eastham J.F., Profumo F., Teneoni A., Gianolio G., Linear drive in industrial application: state of the art and open problems, ICEM, Brugge, Belgium, August 25–28, 2002, paper no. 620.
3. Gordon S., Hillery M.T., Development of a high-speed CNC cutting machine using linear motors, Journal of Materials Processing Technology, No. 166, 2005, pp. 321–329.
4. Kim W.-J., Murphy B.C., Development of a novel direct-drive tubular linear brushless permanent-magnet motor, International Journal of Control, Automation and Systems, Vol. 2, No. 3, September 2004, pp. 279–288.
5. Tomczuk B., Zakrzewski K., Waindok A., Field analysis in permanent magnet tubular linear motor (PMTLM) under variable scaled geometries, MiS'06, Soplicowo, Poland, Sep. 17–21, 2006, pp. 305–311.
6. Tomczuk B., Waindok A., Magnetic field calculations of a permanent magnet tubular linear motor (PMTLM), X Conference on Computer Applications in Electrical Engineering, Poznań, Poland, April 18–20, 2005, pp. 89–90.
7. Tomczuk B., Sobol M., A field-network model of a linear oscillating motor and its dynamics characteristics, IEEE Transactions on Magnetics, Vol. 41, No. 8, August 2005, pp. 2362–2367.
8. Tomczuk B., Sobol M., Analysis of tubular linear reluctance motor (TLRM) under various voltage supplying, ICEM, Cracow, Poland, September 5–8, 2004, paper 792.
9. Meeker D., Finite Element Method Magnetics, User's Manual, 2006.

Field Analysis of a Disc-Shaped Dielectric Motor

Paolo Di Barba, Marie Eveline Mognaschi, and Antonio Savini

Abstract A 3D finite element model of the prototype of a disc-shaped induction dielectric motor is developed; accordingly, a field analysis in time-harmonic condition is developed. A parametric model allows to perform repeated analyses in order to evaluate the dependence of the driving torque on the electrical properties of the rotor disc.

1 Introduction

In the past the authors have studied an induction dielectric motor of cylindrical shape [1]. The motor studied in this paper represents the evolution from a cylindrically shaped device to a disc-shaped one, the prototype of which does exist. This improvement represents an important step towards the device miniaturization.

It is known that in induction dielectric motors the driving torque is due to the interaction of a rotating electric field generated by the stator electrodes, with the lossy dielectric material located at the surface of the rotor. In particular, for the disc-shaped motor, the so-called "rotlin field" is considered [2]. In each point of the rotor surface, "rotlin field" is characterized by a constant magnitude in time and a rotating direction. Potential applications of this kind of field in bioengineering have been already investigated [3].

In order to study the behavior of the motor, a 3D model of the prototype is necessary. In fact, the electrodes are laid in the radial direction, while the field coupling between rotor and stator takes place in the axial direction; moreover, because of the disc-shape, the stray field is not negligible. Consequently, the analytical model of the "rotlin field" [2] is no longer realistic.

P. Di Barba, M.E. Mognaschi, and A. Savini
Department of Electrical Engineering, University of Pavia, via Ferrata 1, I27100 Pavia, Italy
paolo.dibarba@unipv.it, eve.mognaschi@unipv.it, savini@unipv.it

P. Di Barba et al.: *Field Analysis of a Disc-Shaped Dielectric Motor*, Studies in Computational Intelligence (SCI) **119**, 283–287 (2008)
www.springerlink.com

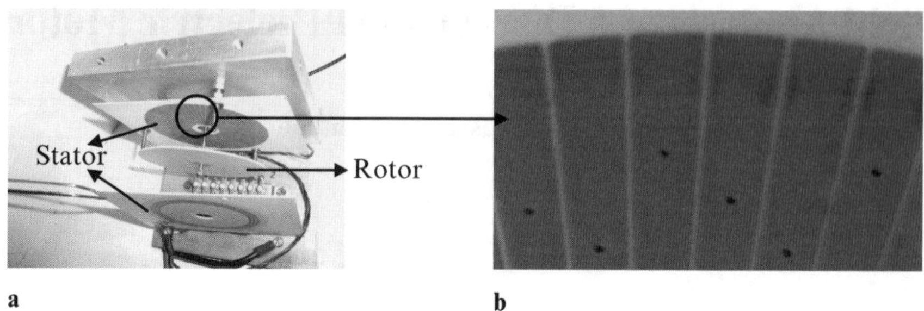

a b

Fig. 1 **a** An exploded view of the motor **b** a detail of the electrodes

2 The Motor

The prototype of the motor under investigation is represented in Fig. 1a:

The device is made of two stator discs and a rotor disc coaxially located. The three discs are made of alumina; on the inner surface of both stator discs a thin film of conducting material is deposited to form the electrodes, while on both sides of the rotor surface a thin film of a lossy dielectric material is laid.

The diameter of each disc is 70 mm and the thickness is about 0.6 mm; moreover, the motor exhibits 72 electrodes on each stator disc. In turn, the shape of each electrode is a circular sector, modified in order to have a distance between two adjacent electrodes constant and equal to 0.3 mm. The 72 electrodes are connected sequentially to a three-phase power supply; the RMS-value phase voltage U_0 is 380 V at a frequency of 50 Hz.

3 Field Analysis

In the frequency domain, the electric field in the region between stator and rotor is governed by the following equations:

$$\nabla \cdot \mathrm{J} = 0, \tag{1}$$

$$\nabla \times \mathrm{E} = 0, \tag{2}$$

subject to the constitutive relationship

$$\mathrm{J} = (\sigma + j\omega\varepsilon)\,\mathrm{E} = \sigma\mathrm{E} + j\omega\mathrm{D}, \tag{3}$$

where J, E, D are the vector phasors of current density, electric field and dielectric induction, respectively; σ is the electric conductivity, ω is the supply frequency and ε is the electric permittivity.

In terms of the scalar phasor U of the potential such that

$$\mathrm{E} = -\nabla \mathrm{U}, \tag{4}$$

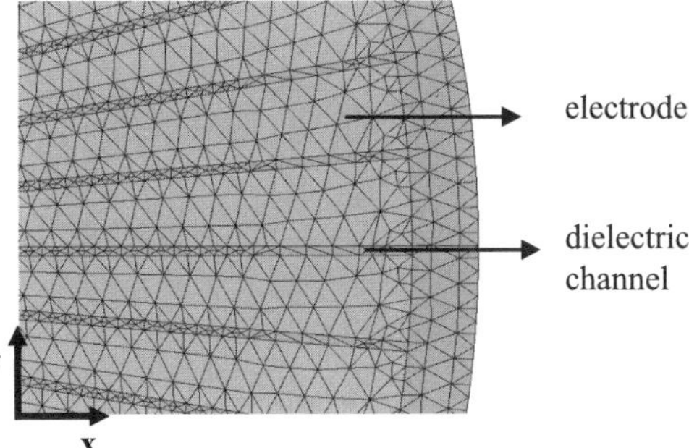

Fig. 2 A detail of the mesh in the plane $z = 0$

the governing equation is

$$\nabla \cdot [(\sigma + j\omega\varepsilon)\,\nabla U] = 0. \tag{5}$$

The problem is subject to the following boundary conditions:

$$U = U_0 \text{ at the electrodes,} \tag{6}$$

$$\frac{\partial U}{\partial n} = 0 \text{ elsewhere,} \tag{7}$$

where n is the normal unit vector.

The electric field in each point of the domain is obtained after solving (4–7) by a finite element scheme. For this purpose, a 3D finite element model was developed. The model is made of about 410,000 tetrahedra, forming five layers in the axial direction. The potential is quadratically interpolated on each element (8 node tetrahedron), so that the whole mesh gives rise to about 94,000 nodes. A detail of the mesh is shown in Fig. 2.

The driving torque acting on the rotor when its speed is zero (i.e. the starting torque) is evaluated by means of the Maxwell electric stress tensor. To this end, a closed cylindrical surface incorporating the rotor disc has been considered as the integrating surface. Second-order effects like static friction and eccentricity were neglected.

4 Results

The results of the field analysis, taken on the $y = 28\,\text{mm}$ plane, are represented in Figs. 3 and 4 in terms of potential lines.

To show the field rotation, in Fig. 3 the potential lines at $t = 0$ and $t = 0.25\,T$, T being the time period, are shown. The material of the rotor has dielectric permittivity equal to 3.3 (paper) and electrical conductivity equal to $10^{-10}\,\text{Sm}^{-1}$.

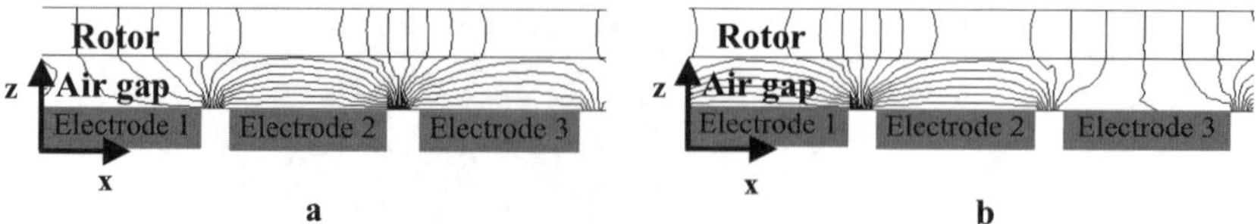

Fig. 3 Potential lines for **a** $t = 0$ and **b** $t = 0.25\,\mathrm{T}$

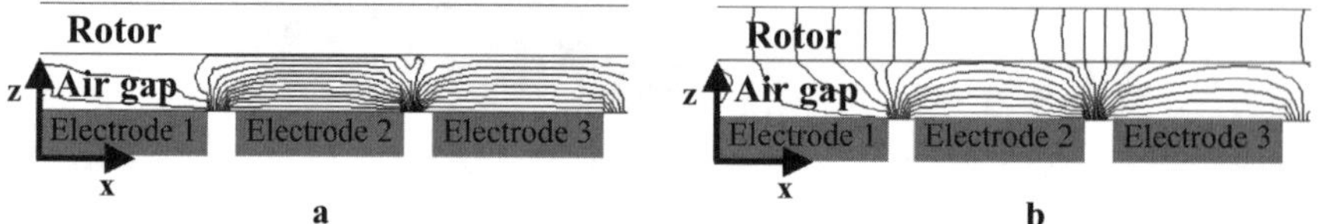

Fig. 4 Potential lines for **a** a perfect conductor rotor and for **b** a perfect insulator rotor

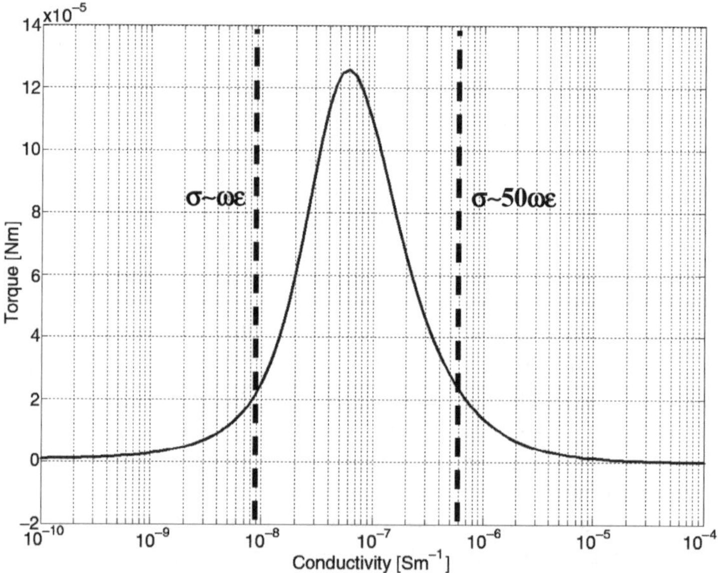

Fig. 5 Motor torque calculated for different values of the rotor conductivity. The gray area highlights the region of best operation of the motor

At the same time instant the potential lines are mapped for two extreme conditions: in Fig. 4a the rotor material is a perfect conductor, while in Fig. 4b the rotor material is a perfect insulator. The corresponding value of torque is practically zero.

In order to assess the numerical procedure of torque calculation based on the Maxwell stress tensor, various integration surfaces are taken into account, each of them incorporating the whole rotor: the error made changing the integration surface is negligible.

In order to investigate the effect of material properties, the torque is calculated for different values of the conductivity of the rotor, fixing the electrical permittivity to 3.3. In Fig. 5 the results are shown.

A region of regular operation of the rotor can be defined as the interval of conductivity values such that the torque is greater than a threshold, e.g. 2×10^{-5} Nm (see Fig. 5).

In the case of electrical permittivity fixed to 3.3, the regular operation region corresponds to a conductivity varying from 1 to 50 times the product of the angular frequency ω of the power supply times the permittivity ε of the rotor. The highest value of the torque, about 1.26×10^{-4} Nm, occurs when $\sigma \sim 22\omega\varepsilon$. The behavior of the curve shown in Fig. 5 confirms that the operation of the motor is based on the presence of a lossy dielectric material.

5 Conclusion

In order to calculate the starting torque of a dielectric motor, a 3D finite element model of the device has been performed. The electrical conductivity of the rotor has been varied and the torque has been updated with a numerical procedure. The results found are helpful to support the design of the motor.

References

1. P. Di Barba, E.R. Mognaschi, M.E. Mognaschi, A. Savini, On a Class of Small Dielectric Motors: Experimental Results and Shape Design, in Electromagnetic Fields in Mechatronics, Electrical and Electronic Engineering, edited by A. Krawcyk, S. Wiak and X.M. López-Fernandez, IOS Press Studies in Applied Electromagnetics and Mechanics, Vol. 27, pp. 306–311, 2006.
2. J.H. Calderwood, E.R. Mognaschi, The Electric Field Created by a Longitudinal Sinusoidal Charge–Density Wave on the Surface of a Plane, Journal of Electrostatics, Vol. 47, pp. 171–182, 1999.
3. J.H. Calderwood, E.R. Mognaschi, The Use of a Rotlin Field in the Dielectrophoretic Separation of Different Types of Bioparticle, IEEE Transactions on Industry Appplication, Vol. 37, No. 6, pp. 1658–1662, 2001.

Fictitious Magnetic Contours as Eddy Currents Exciter

Ilona I. Iatcheva, Slavoljub R. Aleksic, and Rumene D. Stancheva

Abstract Similar to the equivalent electrode method the method of investigation in time changing magnetic field is proposed. The region with concentrated magnetic field is replaced by a number of fictitious magnetic contours. Eddy current exciter is the magnetic flux density time derivative. The solution has been done by the help of electric vector potential. Changing the number of the contours and the appropriate to them magnetic field, the coefficients of the describing linear system of equations are also changed. This procedure stops when at each control point the value of induced current density becomes constant.

1 Introduction

Outgoing from the Pointing's theorem and the energy flow of Poynting's vector through the closed surface the expression for mutual electric energy exchange between magnetic contours is obtained [1,2]. On the bases of the electric vector potential, manifestations of concentrated in magnetic tube time changing magnetic field are considered. Its actions with respect to induced in a conducting medium eddy current and to electric energy interactions between magnetic contours are examined. Up to now investigations are made in a theoretical aspect only [3]. Because of high level accuracy of the contemporary numerical methods numerical instead of experimental confirmation is possible. The goal of the work is to be the first step in the theoretical results checking.

I.I. Iatcheva and R.D. Stancheva
Technical University, Kl.Ohridski 8, 1000 Sofia
iiach@tu-sofia.bg, rds@tu-sofia.bg

S.R. Aleksic
University of Nis, Faculty of Electronic Engineering, Aleksandra Medvedeva 14, 18000 Nis, Serbia
as@elfak.ni.ac.yu

I.I. Iatcheva et al.: *Fictitious Magnetic Contours as Eddy Currents Exciter*, Studies in Computational Intelligence (SCI) **119**, 289–298 (2008)
www.springerlink.com

The paper is devoted to the comparison between eddy current numerical results obtained by finite element method (FEM) and the ones due to the imaginary magnetic contours. The proposed method is applicable for time changing magnetic field. Its essence consists in eddy currents calculation replacing the exciting magnetic field by fictitious magnetic contours. Each of them is supposed coinciding with magnetic tube with discrete cross section. This approach is similar to equivalent electrode method [4–6]. Calculations are implemented for conductive sheet in magnetic system.

2 Eddy Current Determination by the Help of Electric Vector Potential

As an analogue to the magnetic power P determined by the magnetic vector-potential $\overrightarrow{A}$ and the electric current density $\overrightarrow{J}$

$$\mathrm{P} = \frac{1}{2}\frac{\partial}{\partial \mathrm{t}} \iiint_{\upsilon} \overrightarrow{\mathrm{A}}.\overrightarrow{\mathrm{J}} d\upsilon \tag{1}$$

expression for the electric power P_e on the basis of electric vector-potential $\overrightarrow{T}$ is possible to be obtained [1]

$$\begin{aligned} \mathrm{P_e} &= -\iiint_{\upsilon} \overrightarrow{\mathrm{T}}.\mathrm{curl}\,\overrightarrow{\mathrm{E}}\,\mathrm{d}\upsilon \\ &= -\iiint_{\upsilon} (\overrightarrow{\mathrm{T}}_{\mathrm{e}}.\mathrm{curl}\,\overrightarrow{\mathrm{E}} + \overrightarrow{\mathrm{T}}_{\gamma}.\mathrm{curl}\,\overrightarrow{\mathrm{E}} + \overrightarrow{\mathrm{T}}_{\mathrm{D}}.\mathrm{curl}\,\overrightarrow{\mathrm{E}})\mathrm{d}\upsilon. \end{aligned} \tag{2}$$

Accordingly to the subsidiary relation

$$\overrightarrow{\mathrm{J}} = \mathrm{curl}\,\overrightarrow{\mathrm{T}}, \tag{3}$$

the total current density $\overrightarrow{\mathrm{J}}$ components having different nature are closely related to the referring electric vector-potential members. The current density caused by the charges moving in free space or rare gases is not taken into account. The conduction-current density $\overrightarrow{J}_{\gamma}$, the density of the external independent sources $\overrightarrow{J}_e$ and the displacement current density $\overrightarrow{\mathrm{J}}_{\mathrm{D}}$ have been only considered. Therefore the electric vector-potential is accepted composed by the following three terms

$$\overrightarrow{T} = \overrightarrow{T}_{\gamma} + \overrightarrow{T}_{e} + \overrightarrow{T}_{D}. \tag{4}$$

Each component of the electric vector potential determines relative part of the electric power. But it is important to remark that all parts of electric power are generated only by regions where time-changing magnetic field exists and propagates.

Our interest is focused only to the induced eddy currents in conducting medium. It should be taken into account that in low frequencies the excited action is due to $\overrightarrow{T}_\gamma$ itself. At an arbitrary point $\mathrm{P_i}$ the electric vector potential $\overrightarrow{\mathrm{T}}_\gamma$ is determined in a following way. In homogeneous linear medium with conductivity γ at a distance r to the source $(-\frac{\mathrm{d}\overrightarrow{\mathrm{B}}}{\mathrm{dt}}\mathrm{d}\varepsilon)$ the quantity $\overrightarrow{\mathrm{T}}_\gamma$ is found by the integral equation:

$$\overrightarrow{\mathrm{T}}_\gamma^{\mathrm{P_i}} = \frac{\gamma}{4\pi}\iiint\limits_{\upsilon} \frac{\left(-\frac{\mathrm{d}\overrightarrow{\mathrm{B}}}{\mathrm{dt}}\right)}{\mathrm{r}}\mathrm{d}\upsilon = \frac{\gamma}{4\pi}\mathrm{e(t)}\oint\limits_{\Gamma}\frac{\mathrm{d}\overrightarrow{\mathrm{l}}}{\mathrm{r}}. \tag{5}$$

The second term in the equality is referred to linear magnetic contour Γ. The validity of (5) is only for homogeneous conducting medium or in particular case when induced electric field is distributed tangentially to the conducting surface. In this form it could be applied also in the presence of nonlinear magnetic materials. The use of substantial derivative $\frac{\mathrm{d}\overrightarrow{\mathrm{B}}}{\mathrm{dt}} = \frac{\partial\overrightarrow{B}}{\partial t} + (\overrightarrow{v}.\mathrm{grad})\overrightarrow{B}$ makes it possible to determine induced electric field in point $\mathrm{P_i}$ moving by linear velocity $\overrightarrow{v}$ in normal direction to the magnetic field. Taking into account (3) induced through section $s$electric current i is determined from the expression

$$\mathrm{i} = \oiint\limits_{s}\overrightarrow{J}.\mathrm{d}\overrightarrow{s} = \oint\limits_{\Gamma_1}\overrightarrow{T}_\gamma.\mathrm{d}\overrightarrow{l}, \tag{6}$$

where the contour of integration Γ_1 embraces the surface s. After replacing (5) in (6) a following expression for the eddy current i through the section encountered by the contour Γ_1 is obtained

$$i = e(t)G = \mathrm{e(t)}\frac{\gamma}{4\pi}\oint\limits_{\Gamma}\oint\limits_{\Gamma_1}\frac{\mathrm{d}\overrightarrow{\mathrm{l}}.\mathrm{d}\overrightarrow{\mathrm{l}}_1}{\mathrm{r}}. \tag{7}$$

3 Method of Fictitious Magnetic Contours

Concentrated in ferromagnetic body magnetic flux is replaced by the sum of elementary fluxes of the introducing there magnetic contours. In Fig. 1 the disposition of the one of these contours in the planes xy is shown.

If position of the contour in Fig. 2 is described by its vertices: $M_1(m_{ij}, n_{ij}, b_j)$, $M_2(m_{ij}, -n_{ij}, b_j)$, $M_3(p_{ij}, -n_{ij}, b_j)$ and $M_4(p_{ij}, n_{ij}, b_j)$, electric vector potential at a point $M(x_k, y_k, z_k)$ is determined by

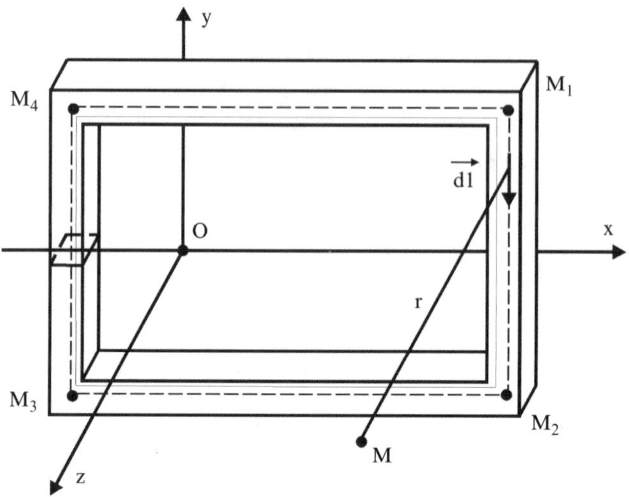

Fig. 1 One of the magnetic contours in the plane xy $(z = b_j)$

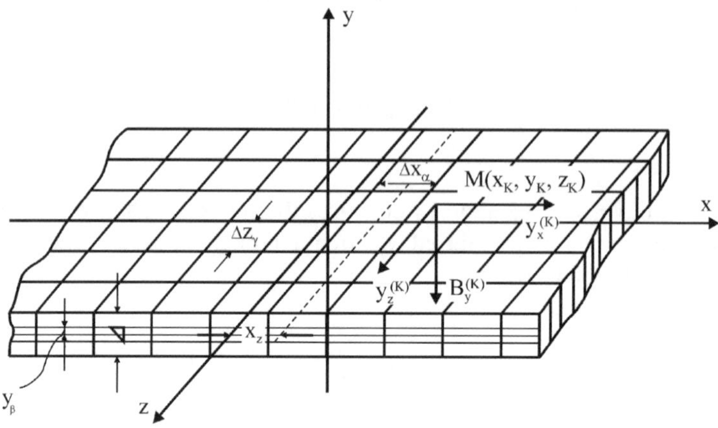

Fig. 2 Calculation of the eddy current in the conducting sheet

$$\begin{aligned}\overrightarrow{T}_{ij}^{k} = \frac{1}{4\pi}\frac{d}{dt}\Phi_{ij}\Bigg\{&\Bigg[arcsh\frac{m_{ij}-x_k}{\sqrt{(y_k+n_{ij})^2+(z_k-b_j)^2}} - arcsh\frac{p_{ij}-x_k}{\sqrt{(y_k+n_{ij})^2+(z_k-b_j)^2}}\\
&+arcsh\frac{p_{ij}-x_k}{\sqrt{(y_k-n_{ij})^2+(z_k-b_j)^2}} - -arcsh\frac{m_{ij}-x_k}{\sqrt{(y_k-n_{ij})^2+(z_k-b_j)^2}}\Bigg]\overrightarrow{x}^0\\
&+\Bigg[arcsh\frac{y_k+n_{ij}}{\sqrt{(x_k-m_{ij})^2+(z_k-b_j)^2}} - arcsh\frac{y_k-n_{ij}}{\sqrt{(x_k-m_{ij})^2+(z_k-b_j)^2}}\\
&+arcsh\frac{y_k-n_{ij}}{\sqrt{(x_k-p_{ij})^2+(z_k-b_j)^2}} - arcsh\frac{y_k+n_{ij}}{\sqrt{(x_k-p_{ij})^2+(z_k-b_j)^2}}\Bigg]\overrightarrow{y}^0\Bigg\}.\end{aligned} \quad (8)$$

Taking into account that $y_k = 0$ at an arbitrary point of the plane xOz, the electric vector potential could be calculated by the following reduced term

$$\vec{T}_{ij}^{k} = \frac{1}{2\pi}\frac{d}{dt}\Phi_{ij}\left[\operatorname{arcsh}\frac{n_{ij}}{\sqrt{(x_k - m_{ij})^2 + (z_k - b_j)^2}} - \operatorname{arcsh}\frac{n_{ij}}{\sqrt{(x_k - p_{ij})^2 + (z_k - b_j)^2}}\right]\vec{y}^{0}. \tag{9}$$

4 Numerical Results and Verification

The proposed method has been applied for a magnetic device, shown in Fig. 3. The system is symmetrical with respect to axis x. Sinusoidal current wI = 1,000. At is flowing in the winding. Different frequencies have been investigated: f=50 Hz, f = 200 Hz, f = 500 Hz and f = 1,000 Hz. Dimensions are as follows: h = 4.5 cm, h_1 = 6.5 cm, a = 4 cm, a_1 = 8 cm. The height of the copper sheet is $\Delta = 2$ cm and the distance to the system is $\delta = 1$ cm.

Eddy currents in 17 control points along X axis at level y = 0 (middle of the conducting sheet) have been calculated using the approach with replacing the exciting magnetic field by fictitious magnetic contours. The comparison has been done between results for eddy current distribution obtained by finite element method (FEM) and those calculated by method of fictitious magnetic contours.

The electromagnetic field problem has been considered as two-dimensional and time-harmonic with eddy currents in the conductive media.

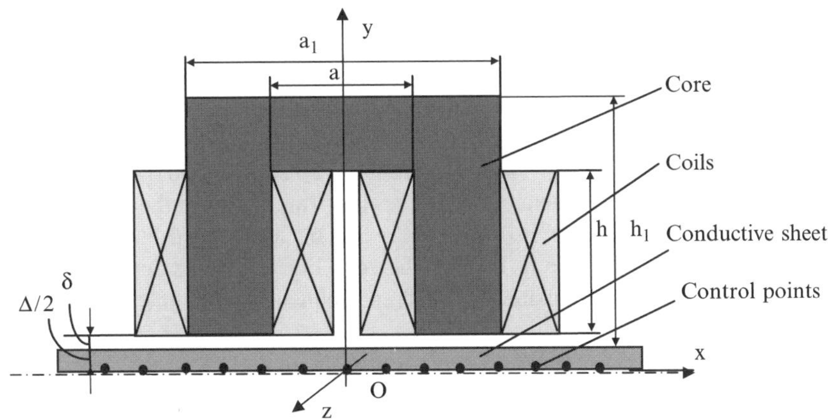

Fig. 3 Geometry of the magnetic system (*upper half*) with control points

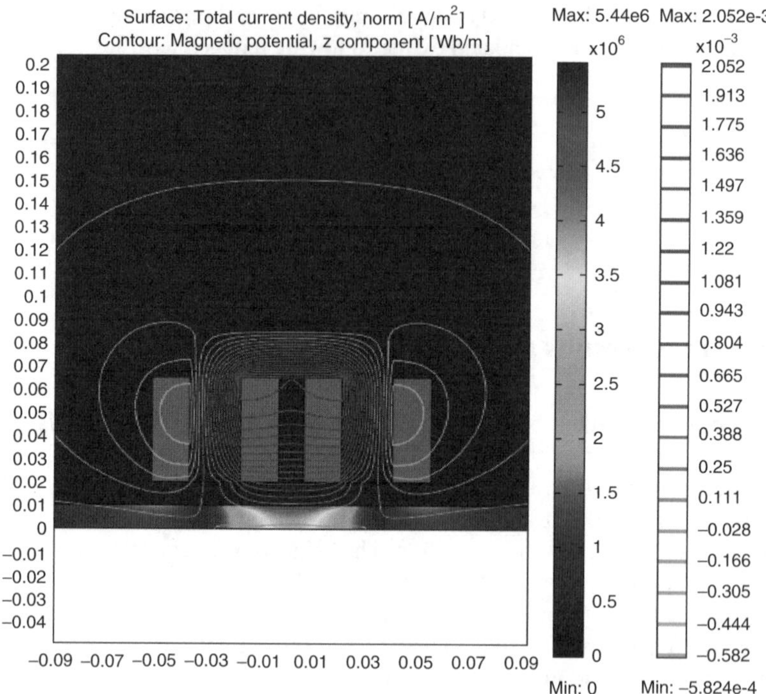

Fig. 4 Magnetic field and current density distribution for frequency $f = 50\,\text{Hz}$

4.1 FEM Analysis

Applying COMSOL 3.3 FEM package [7] the electromagnetic problem has been solve and eddy currents distribution has been obtained for different current frequencies. Special attention has been paid to detailed analysis of the eddy currents in the conductive sheet, where 17 control points have been observed.

Magnetic field and current density distribution for the investigated system are shown in Fig. 4 for frequency $f = 50\,\text{Hz}$ and in Fig. 5 for $f = 500\,\text{Hz}$, respectively. Current distribution for frequency $f = 50\,\text{Hz}$ along the X axes in the middle of the copper sheet ($y = 0$, $x = 0$–$16\,\text{cm}$) is shown in Fig. 6. Current distribution along the height of the sheet ($x = 0$, $y = 0$–$1\,\text{cm}$) is shown in Fig. 7. Current density values in control points have been compared for different frequencies in Fig. 8.

4.2 Analysis with Method of Fictitious Magnetic Contours

Magnetic field and current density distribution for the considered system are investigated using the above described method of fictitious magnetic contours. Eddy currents in control points along X axis have been calculated using the method with

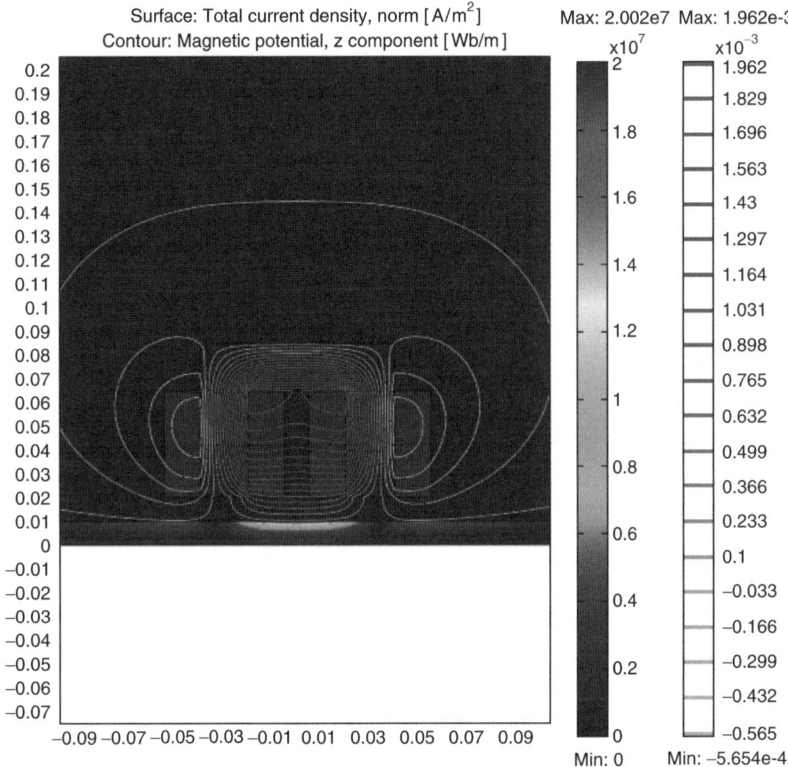

Fig. 5 Magnetic field and current density distribution for frequency $f = 500\,\text{Hz}$

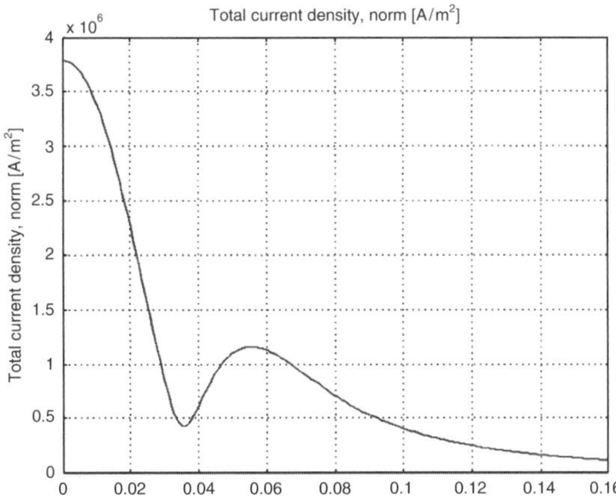

Fig. 6 Current density along the X axes in the middle of the copper sheet ($y = 0,\ x = 0\text{–}16\,\text{cm}$), $f = 50\,\text{Hz}$

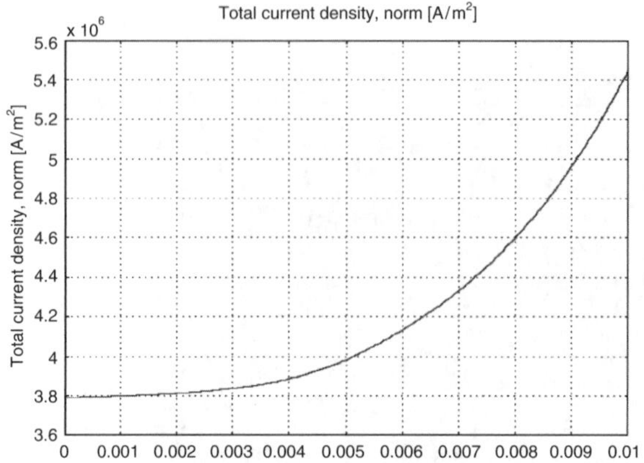

Fig. 7 Current density along the height of the sheet ($x = 0$, $y = 0$–$1\,\text{cm}$), $f = 50\,\text{Hz}$

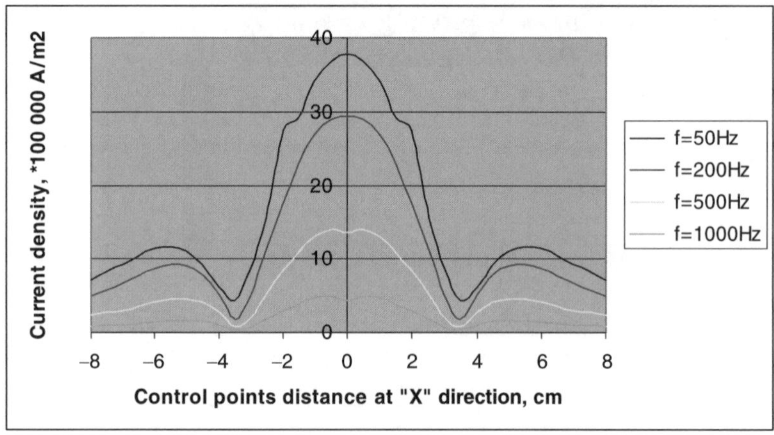

Fig. 8 Current density in control points for different frequencies obtained by FEM

replacing the exciting magnetic field by fictitious magnetic contours. The problem has been solved for different values of current frequencies: $f = 50\,\text{Hz}$, $f = 200\,\text{Hz}$, $f = 500\,\text{Hz}$ and $f = 1{,}000\,\text{Hz}$. The obtained results have been presented in Fig. 9. The comparison between obtained current distribution in observed points with FEM results have been shown in Fig. 10 for frequency $f = 50\,\text{Hz}$ and in Fig. 11 for frequency $f = 500\,\text{Hz}$. The results of comparison show satisfactory agreement.

5 Conclusions

The method based on the imaginary magnetic contours, applicable for analysis of time changing magnetic field is proposed. The method has been applied for a magnetic system and eddy currents in 17 control points in the conductive sheet have

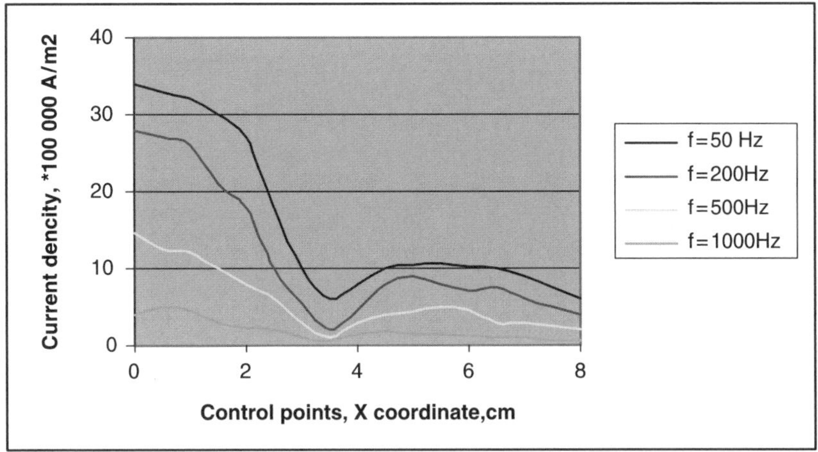

Fig. 9 Current density in control points for different frequencies obtained by using magnetic contours

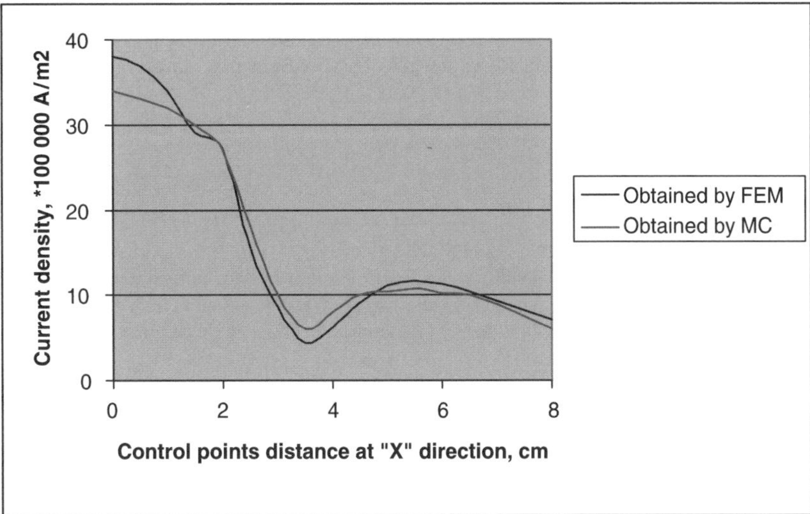

Fig. 10 Comparison between eddy currents in control points, obtained by magnetic contours and by FEM for frequency $f = 50\,\text{Hz}$

been calculated using fictitious magnetic contours instead exciting magnetic field. The problem has been solved for different frequencies. The obtained results have been compared with results obtained by FEM and the results show satisfactory agreement.

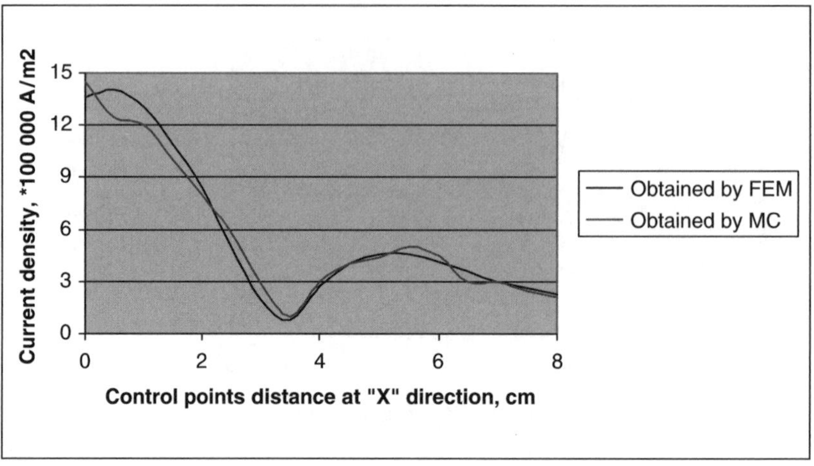

Fig. 11 Comparison between eddy currents in control points, obtained by magnetic contours and by FEM for frequency $f = 500\,\text{Hz}$

Acknowledgements The study is a part of the work carried out on the project "Couple and inverse problems in electrical machines" supported by NFSI of Bulgarian Ministry of Education and Science.

References

1. R.D. Stancheva, Energy conversion on the basis of electric vector-potential, Conference Proceedings, VII, XII Int. Symp. Theoret. Electr. Engng, Warsaw, Poland, 2003, pp. 273–276.
2. R. Stancheva, S. Papazov, Self and Mutual Parameters of Magnetic Contours, Izv. TU, vol. 41, no. 7, 1986, pp. 33–44.
3. M.P. Zlatev, Determination des courants de foucault an moyen du potentiel vector electric, Paper Bull. Bulg. Academy et Science, vol. 21, 1960.
4. D.M. Velickovic, Metod ekvivalentne elektrode, XXX Conference ETRAN, Herceg Novi, 1986, pp. 173–180.
5. S.R. Aleksic, Z.Z. Cvetkovic and B.R. Nicolic, Shallow-buried ring grounding electrodes, Seventh International Conference on Applied Electromagnetics, Nis, Serbia & Montenegro, 2005, pp. 11–12.
6. S.R. Aleksic, S.S. Ilic and M.T. Peric, Capacitance calculation of Saturn capacitor, Seventh International Conference on Applied Electromagnetics, Nis, Serbia & Montenegro, 2005, pp. 13–14.
7. COMSOL Version 3.3, AC/DC Module User's Guide.

Non-Destructive Testing for Cracks in Special Conductive Materials

Jarmila Dedkova

Abstract In this paper we propose new approaches for recovering conductivity distribution of special structures called honeycombs using electric impedance tomography (EIT). There is presented that algorithms based on stochastic methods are suitable for the detection of some cracks in the conductive honeycombs. There are compared numerical results obtained using stochastic method and using widely known deterministic methods.

1 Introduction

Electrical impedance tomography is a soft-field tomographic modality, where images of the electrical conductivity distribution in a volume can be reconstructed from voltage measurement captured on its boundaries. Usually, a set of voltage measurements is acquired from the boundaries of an investigated volume, whilst this is subjected to a sequence of low-frequency current patterns. In principle, measuring both the amplitude and the phase angle of the voltage can result in images of the electric conductivity and permittivity (impedivity) in the interior of a body. Alternating current patterns are preferred to DC to avoid polarization effects. In the usual frequency range (below 1 MHz) the field can be considered a steady current field, which is governed by the Laplace equation. It is well known that while the forward problem is well-posed, the inverse problem is highly ill-posed. Various numerical techniques based on deterministic or stochastic methods with different advantages have been developed to solve this problem. The aim is to reconstruct, as accurately and fast as possible, the impedivity or conductivity distribution in two or three dimensional models.

Jarmila Dedkova
FEEC BUT, Kolejni 4, Brno, Czech Republic
dedkova@feec.vutbr.cz

J. Dedkova: *Non-Destructive Testing for Cracks in Special Conductive Materials*, Studies in Computational Intelligence (SCI) **119**, 299–303 (2008)
www.springerlink.com

2 Basic Principle

The theoretical background of EIT is given in [1]. The forward EIT calculation yields an estimation of the electric potential field in the interior of the volume under some Neumann and Dirichlet boundary conditions. The finite element method (FEM) in two or three dimensions is exploited for the forward problem with current sources. The honeycombs are special structure which can be described by surface conductivity only (Fig. 1). The theoretical background of EIT with surface conductivities is given in [2]. Here we assume volume conductivity $\sigma = 0$ and only a surface conductivity σ_s defined on finite element surfaces is to be determined. Furthermore we assume the constant distribution of surface conductivity σ_s given by its edge values. To recover these values we use least square method. We have to minimize the primal objective function $\Psi(\sigma_s)$

$$\Psi(\sigma_s) = \frac{1}{2}\sum \|U_{\mathrm{FEM}}(\sigma_s) - U_{\mathrm{MEAS}}\|^2, \tag{1}$$

where U_{FEM} is the vector iteratively calculated using FEM, U_{MEAS} is the vector of measured nodal voltages. The optimization necessitates algorithms that impose regularization and some prior information constraint [3]. The regularization techniques vary in their complexity. Also we proposed the methods based on stochastic approach, which are expected to be used for the non-destructive testing conductive materials.

There is widely known deterministic methods Tikhonov regularization to be used for the acquirement of more accurate reconstruction results [3]. The surface conductivity is defined from the nodal values of the finite element grid. To recover these values we use LS with a modification of the Tikhonov regularization method (TRM) described in [4]. We have to minimize the primal objective function $\Psi(\sigma_s)$

$$\Psi(\sigma_s) = \frac{1}{2}\sum \|U_{\mathrm{M}} - U_{\mathrm{FEM}}(\sigma_s)\|^2 + \alpha \|L\sigma_s\|^2. \tag{2}$$

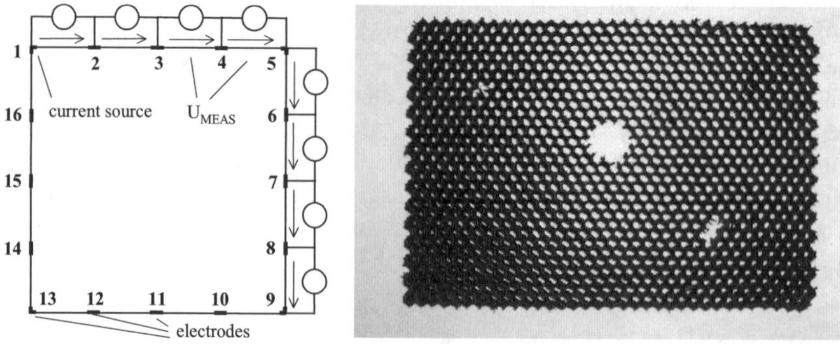

Fig. 1 An arrangement of 2D model and some cracks in honeycombs

Here are the regularization parameter α and a suitable regularization matrix L, connecting adjacent edges of the different conductivity values.

Furthermore, we compare the recovered results obtained by this algorithm and the controlled selection of non-homogeneities (CSN). The optimization of the primal objective function (1) based on the controlled selection of non-homogeneities (CSN) is new technique for reconstruction of conductivity distribution. The basic principle of the CSN is very simple. Let us consider a homogenous region with NE edges of the FEM mesh. We would find all cracks inside this conductive region with constant surface conductivity σ_S; it means that we have to find the number of edges with zero conductivity and their positions. We test all suitable permutations to minimize the objective function (1).

Global optimizing evolutionary algorithms, such as genetic algorithms, have been recently applied to the EIT problem [5]. Some results of genetic algorithm research are described in [6]. Compared to the genetic algorithm, the differential evolution algorithm (DEA) is a relatively new heuristic approach to minimizing non-linear and non-differentiable functions in a real and continuous space. DEA converges faster and with more certainty than many other global optimization methods according to various numerical experiments. It requires only a few control parameters and it is robust and simple in use.

3 Simulation Results and Comparison

The following example describes the application of the above-mentioned techniques for the recovering a collection of linear cracks in a homogeneous electrical conductor from boundary measurements of voltages induced by specified current fluxes. To recover σ_s distributions the LS method with a different type of the regularization's way was used. Furthermore, we compare the results obtained by CSN and DEA together with the results which were recovered by the modifications of the TRM. To evaluate the quality of simulation results, the total error *Err* of the recovered σs distribution is defined as

$$Err = \sqrt{\frac{\sum\limits_{i=1}^{NE} (\sigma_s(i) - \sigma_{s_{orig}}(i))^2}{\sum\limits_{i=1}^{NE} (\sigma_{s_{orig}}(i))^2}} 100\%. \tag{3}$$

Here surface conductivity σ_{sorig} (in S) is the original value, σ_s is the value recovered by EIT. The model of 2D arrangement for a numerical experiment is given in Fig. 2 (on the left). The total number of edges is 384; the number of nodes is 272. We assume a homogeneous object with $\sigma_s = 100\,\text{S}$ on all edges except for the chosen ones, where the values of $\sigma_s = 0\,\text{S}$ (bold marked edges). These edges can represent some cracks. The crack's distribution is given by two crowds of six and three edges.

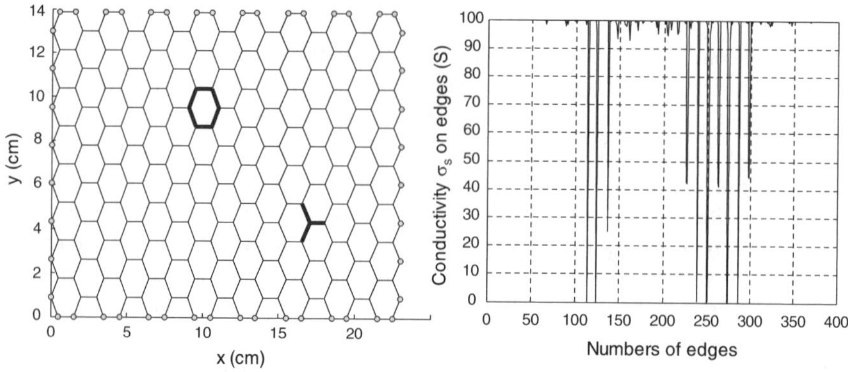

Fig. 2 The cracks distribution and σ_s on edges obtained using TRM

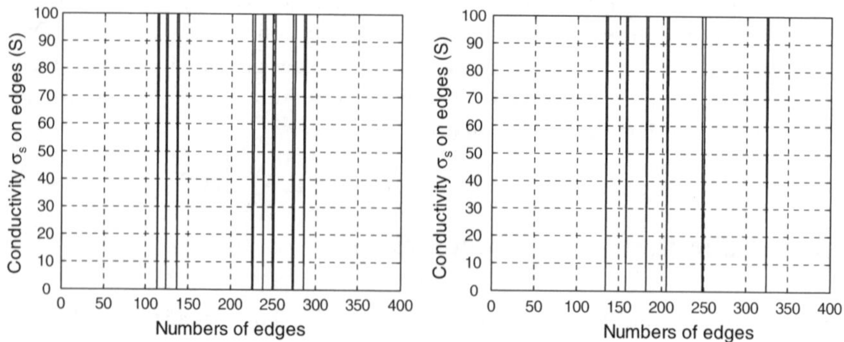

Fig. 3 Conductivity σ_s on edges obtained using CSN and DEA

Table 1 Comparison of recovered results

Method	Ψ	α	*Err* (%)	*t* (min)
CSN	0	–	0	45
DEA	2×10^{-10}	–	10	5
TRM	5×10^{-7}	2×10^{-22}	3	2

The results of numerical experiments you can see in Figs. 2 and 3. The final values of primal objective function $\Psi(\sigma_s)$, regularization parameter α for TRM, total error *Err*, and total time *t* are given in Table 1. You can see that the best results for reconstruction of arbitrary crack's distribution are obtained using CSN algorithm.

4 Conclusion

In this paper a new approach to the reconstruction of non-homogeneities using stochastic algorithms has been presented. There is shown one of possibilities of practical utilization of the image reconstruction based on electrical impedance tomography, which can be used to test for cracks in material structure like honeycombs. Many numerical experiments performed during the above-described algorithms have resulted in the conclusion that the application of the CSN reconstruction algorithms has an advantage over the TRM in better accuracy and stability of the reconstruction process. On the other hand the CSN is very time-consuming techniques. The DEA is not so much suitable method for the testing conductive materials for cracks. The results stated above as well as many other examples were obtained using a program written in MATLAB by author.

Acknowledgements The research described in the paper was financially supported by the research program MSM 0021630513.

References

1. M. Cheney, D. Isaacson, and J.C. Newell, Electrical impedance tomography, SIAM Rev., vol. 41, no. 1, 1999, pp. 85–101.
2. L. Dedek and J. Dedkova, Evaluation of thin layer conductivity in a volume using electrical impedance tomography, Proceedings BIOSIGNAL 2004, Brno VUTIUM, p. 3.
3. E. Somersalo, M. Cheney, and D. Isaacson, Existence and uniqueness for electrode models for electric current computed tomography, SIAM J. Appl. Math., vol. 52, 1992, pp. 1023–1040.
4. M. Vauhkonen, D. Vadász, P.A. Karjalainen, E. Somersalo, and J.P. Kaipio, Tikhonov regularization and prior information in electrical impedance tomography, IEEE Trans. Med. Eng., vol. 17, 1998, pp. 285–293.
5. R. Olmi, M. Bini, and S. Priori, A genetic algorithm approach to image reconstruction in electrical impedance tomography, IEEE Trans. Evol. Comp., vol. 4, 2000, pp. 83–88.
6. Z. Michalewicz, Genetic Algorithms+Data Structure=Evolution Programs. 2nd ed. Springer Verlag, Berlin Heidelberg New York, 1994.